Advances in 21st Century Human Settlements

Series editor

Bharat Dahiya, Chulalongkorn University, Bangkok, Thailand

More information about this series at http://www.springer.com/series/13196

Puay Yok Tan · Chi Yung Jim
Editors

Greening Cities

Forms and Functions

Springer

Editors
Puay Yok Tan
National University of Singapore
Singapore
Singapore

Chi Yung Jim
The University of Hong Kong
Hong Kong
China

ISSN 2198-2546 ISSN 2198-2554 (electronic)
Advances in 21st Century Human Settlements
ISBN 978-981-10-4111-2 ISBN 978-981-10-4113-6 (eBook)
DOI 10.1007/978-981-10-4113-6

Library of Congress Control Number: 2017933066

Printed on acid-free paper

This Springer imprint is published by Springer Nature
The registered company is Springer Nature Singapore Pte Ltd.
The registered company address is: 152 Beach Road, #21-01/04 Gateway East, Singapore 189721, Singapore

Foreword I

Towns and cities have become the dominant human habitat. A species that evolved to run and hunt on grassy plains now spends a majority of its time inside structures that modify environmental conditions and largely isolate humans from nature. Yet among those structures, from the time of the earliest cities, there have usually been gardens and vegetable plots. Relaxing or working in such green spaces has long been considered a pleasant and reviving part of urban life.

Historically, the green spaces were often just for the elites. They were parts of palace grounds or hidden within the walls of the most prestigious residences. Sometimes, as cities grew, they incorporated rural commons. Some remained pieces of countryside to which all local inhabitants had access, others were made available later. For example, in Scotland in 1450, Bishop Turnbull gifted the common lands of Glasgow Green to the townspeople. Many walled towns had spaces for growing crops within their walls, within the curtilages of individual properties or as a shared space where individuals had vegetable plots. Domestic animals were kept within cities and horses and mules provided transport. Such situations had both advantages and disadvantages for human health and well-being.

Industrialisation and the accompanying rural to urban migration began to produce rapid urban growth in the late eighteenth century. So rapid was the movement of people that housing was built rapidly and cheaply with inadequate sanitation, water supply and concern for health. Cities such as Georgian London revealed great environmental contrasts between the spacious, leafy squares in wealthy housing areas, and the tenements of the poor where human wastes piled up in the streets and drained into local streams. Today, the informal settlements of most low latitude cities in many respects replicate those late eighteenth century European contrasts between affluent and impoverished areas. Urban greenspace remains inequitably distributed across cities, wealthy areas enjoying much greater vegetation cover.

Into the industrial cities came public health pioneers and the city visionaries who persuaded powerful city governments to improve living conditions by better water supplies, sanitation and housing standards. The benefits of clean air were extolled and public parks began to be created. In the early 17th century, a former royal deer park in London, Hyde Park, was opened to the public. In Munich, the royal Hofgarten became accessible to the public in 1780, and a new public park, now known as the English Garden, was opened in 1792. Later, former royal hunting grounds, such as St. James Park in London, the Tiergarten in Berlin and the Bois de Boulogne in Paris became public open spaces.

Partly inspired by the efforts of such public health reformers such as James Kay and Edwin Shuttleworth, prominent philanthropic local residents and councillors supported the provision of parks. In British industrial cities, Princes Park was opened in Liverpool in in 1843, followed by Phillips and Queen's Parks in Manchester and Peel Park in Salford in 1846, and Birkenhead Park, across the Mersey from Liverpool, in 1847. Formal legislation providing for local governments to acquire land "for the purpose of being used as public walks or pleasure grounds" was included in the British Public Health Act of 1875. By such gradual steps, fragments of urban greenspace acquired legal status and provided the first components of urban green infrastructure.

Remnants of ancient woodland, from Kenwood in London to Bukit Timah in Singapore, encapsulated in expanding cities became key elements of the green infrastructure. River floodplains were frequently kept free of development to act as floodways. Greenspaces developed along utility line easements, railways and canals, which, together with mature street trees acted as migration corridors for certain species. The mosaic of urban gardens, public and private, became host to a diversity of plants, native and exotic, introduced and invasive, that in turn supported a food chain from insects to birds and urban mammals. Occasional disturbance, from urban land use change and dereliction to wartime bombing and fires, created new ecological niches which added to urban biodiversity. Some of the opportunities for urban wildlife thus created were exploited by committed enthusiasts to create urban nature reserves where city children could experience nature close to their homes.

With the concern about the human impact on the environment rising after the 1960s, the value of urban nature became increasingly recognised, culminating in many countries with the development of urban nature conservation strategies and local biodiversity action plans by many local governments. As the evidence of climate change became widely accepted, urban strategies to both mitigate and adapt to climate change incorporated the use of plants to reduce the urban heat island effect, remove some air pollutants, and retard the movement of storm-water became widely adopted. People concerned with physical and mental health and general human well-being also advocated wider use of varied forms of greenspace for relaxation and recreation. Comparative studies of birds, mammals and plant species in cities demonstrated how species adapted to the urban environment and took advantage of niches within the built environment. Such knowledge is brought together in this book with a particular emphasis on the ecological underpinnings of urban greenspace, the varied functions of urban greenspace from an ecosystem services perspective and the forms of urban greenspace, from blue-greenspaces associated with water bodies to green walls and green roofs.

Urban greenery is essentially either a human creation or a human modified form of natural vegetation. It is greatly affected by the extent of care or neglect in the management of particular greenspaces. The same greenspace may be used in different ways for different purposes by varied individuals and social groups. Some potential uses may conflict with one another. The components of urban green infrastructure therefore have to be seen as multi-modal social spaces, with planning for opportunities of many different activities, separated sometimes into different localities so that incompatible greenspace uses do not clash. Other forms of uses involve accepting compromises, such as when a football pitch within a grassed flood basin becomes temporarily inundated, or when access is restricted to protect nesting birds. Urban farmers and gardeners may resent some insect and bird species that exploit their fruit and vegetables. Street trees provoke arguments between those who desire tree preservation at all costs and those concerned about root damage to pavements and walls, falling branches and potential wind throw hazards. These all-important community contrasts and diverse attitudes to nature are given due attention here, particularly in terms of impacts on household well-being, income and food security.

Urban food-growing takes many forms, from a few herbs in a window box, to the cultivation of vegetables as subsistence food on any accessible vacant land, even roadside verges, river banks and floodplain wetlands. There is an enormous difference between hobby gardening in affluent suburbs and urban agriculture practised by poor households in cities such as Harare, Zimbabwe, where around 90% of the leafy vegetables consumed by disadvantaged families are home grown on urban plots that have little security of tenure or occupation. The opportunities for increasing urban food production are great, but they need to be incorporated into planning and land use management systems. Readers will find ideas about how to improve urban greenspace planning throughout this book. Understanding the local situation, land tenure systems, traditional practices and political dilemmas is always required.

Landscape tastes vary. Urban woodland may enchant some people and frighten others. The risks associated with dark shady spaces are very real. Urban greenspaces are so diverse and occur in such varied situations that care has to be taken to plan urban greening in ways that suit particular places, communities and social norms. The social-economic dimensions of urban greenspace planning and management are integrated into individual chapters, with examples from around the world to show how solutions sometimes are significantly different.

Many people are now engaged in valuing the ecosystem services provided by urban vegetation and greenspace. This economic approach helps to reinforce the health, human well-being, environmental improvement, climate change adaption, biodiversity, and nature conservation arguments for improving urban green infrastructure. Caution has to be applied, however, in the way these arguments are pitched. The evidence in this book suggests the value of urban wetlands in storm runoff management, but also considers ecosystem disservices, such as the way invasive species, for example Himalayan balsam, water hyacinth and water lettuce, can occupy streams, lakes and reservoirs and impede water flows and reduce water storage capacities. This balanced approach is important as overstating the benefits of urban greenery may be as dangerous as not advocating its preservation, enhancement and extension. This book provides a wealth of insights into opportunities, mechanisms and procedures for developing innovative ways of urban greening to enhance the beauty of cities and to make them more sustainable, more resilient to environmental change and healthier, more exciting places in which to live.

18 March 2017

Ian Douglas
The University of Manchester
Manchester, England

Foreword II

In greening cyborg cities, be an *ecophronetic* scholar-practitioner

> We must learn to green the earth, to restore the earth, and to heal the earth.
>
> Ian L. McHarg (1996, p. 374)

"I would love to be here when this process (of greening, restoring, and healing the earth) is apace." Wrote the late American landscape planner and educator Ian L. McHarg (1920–2001) in his 1996 autobiography *A quest for life*. "In my mind's eye I see myself with a group of scientists, looking at the earth from space, viewing the shrinking deserts, the burgeoning forests, the clear atmosphere, the virgin oceans, smiling at the recovery, anticipating the day when a successor will announce, 'the earth is healed, the earth is well'." (p. 375)

Two decades later in 2017, McHarg should have been far less impressed by what he would have seen on the earth at large. But euphoric he would most definitely be to have learnt, from reading this striking monograph, that not only does the process of greening the earth continue to gain new momentum in the Anthropocene, but also the greening troops have now been advancing to the brutal battleground of cyborg cities, and combatting triumphantly along all frontiers of *greening cities*.

As a student of McHarg's in the 1980s whose academic aspiration has been ever since inspired and professional path illuminated by his ideal of design-with-nature, I share the same excitement; on behalf of my comrades inside the front trench on the battleground near Shanghai Disneyland Resort in Pudong, one of satellite cyborg cities of Shanghai, China, I congratulate the editors, Professors Puay Yok Tan and C.Y. Jim, and the authors for the publication of this delicate collection of important scholarly works on the ecological practice of greening cities.

Yet I become even more impressed and delighted when viewing this monograph through the lens of *ecophronetic* practice research, a perspective that helps bring to light some unique and admirable qualities demonstrated by the editors and authors.

What is *ecophronetic* practice research? How relevant is it to both the practice and scholarship of ecological planning, design, construction, restoration, and management (For brevity, hereafter, I shall use the umbrella term *ecological practice* to represent the whole of these five distinct yet interrelated human activities)? As the founding director of the Center for *Ecophronetic* Practice Research at Tongji University, Shanghai, China, I have been constantly asked to respond to these and other pertinent questions. My response has always been that *ecophronetic* practice research is ecological practice research at its best that exhibits two distinctive characteristics: the researcher mindfully enacts the role of *scholar*-practitioner (instead of that of spectator) who is committed to the dual responsibility of producing knowledge and influencing (instead of informing) practice; and the research process is being *ecophronetical*—inspired and informed by ecological practical wisdom, *ecophronesis*. I believe that the editors and authors demonstrate these qualities cogently in this book.

In the varied landscape of social practice (for example, education, law, medicine, and one or any combination of the five activities in ecological practice), according to American philosopher and planning theorist Donald Schön, scholars are confronted with a choice between standing on the high ground of theory and descending to the swampy lowlands of practice (Schön 2001, p. 191). Problems in the lowlands are wicked and thus less capable of, or even resistant to, scientific or technical solution, while those on the high ground are tame, or can arguably be perceived to be so, lending themselves to solution through the use of scientific theory and technique (For a recent and succinct account on wicked and tame problems, see Xiang 2013). In making this choice, scholars face the irony that "the problems of the high ground tend to be relatively unimportant to individuals or to society at large, however great their technical interest may be, while in the swamp lie the problems of greatest human concern." (*Ibid.*) Scholar-practitioners, as defined by American management scholar Ed Schein (cited by Wasserman and Kram 2009, p. 12), are those scholars who audaciously choose to descend to the lowlands and are "dedicated to generating new knowledge that is useful to practitioners."

What does this choice of becoming a scholar-practitioner mean to a scholar? It means that she needs to embrace the following challenges-cum-opportunities: acting as a scholar of practice research *in* and *for* practice, rather than a scholar of scientific or applied research in and for science or applied science; assuming the dual responsibility of producing useful knowledge that may not even be novel in the conventional sense, and influencing, not just informing, practice; bridging the theory-practice gap in research; and overcoming the rigor-relevance hiatus in scholarship (For a recent and succinct account on the theory-practice or rigor-relevance gap, see pp. 338–342 in Sandberg and Tsoukas 2011). In the lowlands of ecological practice, the choice also means that a scholar-practitioner faces the peculiar challenge-cum-opportunity of being responsible for all living and non-living beings on the earth. Ecological practice is the action and process through which humans attempt to bring about a secure and harmonious socio-ecological condition that serves the basic human need for survival and flourishing. Unlike in other kinds of social practice, such as education, mechanical engineering, medicine, and law, where practitioners primarily deal with human affairs, practitioners in ecological practice concern themselves primarily and explicitly with the relationship between human and nature on top of social relationships (Steiner 2016, pp. 1–4, p.173; Xiang 2016, p.55). Because of this unique characteristic, in conducting ecological practice research, a scholar-practitioner needs to not only hold the belief that it is in human beings' self-interest—ethical, moral as well as material—to respect and appreciate the intrinsic value of all living and non-living beings on the earth, but also to employ this very *human-beings' enlightened self-interest* as a benchmark for judging what is the right knowledge to develop for the right practice.

The second characteristic of *ecophronetic* practice research is that the research process is being *ecophronetical*—inspired and informed by ecological practical wisdom, *ecophronesis*. As a reinvented concept of neo-Aristotelian *phronesis* (practical wisdom) within the realm of ecological practice, *ecophronesis* is the master skill par excellence of moral improvisation to make, and act well upon, right choices in any given circumstance of ecological practice (Xiang 2016, p. 55). The term *ecophronesis* is coined as an *ex post* recognition of and a revered tribute to those human beings who developed the master skill par excellence through immersed and mindful ecological practice. With the blessing of the skill, these individuals of ecological practical wisdom, *ecophronimos*, that is, achieved a paramount level of "doing real and permanent good in this world" (Xiang 2014, p. 65), exemplifying the stellar quality of *ecophronetic* practice. Among those celebrated on the exalted roster of *ecophronimos* are Li Bing and his colleagues of many generations who built and sustained the 2300 year old Dujiangyan irrigation system in Sichuan, China (Needham et al. 1971, p. 288; Xiang 2014, pp.65–66), and McHarg and his colleagues who planned, designed, and constructed the ecological town of The Woodlands, Texas, the United States near half-of-a-century ago (McHarg 1996, pp. 256–264; Yang and Li 2016).

As an invaluable intellectual heritage and character asset, *ecophronesis* enlightens scholar-practitioners in the lowlands of ecological practice who strive to overcome the above mentioned challenges. For example, a distinctive virtue shared by all *ecophronimos* is their ability to sustain a carefully nuanced balance between the act of following the logic of ecological practice on the one hand, and that of embracing the logic of ecological science on the other. As such, never exists to them the arguably unbridgeable hiatus between scientific rigor and practical relevance that has been a persistent concern in both circles of ecological practice and science in the modern-day world. Equally illuminating, for another example, is the way *ecophronimos* produce useful knowledge. Solely motivated by and entirely devoted to practical interest, these individuals are capable of sorting out the science, techniques, and tacit knowledge of all kinds that are suitable for the issues most relevant to practice, and developing a new and useful body of knowledge that is *pertinent*, *actionable* and *efficacious* in specific instance of ecological practice. This eclectic way of ecological practice research is not only utterly different from that of ecological science research, but also notably in sharp contrast to that of applied ecological research. Indeed, in applied ecological research, ecological practice is often regarded as an "applied" version of the knowledge, methods, and principles of a specific branch of ecological science, and thus treated as a practical demonstration of the pertinence of scientific principles.

Celebrating and embracing the intellectual heritage of *ecophronesis* and character asset of *ecophronimos*, scholar-practitioners of *ecophronetic* practice research strive to emulate the way *ecophronimos* produced useful knowledge in specific circumstances of ecological practice. As such, they attempt to answer two sets of questions:

If we identify ourselves as *scholar-practitioners* dedicated to the dual responsibility of developing knowledge and influencing practice,

> How do we enact our role to generate useful—*pertinent*, *actionable*, and *efficacious*—knowledge practitioners will celebrate? How do we know the knowledge we develop is useful? How should this process of ecological practice research be implemented?

Before emulating, we need to know the way *ecophronimos*, who might or might not identify themselves as scholar-practitioners, conducted practice research. More specifically,

> How did those who are recognized as *ecophronimos* navigate the swampy lowlands of ecological practice without being struck by the theory-practice or rigor-relevance torpedoes? What light, that is, *ecophronetical* light, do their effective experience and exemplary practice bring to the contemporary ecological practice in general, and ecological practice research in particular? In the *ecophronetical* light, how did they produce useful knowledge in any given instance of ecological practice? How should we as *scholar-practitioners* in the contemporary ecological practice emulate their way of practice research to produce useful knowledge?

I believe that this *ecophronetic* mode of ecological practice research, still in its infancy as compared to its elder siblings—ecological science research, applied ecological research, and even action research, is most promising and merits attention from the community of scholar-practitioners in the lowlands of ecological practice. As a matter of fact, many colleagues on the scholar-practitioner continuum have already been, and may well continue to be, working under this mode without knowing it or without calling it as such (For a brief account on the continuum, see Wasserman and Kram 2009, p. 14). Among ample examples for the latter case, I found with great delight, is the delicate collection of scholarly works in this book.

I therefore hope that in addition to the much needed useful knowledge about greening cities Professors Puay Yok Tan and C.Y. Jim and other authors generously shared and eloquently articulated, in this book the readers will also be able to recognize, by means of the lens of *ecophronetic* practice research above outlined, some virtuous qualities the cohort of scholar-practitioners cogently demonstrated. These include, but are certainly not limited to, the audacity to act in the lowlands as a *scholar-practitioner* who does practice research *in* and *for* the important practice of greening cities, rather than on the high ground as a scholar of scientific or applied research in and for science or applied science; the commitment to the dual responsibility of producing useful knowledge and influencing (instead of informing) practice; the ethical belief in human-beings' enlightened self-interest as a benchmark for judging what is right in ecological practice and science; the resolve to bridge the theory-practice gap in research; the determination to overcome the rigor-relevance hiatus in scholarship; and the capability of conducting transdisciplinary research to embrace diverse perspectives. With a gentle caveat that this way of inquiry aims to examine and advocate *ecophronetic* practice research as a distinctive mode of practice research drawing upon the experience and examples of *ecophronetic* individuals, rather than promoting the individuals themselves, I trust that the readers will, like me, be motivated to celebrate these and other admirable exemplary qualities they identified themselves through the lens of *ecophronetic* practice research, and to emulate in their own ecological practice research.

Finally, to the *ecophronimo* Ian L. McHarg whose mind's eye has been watching from space the greening operations on the earth, what do we have to say? On behalf of all the comrades from around the world who are immersed in the operation *greening cities*, I shall send the following message—

> Continuing in fractured cyborg cities,
> Is our brutal battle for greening city;
> Block by block, building by building,
> Our troops advance steady;
> Foot by foot, inch by inch,
> Is greenery burgeoning horizontal and vertically;
> To accomplish the noble mission,
> *Ecophronetic* scholar-practitioners are whom we strive to be!

18 March 2017

Wei-Ning Xiang
Tongji University, Shanghai, China;
University of North Carolina, Charlotte, USA;
Holly Hill Farm, Charlotte, USA
e-mail: wnxiang@tongji.edu.cn; wxiang@uncc.edu

References

McHarg IL (1996) A quest for life: An autobiography. New York, John Wiley & Sons

Needham J, Wang L, Lu, G-D (1971). Science and civilization in China. Volume 4: Physics and physical technology, Part III: Civil engineering and nautics. Cambridge, UK: Cambridge University Press

Sandberg J, Tsoukas H (2011) Grasping the logic of practice: Theorizing through practical rationality. Acad Manag Rev 36(2):338–360

Schön D (2001) The crisis of professional knowledge and the pursuit of an epistemology of practice. Chapter 13, in Competence in the Learning Society, Raven J, Stephenson J (eds) Reproduced on the HE Academy website by kind permission of Peter Lang Publishing, Inc, pp. 185–207. Retrieved online from http://wwwnew1.heacademy.ac.uk/assets/documents/resources/heca/heca_cl13.pdf

Steiner FR (2016) Human ecology: How nature and culture shape our world. Washington, D.C: The Island Press (Forwarded by Richard T.T. Forman)

Wasserman IC, Kram KE (2009) Enacting the scholar-practitioner role: An exploration of narratives. J Appl Behav Sci 45(1):12–38

Xiang W-N (2016) *Ecophronesis*: The ecological practical wisdom for and from ecological practice. Landscape and Urban Plan 155:53–60

Xiang W-N (2014) Doing real and permanent good in landscape and urban planning: Ecological wisdom for urban sustainability. Landscape Urban Plan 121:65–69

Xiang W-N (2013) Working wth wicked problems in socio-ecological systems: Awareness, acceptance, and adaptation. Landscape Urban Plan 110(1):1–4

Yang B, Li S (2016) Design with nature: Ian McHarg's ecological wisdom as actionable and practical knowledge. Landscape Urban Plan 155:21–32

About the Book

Cities are considered one of humankinds most spectacular creations. They weren't created by a single person at one location nor are they revered by all their inhabitants. Due to the unprecedented growth of urban areas around the world, there is currently widespread agreement amongst planners, landscape designers, architects, engineers, scientists and the public of the need to create sustainable and resilient cities in the future. One of the fundamental elements required to create such cities is the preservation, creation and management of urban vegetation or green space. Puay Yok Tan and Chi Yung Jim have assembled an outstanding group of experts and have produced one of the most comprehensive volumes on the current state of knowledge on urban greening and ecology. Chapters provide extensive analysis and discussion of the benefits, techniques, approaches and challenges of urban greening including climate amelioration, hydrology, design, biodiversity, agriculture, ecosystem services and social-cultural services as well as some specific types of green space such as heritage landscapes, remnant vegetation, green roofs and blue-green infrastructure. The authors have done an excellent job at integrating scientific knowledge with practical applications which is critical if we are to successfully create healthy, liveable cities in the future. In addition, they have provided urban practitioners and students with valuable insights into the current gaps in our knowledge which will greatly assist in the identification of new research directions, design approaches and management practices. I believe the authors have done an outstanding job in producing a volume that significantly enhances our knowledge of urban greening and ecology. Due to the importance of maintaining green cities this should be a must read for anyone involved in the study, design, creation and management of cities.

Mark J. McDonnell
Editor-in-Chief, Journal of Urban Ecology
Oxford University Press, Oxford, UK

Contents

Contributors

Robert Brears Mitidaption, Christchurch, New Zealand

Wendy Y. Chen University of Hong Kong, Pokfulam, Hong Kong

Shinuo Deng AECOM Asia Company Limited, Sha Tin, Hong Kong

Evyatar Erell Ben-Gurion University of the Negev, Beersheba, Israel

Yuanqiu Feng National University of Singapore, Singapore, Singapore

Abdul Rahim Hamid National University of Singapore, Singapore, Singapore

Jeffrey Hou University of Washington, Seattle, WA, USA

Shalene Jha University of Texas at Austin, Austin, TX, USA

Chi Yung Jim University of Hong Kong, Pokfulam, Hong Kong

John G. Kelcey Ceckovice, Bor U Tachova, Czech Republic

Cybil Kho National Parks Board, Singapore, Singapore

Kuei-Hsien Liao National University of Singapore, Singapore, Singapore

Heidi Liere Reed College, Portland, OR, USA

Brenda B. Lin CSIRO Land & Water, Aspendale, VIC, Australia

Stacy M. Philpott University of California, Santa Cruz, CA, USA

Swinal Samant National University of Singapore, Singapore, Singapore

Angelia Sia National Parks Board, Singapore, Singapore

Puay Yok Tan National University of Singapore, Singapore, Singapore

Christine A. Vogt Arizona State University, Phoenix, AZ, USA

Peter Werner Institute for Housing and Environment, Darmstadt, Germany

Chapter 1
Introduction to Green City Idea and Ideal

Chi Yung Jim and Puay Yok Tan

1.1 The Green City Imperative

To different people, the idea of green city can conjure up a plethora of connotations and imaginations. They represent a spectrum of human responses to the form and substance of nature embedded in and enveloping cities. Some laypersons may consider urban parks or urban green spaces in general as venues for outdoor recreational pursuits in the company of some vegetative elements. Some would be attracted by the ornamental or decorative traits providing visual divergence from the preponderant artificial structures and surfaces of the built environment. Others may go beyond the skin-deep response to explore, appreciate and relish the diverse and high-order ecosystem services.

The favourite type of greenery preferred by citizens could vary greatly. It covers bequests from the pre-urbanization era with rich natural contents such as remnant enclaves of woodlands that have been conserved by default or by design in cities. They denote urban green spaces with the most complex ecosystem diversities, structures and functions. In the eyes of the beholders, the wild forest landscape could be earnestly preferred. However, it could be feared and shunned by some citizens who may opt for neatly manicured parks with nature presented formally and orderly on human terms. At the other end of the spectrum, it could be highly regimented vegetation composed of common horticultural species planted in urban parks or at roadsides.

The innate urge to get close to greenery and water, probably deeply etched in the human psyche, may drive the desire to preserve and rear them. When our ancestors

C.Y. Jim (✉)
University of Hong Kong, Pokfulam, Hong Kong
e-mail: hragjcy@hku.hk

P.Y. Tan
National University of Singapore, Singapore, Singapore
e-mail: akitpy@nus.edu.sg

P.Y. Tan and C.Y. Jim (eds.), *Greening Cities: Forms and Functions*,
Advances in 21st Century Human Settlements,
DOI 10.1007/978-981-10-4113-6_1

invented the institution of town and city, the escape from the vagaries of unforgiving and capricious nature would be considered a pleasant blessing or even good riddance and good fortune. As cities grew bigger and denser, the harshness induced by nature deficit began to be keenly felt and forlornly lamented. As the connection with nature became more tenuous in cities, the desire to restore nature began to grow stronger. Humans have followed circuitously and clumsily a circular route in our relationship with nature, from depending on nature to escaping from or even excluding nature, and then reverting to embracing and treasuring nature.

The development of ecological science and its increasing application to the urban sphere has enriched our knowledge base regarding the multiple and pertinent benefits. We have learnt the intricate ins and outs, and arts and sciences of urban green space design, use, management and conservation. Many cities have emerged from the fuzzy understanding of greening benefits in the past to awakening and enlightenment. The broader and deeper appreciation of ecosystem services has nurtured, mobilized and reinforced a latent force to transform the traditional mode of city development. In the course of urban planning, the hitherto often side-lined green infrastructure has been increasingly mainstreamed and considered in tandem with the grey infrastructure and other hardware paraphernalia of cities.

The benefits brought by the urban green infrastructure have been focused traditionally on the environmental and then ecological aspects which form the basis to generate and sustain other services. In recent years, attention has been extended to the social and economic dimensions, and deeper exploration of ecological ingredients and contributions. The research findings have been gradually translated and transferred to the policy and practice provinces. Meanwhile, researchers have continued to move into new directions to expand the knowledge frontier. The following paragraphs summarize the key growth hotspots in recent years related to the city greening theme.

1.1.1 Spatial-Ecological Planning of Urban Green Space

- From the establishment of individual green space sites per se to the city-scale spatial-ecological planning especially with reference to sustainability and liveability issues (Caspersen and Olafsson 2010).
- From non-spatial and non-ecological planning of urban green space to firm grasp and application of mainstream ecological, urban-ecological, and landscape-ecological principles.
- From isolated green patches to connectivity with the help of stepping stones and habitat corridors, and permeating ecological network of greenways and blueways (Ignatieva et al. 2011).
- From casual or chance inheritance of natural enclaves by default to well-planned conservation of high-calibre remnant nature by design, as well as creation of complex natural ecosystems at green or brown fields.

- From an emphasis on vegetation to seamless integration with water features to form the green-cum-blue infrastructure with applications in stormwater management in terms of both quantity and quality (District of Columbia 2014).
- From landscape planting to urban farming and allotment and community gardening realms that can fuse natural with productive and amenity functions.
- From the emphasis on formal green space to increasing awareness and attention to informal urban natural areas.
- From ground-level greening to elevated skyrise greening expressed as green roofs, green walls and sky terraces.
- From single organism, factor or process to the integrated ecosystem approach adopting the multidisciplinary perspective.
- From simplistic off-site habitat compensation or offset to full re-creation of ecological components, interactions and processes (Quétier and Lavorel 2011).
- From urban densification dominated by the grey infrastructure to embracing the green imperative in a co-ordinated manner under a nature-friendly planning and development regime (Jim and Chan 2016).
- From the redevelopment of old districts without improvement in green space provision to visionary allocation of adequate land for high-quality green sites with the help of development right transfer.
- From little attention to soil coverage in cities to increasing studies on the role of permeable land surface in urban vegetation performance and as a surrogate indicator of environmental quality (Artmann and Breuste 2015).
- From piecemeal transfer of research findings to the comprehensive and integrated knowledge exchange in the package of nature-based solutions (European Commission 2015).

1.1.2 Design and Management of Urban Greenery

- From the traditional horticultural-landscape design to the innovative ecological-naturalistic alternative aiming at emulating the local or regional ecology (Jim and Chen 2003).
- From routine provision of limited species richness to meticulous design to plant and attract more species to achieve high biodiversity (Secretariat of the Convention on Biological Diversity 2012).
- From ingrained preference for domination by a common cohort of exotic species to cultivating and accommodating native flora and fauna.
- From realizing the individual benefits of urban green space to the holistic understanding of the whole package of ecosystem services.
- From the contribution of green space to the quality of the environment to the quality of life, including passive health benefits and active enhancement of physical and mental health.

- From the welfare of people to altruistic concerns with other floristic and faunal companions and their interplays and interactions with humans.
- From the amelioration of the urban microclimate and urban heat island effect to comprehensive rendering of cities for climate-change adaptation, proofing, readiness and resilience (Brown et al. 2015).
- From the preoccupation with green space coverage to the elaborate assessment of ecological value (e.g. biotope area factor or green plot ratio) (Landschaft Planen and Bauen 1990).
- From the routine care of heritage trees to recognition of multiple contributions and specialized and dedicated treatment as ecological as well as cultural treasure.
- From a narrow focus on ecosystem services to the realization and understanding of disservices and the related trade-offs.
- From general knowledge on the contributions of site and vegetation factors to cooling and air cleansing services to detailed studies of specific causative attributes and processes.
- From the neglect of soil for urban vegetation to increasing examination of urban soil as a critical constraint to plant growth in urban areas.
- From inadequate understanding of ecological restoration to development of practical field techniques to restore degraded urban ecosystems (Simmons et al. 2016).
- From taming or regimentation of the wildscape to preservation of nature and re-wilding of degraded semi-natural or ruderal enclaves (Connop et al. 2016).

1.1.3 Attention to the Socio-economic Dimension

- From meeting the basic needs for recreation and leisure to creating and cultivating high-order multimodal social space.
- From the treatment of park visitors as a monolithic group to exploring their socio-demographic variations and social inclusiveness needs as a key design component aiming at fostering community attachment, integration and cohesion.
- From the indifference towards differential usage of urban green space by different socio-economic groups to heightened awareness of spatial inequality in provision and associated implications and consequences of environmental injustice (Tan and Samsudin 2017).
- From patronizing park design and management dominated by the administration and professionals to active involvement, participation and engagement of the community.
- From official estimate of conventional recreational demands in guiding park design to objective evaluation of citizen expectations and preferences for park facilities and ambience (Southon et al. 2017).

- From insouciance towards the perception of safety in parks to understanding the role of park design and management in the fear and occurrence of crime.
- From limited understanding of the health effects of urban green space to in-depth investigations of the underlying factors and mechanisms in relation to both well-being restoration and enhancement (Cleary et al. 2017).
- From preliminary assessment of the perception of urban green space in terms of design, management and ecosystem services to detailed evaluation yielding findings to inform policies and planning (de la Barrera et al. 2016).
- From direct measurement of Euclidean distance in park accessibility studies to in-depth evaluation of the perception of accessibility, tangible and intangible barriers, and socio-demographic profiles of catchment areas.
- From loose studies of various benefits of urban green space to economic and monetary valuation of the wide spectrum of ecosystem and other services based on hedonic pricing, contingent valuation, travel cost and other methods (Bertram and Larondelle 2017).

1.2 Objectives and Organization of the Book

This book has been conceived to analyse and condense the vast contributions in the scientific and professional literature especially in recent years in the burgeoning field of greening cities. The universal subject matter provides a fertile common ground for ideas from diverse disciplines to incubate, interact, hybridize and flourish. The themes that can be encompassed under this umbrella are pretty comprehensive if not eclectic. The large pool of research findings scattered in scholarly journals and reports may not be easily accessible to researchers and practitioners and converted into concepts as well as useful hints and skills. We have selected a subset based on relevance to the liveability and sustainability aspects under the ambit of urban development. They cover conceptual and empirical studies, as well as the knowledge exchange endeavours to transform research findings into policies and practices.

The ecological core of the subject matter provides the pertinent strands to connect with the social, cultural and economic spheres to achieve integration. We aim at mobilizing the innate strength of urban ecology to understand the intricacies of urban nature, and to enhance the planning and development of the urban green infrastructure. In organizing and advancing the knowledge domain, we have tried to strike a balance between using nature to promote urban health as well as conserving nature to ensure that the ecosystem services will not be impaired. We hope that the book can furnish a ready reference for researchers, practitioners and policy-makers to grasp the critical issues, and to trigger further studies and applications to fulfil the quest of high-order green cities.

Following this introductory remarks (this chapter), the flow of the book adopts a logical sequence. Part I begins with an in-depth survey of the key ecological principles that constitute an overarching framework for urban greening inquiries and the application of urban ecology to the design and planning of different facets of the urban green infrastructure (Chap. 2). As the past is a key to the comprehension of the present and the planning for the future, the historical root of the origin and rationale for human endeavours to green the urban domain could enhance understanding of where we are and where we should go henceforth (Chap. 3).

Part II shifts the attention to the diverse range of services – the functions – provided by urban greening. It starts with the favourable impacts on the abiotic component, the microclimate, with a focus on the cooling effect and the related human thermal sensation and thermal comfort (Chap. 4). The air temperature amelioration and other functions are connected with the wholesome benefits of urban vegetation on physical and mental health as well as high-order social values (Chap. 5). Readers are taken to a special mode of urban greening, community gardens, which can play a handsome basket of social, cultural and economic roles (Chap. 6). The vista is opened to the important contributions of urban greening to biodiversity with respect to the enabling and suppressing factors and the strategies for its enhancement (Chap. 7). This is followed by the practice of urban agriculture, offering a broad complement of services but encountering some disservices and challenges (Chap. 8). The multiple ecosystem services of the heterogeneous modes and expressions of urban greening can be knitted together and elucidated in a condensed rendition to enhance understanding and identify research directions (Chap. 9).

In Part III, the multiple forms of urban green spaces are explored. Water in cities in tandem with greenery offer a unifying theme and tool to bring benefits in the form of the combined blue-and-green infrastructure. It can be designed to provide sustainable stormwater management by regulating water storage, treatment and release, to be discharged at prescribed outlets at predetermined flow rate (Chap. 10). Many compact cities with inadequate ground-level green space could extend the greening efforts to the largely bare rooftops to contribute to ecosystem services and amenities space (Chap. 11). The apparently disparate spatial patterns and benefits of the urban green infrastructure deserve a cohesive assessment. The usually fragmented green space patches, embedded within the ecologically-deprived urban matrix, could be connected by habitat corridors to form a spatially and functionally cohesive ecologically network (Chap. 12). A unique cohort of urban doyens, heritage trees, is studied in relation to their cultural-cum-historical significance, keystone ecological roles, together with recommendations for their protection and conservation (Chap. 13). A special member of the urban ecosystem is expounded and expanded by examining the multivariate functions and factors of urban woodland existence, and the innovative methods for their conservation and creation (Chap. 14). For cities endowed with a waterfront, efforts can be directed to optimizing the planning and development regime to bring benefits to both nature and people through creating and rehabilitating the use of the land and its adjacent waters (Chap. 15). The concluding remarks sum up the contributions of this monograph to the state-of-art of this beloved multidisciplinary subject matter with universal relevance and appeal (Chap. 16).

1.3 Synopsis of the Chapters

In Chap. 1 Introduction, Chi Yung Jim and Puay Yok Tan evaluate the green city imperative by tracking the recent development hotspots of the multidisciplinary field and contrasting them with the past research scopes. The objectives and organization of the book are then elucidated, followed by synopsis of the individual chapters.

In Chap. 2 Perspectives of Greening of Cities through an Ecological Lens, Puay Yok Tan evaluates the importance of adopting ecological principles in the design, planning and management of urban green spaces (UGS). The transfer of ecological knowledge to cities has evolved through three phases. It began with *ecology in cities* which studies operation of traditional ecological processes in natural analogue habitats in urban areas. The subsequent shift to *ecology of cities* recognizes the independence and uniqueness of the urban ecosystem. The latest episode *ecology for cities* employs the relevant knowledge and skills to resolve and refine multiple UGS challenges. Specific ecological principles and strategies have been elucidated with a view to translating them into practices to meet high-order and discerning expectations.

In Chap. 3 Imperatives for Greening Cities: A Historical Perspective, Yuanqiu Feng and Puay Yok Tan survey the evolution of the green city idea and ideal as part and parcel of human cultural advancement in conjunction with urban development history. Ranging from the spiritual to the pragmatic and from the tangible to the intangible, the multiple motivations for humans to insert and maintain greenery in dwellings and settlements have been articulated in details. The stimuli and impetuses to nurture vegetation in cities have stood the test of time, with shared commonality across geographical and temporal divides. The co-existence of nature and city, well rooted in antiquity, has remained highly relevant and become more earnestly sought and necessary in response to the rising tide of urbanization and urban densification.

In Chap. 4 Urban Greening and Microclimate Modification, Evyatar Erell focuses on the important utility of vegetation in moulding a wide spectrum of climatic parameters under the shadow of the built environment. From the energy balance viewpoint, the intricate interactions between vegetation parameters and air temperature have been deciphered. A functional classification of urban vegetation provides the basis to assess and understand differential thermal benefits across urban land cover types. The mitigation of the urban heat island effect has been analysed in conjunction with human thermal comfort in the outdoor milieu based on well-established indices. The analysis offers the conceptual bedrock to inform different modes of integrating greenery into the urban fabric in the course of urban planning and design to achieve microclimatic benefits.

In Chap. 5 Urban Greening and its Role in Fostering Human Well-being, Christine A. Vogt, Cybil Kho and Angelia Sia provide a comprehensive review of the health benefits of urban vegetation. A historical survey tracks the provision of public green space in response to the plight of factory workers who had to live in abysmal urban dwellings and neighbourhoods deprived of access to salubrious nature. The pioneering institution of urban parks, as a novel municipal amenity initiated in the 1830s in Europe and America, would thereafter be adopted by most cities around the world to become a *de rigueur* of urban planning. Besides improvement to physical and mental health, relevant public policies could encompass the high-level and multiplier of social benefits.

In Chap. 6 Urban Community Gardens as Multimodal Social Spaces, Jeffrey Hou explains the participation of citizens in the urban greening endeavour in the form of community gardens. The origin of urban gardening has been traced to provide the historical perspective. Instead of passive receivers of urban green spaces provided by the government, people can create and relish in the collective cultivated space which can play multiple social, cultural and economic and roles. They could serve specifically and synergistically as convivial space, cultural space, inclusive space, restorative space, democratic space, and resilient space. By active engagement of citizens, community gardens could offer a new dimension to urban greening, and deserves to be incorporated into associated policies, plans and strategies.

In Chap. 7 Urban Green and Biodiversity, Peter Werner and John G. Kelcey attempt a broad assessment of the varied contributions of urban greening to biodiversity. The highly variable combinations of biogeographical and ecological conditions for the existence of flora and fauna in cities could account for the notable intra- and inter-urban spatial variations. Dynamic modifications in the urban fabric through time would contribute to continual ecological changes and divergence. The attitudes and actions of citizens and governments can influence the ratio between native and non-native species. Analysing the biotic patterns at three scales, namely city in the region, urban matrix and green patches, furnishes the framework to the understanding of underlying relationships, and to inform tailor-made management aiming at preservation and enhancement of urban biodiversity.

In Chap. 8 Urban Agriculture as a Productive Green Infrastructure for Environmental and Social Well-being, Brenda B. Lin, Stacy M. Philpott, Shalene Jha and Heidi Liere explore the expressions of urban agriculture in diverse forms and patterns. The multiple benefits could extend from the environmental to ecological and social domains, with implications for urban sustainability and liveability. Expansion of this unique sector of productive urban greenery has been stifled by various challenges, including keen competition for limited urban land resource, and shortage of sites with suitable environmental conditions for crop growth. Measures should be taken to guard against disservices of urban farmlands, such as spreading pests, pathogens and weeds into natural areas, providing vector breeding grounds for human diseases, and inducing gentrification which may displace low-income groups.

In Chap. 9 Urban Nature and Urban Ecosystem Services, Wendy Y. Chen shifts our attention to the important natural functions offered by our green companions in the built environment. The heterogeneity of urban habitats, including remnant natural, ruderal or artificial, could nurture a rich diversity of flora and fauna in cities to rival the biodiversity of the extra-urban realm. Besides studies of their inherent biology and ecology, the broad range of ecosystem services has increasingly been explored. They include biophysical, economic and social benefits operating under anthropogenic influences. The diverse services, organized in a classification scheme, can be subsumed under provisioning, regulating, cultural and supporting categories. Recognition of disservices and identification of knowledge gaps and challenges can enhance understanding and provide a framework for future research.

In Chap. 10 Blue-Green Infrastructure: New Frontier for Sustainable Urban Stormwater Management, Kuei-Hsien Liao, Shinuo Deng, and Puay Yok Tan concentrate on a special benefit, stormwater management, brought by urban green and blue ingredients. The quantity and quality of water can be modified and regulated through various biophysical processes, including detention, storage, infiltration and biological uptake of pollutants. They can be expressed in diverse forms such as rain garden, bioswale, constructed wetland, retention and detention basin and green roof. Some pertinent ecosystem services are provided by such facilities, notably flood risk mitigation, water quality treatment, thermal reduction, and urban biodiversity enhancement. Different cities adopt mitigation measures in response to a plethora of push and pull factors. The successful implementation in four cities, namely Portland, New York City, Singapore and Zhenjiang, illustrates the best management practices for urban stormwater management.

In Chap. 11 Highrise Greenery: Ancient Invention with New Lease of Life, Chi Yung Jim takes us to the rooftops of buildings which often remain bare and devoid of nature. This wasted resource can provide ample chances to usher nature into dense built-up areas which otherwise are deficient in ground-level green space. The ancient origin of green roof as human response to unforgiving climate offers the prototype to inspire modern highrise greening design. Exemplary vegetated roofs in classical, pre-industrial and industrial periods present pioneering attempts to innovate in tandem with scientific and cultural advances. The technological and material revamp in the 1960s in Germany laid the foundation of modern green roof which has since been liberally adopted in many cities. Despite the recent overhaul, the basic principles and functions have remained similar. Future developments could focus on tailor-made, cost-effective and environmentally-friendly dimensions.

In Chap. 12 Urban Ecological Networks for Biodiversity Conservation in Cities, Abdul Rahim Hamid and Puay Yok Tan investigate the issue of habitat connectivity in cities and application to urban biodiversity enhancement. Analysing the concept of ecological network from the multidisciplinary perspective provides the basis to understand the patterns and distribution of urban green spaces, and their ability to support diverse ecosystems. The built-up matrix and constituent biogeographical barriers, often enveloping the isolated natural pockets, present obstacles and friction to biotic movement within urban areas and with the surrounding countryside. A comprehensive package of methods has been proposed to design ecological networks and monitor their performance, with the help of landscape metrics and indicators. Mapping and characterizing the spatial structure of the green infrastructure marks the initial step towards an improved green-space configuration in the context of spatial ecological planning.

In Chap. 13 Urban Heritage Trees: Natural-cultural Significance Informing Management and Conservation, Chi Yung Jim considers a special component of the urban forest, the respected heritage trees which often conjure up scientific as well as sentimental rejoinders. Large, old, magnificent and sturdy trees would invariably draw the attention of people, who may elevate the appreciation to the level of respect, adoration and veneration. Some 60 epithets have been harvested from the literature to symbolize their cultural significance by different peoples through the ages. As keystone organisms, they provide critical and often unique ecological services to numerous companion creatures. Assessing their economic value could reinforce the conservation message and support. Understanding their survival needs and eclectic contributions in the natural and cultural dimensions could provide the basis to chart an assured protection and conservation regime.

In Chap. 14 Conservation and Creation of Urban Woodlands, Chi Yung Jim inspects the ecological, environmental, social and cultural significance of woodlands embedded in built-up areas from historical to modern times. As the most complex ecosystems in cities usually with high nature content and biodiversity, they are often widely disturbed or threatened yet earnestly wanted by both people and wild life. Protection of existing woodlands could begin with thorough understanding of their intricate ecology, especially the factors and processes associated with their progression, stagnation or decline. Sustainable management would depend on respecting, facilitating or recreating natural conditions. Degraded woodlands could benefit from specific restoration and afforestation measures attending to both organism and soil components. New woodlands can be proactively created on suitable green, brown and grey (rooftop) fields to expand their coverage and services using innovative approaches.

In Chap. 15 Urban Waterfront Revivals of the Future, Swinal Samant and Robert Brears examine the distinctive city-water interface, looking at development and revitalization strategies to provide purpose and identity that are specific if not unique to individual cities. The keenly contested waterfront land for different uses presents obstacles, challenging the quest for a well-balanced and sensitive governance. The ecological approach calls for integrated water-edge planning in conjunction with the green and blue infrastructure and protection of critical natural landforms and ecosystems. The waterfront zone devoted to urban-park use can be linked to the city-wide green space network to optimise its accessibility, outdoor recreational utility and social functions. Besides drawing examples from different cities, the cases of HafenCity of Hamburg and Waterfront Toronto are evaluated to illustrate the successful implementation of innovative and unique measures.

References

Artmann M, Breuste J (2015) Cities built for and by residents: Soil sealing management in the eyes of urban dwellers in Germany. J Urban Planning and Development 141(3). doi:10.1061/(ASCE)UP.1943-5444.0000252

Bertram C, Larondelle N (2017) Going to the woods is going home: recreational benefits of a larger urban forest site: a travel cost analysis for Berlin, Germany. Ecol Econ 132:255–263

Brown RD, Vanos J, Kenny N, Lenzholzer (2015) Designing urban parks that ameliorate the effects of climate change. Landscape Urban Plann 138:118–131

Caspersen OH, Olafsson A (2010) Recreational mapping and planning for enlargement of the green structure in greater Copenhagen. Urban Forest Urban Greening 9:101–112

Cleary A, Fielding KS, Bell SL, Murray Z, Roiko A (2017) Exploring potential mechanisms involved in the relationship between eudaimonic wellbeing and nature connection. Landscape Urban Plann 158:119–128

Connop S, Vandergert P, Eisenberg B, Collier MJ, Nash C, Clough J, Newport D (2016) Renaturing cities using a regionally-focused biodiversity-ledmultifunctional benefits approach to urban green infrastructure. Environ Sci Policy 62:99–111

De la Barrera F, Reyes-Paecke S, Harris J, Bascunán D, Farías JM (2016) People's perception influences on the use of green spaces insocio-economically differentiated neighborhoods. Urban Forest Urban Greening 20:254–264

District of Columbia (2014) Green infrastructure standards. Department of Transportation, Washington, DC

European Commission (2015) Nature-based solutions & re-naturing cities. Final report of the Horizon 2020 Expert Group on nature-based solutions & re-naturing cities. Publications Office of the European Union, Luxembourg

Ignatieva M, Stewart GH, Meurk C (2011) Planning and design of ecological networks in urban areas. Landscape Ecol Eng 7:17–25

Jim CY, Chan MWH (2016) Urban greenspace delivery in Hong Kong: spatial-institutional limitations and solutions. Urban Forest Urban Greening 18:65–85

Jim CY, Chen SS (2003) Comprehensive greenspace planning based on landscape ecology principles in compact Nanjing city, China. Landscape Urban Plann 65:95–116

Landschaft Planen & Bauen (1990) The Biotope Area Factor as an ecological parameter: Principles for its determination and identification of the target. Senate Department for Urban Development and the Environment, Berlin

Quétier F, Lavorel S (2011) Assessing ecological equivalence in biodiversity offset schemes: key issues and solutions. Biol Conserv 144:2991–2999

Secretariat of the Convention on Biological Diversity (2012) Cities and biodiversity outlook. Montreal

Simmons BL, Hallett RA, Sonti NF, Auyeung DSN, Lu JWT (2016) Long-term outcomes of forest restoration in an urban park. Restor Ecol 24:109–118

Southon GE, Jorgensen A, Dunnett N, Hoyle H, Evans KL (2017) Biodiverse perennial meadows have aesthetic value and increase residents' perceptions of site quality in urban green-space. Landscape Urban Plann 158:105–118

Tan PY, Samsudin R (2017) Effects of spatial scale on assessment of spatial equity of urban park provision. Landscape Urban Plann 158:139–154

Part I
Urban Greening as a Component of Urban Development

Chapter 2
Perspectives on Greening of Cities Through an Ecological Lens

Puay Yok Tan

Abstract The increasing focus on urban green spaces (UGS) leads to them becoming an important component of the physical makeup of cities. However, it is useful to be mindful that UGS implementation competes for precious land resources in cities, incur carbon and energy footprint, and can have long payback periods for net benefits to be achieved. The net benefits provided by UGS are thus not assured by its mere presence; functional benefits need to be achieved through deliberate design. In particular, it is suggested here that combining design with an understanding of urban ecological knowledge is a useful approach to increase the ecological functions of UGS. A conceptual model using coupled human-ecological function is described to explain how increasing ecological functions of UGS to reduce resource consumption, restore ecological processes and functions, and reduce waste generation can shift the coupled human-ecological function for both humans and the environment. Four principles distilled from conceptual advances in urban ecology and landscape ecology are proposed as a means to bridge scientific knowledge and UGS implementation through design: (1) spatial patterns of UGS across different scales influence the ecological functions of UGS; (2) heterogeneity of UGS determines its resilience to disturbances; (3) urban ecosystems are dynamic; and (4) ecological processes remain important in cities. More application-focused strategies are in turn, derived from these principles, and how these can be applied to UGS are highlighted. It is also suggested that while current scientific knowledge still limits the application of ecological principles in many aspects of UGS design and management, the increasing emphasis on UGS in cities provides good learning opportunities for scientists, practitioners and policy makers to work in concert to enhance the ecological functions of UGS.

Keywords Urban greening · Urban ecology · Ecological function · Ecological design · Ecological principles · Coupled human-ecological function

P.Y. Tan (✉)
National University of Singapore, Singapore, Singapore
e-mail: akitpy@nus.edu.sg

P.Y. Tan and C.Y. Jim (eds.), *Greening Cities: Forms and Functions*,
Advances in 21st Century Human Settlements,
DOI 10.1007/978-981-10-4113-6_2

2.1 From Being Green to Becoming Ecological

The effort of cities to introduce more urban green spaces (UGS) can be seen in numerous strategic and sustainable development initiatives of cities worldwide. Such a focus on greening cities can also be observed in the physical increases in green spaces within cities over the past few decades. For instance, Zhao et al. (2013) reported that more than 90% of 286 Chinese cities in their study increased in average green space coverage from 17 to 27% between 1989 and 2009. The authors suggested that this could in part, be attributed to the increasing economic wealth experienced in many cities over this period. Kabisch and Haase (2013) in their assessment of 202 European cities also reported that in Western and Central Europe, there was a marked increase in urban green spaces between 2000 and 2006. In these European cities, urban green spaces[1] occupy almost 15–30% of the city area, depending on urban data set used. Between 1991 and 2006, 12 out of 13 cities in England in the study of Dallimer et al. (2011) also had increased green spaces, with green space occupying an average of 24% of the city area. In the American rust belt, shrinking cities also experience increasing number of vacant lots, some of which have become woodlands due to spontaneous regrowth, and others are proposed to be incorporated into green network in the cities (Burkholder 2012; Frazier and Bagchi-Sen 2015). These studies demonstrated two key points: that green spaces respond to urban policies (such as densification, open space provision, and brownfield development) and economic forces, and are hence dynamic. In addition, as shown over a large sample of cities, UGS can occupy as much as 30% of the city area.

Especially in the context of dense, high-population densities with land constraints, land occupied by such green spaces are not insignificant compared to other land uses. In Singapore, for example, total amount of green spaces dedicated to parks, nature reserves, street planting verges, rooftop greenery, and interstitial green spaces between buildings in residential estates and commercial land together occupy about 30% of land in Singapore. This is more than twice the amount of planned residential land which houses 83% of the total population (about 4.6 million), and two and half times the amount of land set aside for industry and commerce.[2] In Hong Kong, land for open space which includes park and garden, playfield, pavilion, etc. constitutes about 60%[3] of total land area set aside for residential use, which houses about 7 million people.

The comparison of land uses pointed out here is not to suggest that land set aside for urban green spaces is disproportionately higher or wasteful compared to other

[1]Note that 'urban green space' in this study was defined as green spaces ≥ 25 ha, i.e., excluding green open spaces between buildings, sports and recreational facilities.

[2]Data from Concept Plan Review 2000 and Tan et al. (2013). Urban green spaces considered exclude green field sites zone for other land uses.

[3]Source of information: Land Use Utilization (2014) from Planning Department, Hong Kong Special Administrative Region.

land uses, nor to suggest that urban green spaces should always have an instrumental value. Rather, it is to highlight that uptake of UGS is not insignificant compared to other land uses, and therefore, what are the benefits that can be derived from the land dedicated for UGS is a responsible question to be asked by urban planners and designers. After all, all UGS, regardless of forms and scale, consumes resources for construction and maintenance, i.e., there is a carbon and energy footprint associated with their provision. It is also necessary too to consider that urban greenery implementations as shown in life cycle assessment, can have long payback periods of 20–30 years even under optimally designed conditions, such as in the case of green walls (Pan and Chu 2016) and water sensitive urban design elements (De Sousa et al. 2012; Flynn and Traver 2013). Other forms of UGS which have not been designed for specific functions could even have longer payback periods. In cities which undergo short real estate redevelopment cycles, urban greenery installations may not even exist long enough for payback period to be reached and net benefits of the installations to be achieved. Yet, it is common to see in cities implementation of UGS that have seemingly failed to consider the costs aspects of installation and maintenance, and large scale installations that have even questionable aesthetics benefits. An example is shown in Fig. 2.1 (see also examples of poorly designed UGS in Jim and Chan 2016). Quigley (2011), in

Fig. 2.1 A towering green wall covering the stairwell of a residential building in Singapore (*left*). However, while ambitious as a design feature, the creepers on trellis can reduce daylight and ventilation in the stairwell, and the green wall is only visible on exterior of the building. The visual attractiveness is questionable, ecological functions are limited, and maintainability of the system is low (*right*) (*Photo credit* P.Y. Tan)

referring to the potential of urban landscapes to function as habitats if appropriately designed, criticised many urban landscapes as 'Potemkin gardens', which are created for 'two-dimensional visual effect', and which lack 'connectivity, function or self-sustainability'. Lim and Lu (2016) also remarked that for the large capital investment that has been put in for the 60 completed water sensitive urban design projects in Singapore for the past decade, efforts should be directed at realizing the hydrological and ecological potential of such projects, as otherwise, many of these are mere 'gardening efforts' around urban infrastructures. These accounts highlight an observation made more than thirty years ago that 'nature has been seen as a superficial embellishment, as a luxury, rather than an essential force which permeates the city' (Spirn 1984: 5).

The key point emphasized here is that due to land uptake, energy and material costs of UGS, making urban areas visually green is useful but inadequate; greening should be a means to deliver more functional benefits for human well-being and environmental quality. One relevant perspective suggested here is that urban greening should become more 'ecological'. What does being more ecological entail and what are the broad strategies that could guide the design and implementation of UGS to achieve this objectives? Given that UGS takes many forms, from city-scale green infrastructure, to community scale neighbourhood green spaces, and to building scale installations of green roofs or green walls, these strategies should also necessarily consider the effects of scale and specificity of sites. Using key concepts of urban ecology and landscape ecology that have shaped our understanding of cities as socio-ecological systems, this chapter describes a conceptual model to frame the notion of greening as a means to increase ecological functions in cities and the key approaches for achieving this aim.

2.2 Greening to Increase Ecological Functions in Cities: A Conceptual Model

Making urban greening more 'ecological' as used here simply denotes deriving more ecological functions from UGS through deliberate design. There are two parts to this statement. The first on ecological functions draws reference to the rapidly expanding literature that UGS is capable of providing ecosystem services, or functional benefits through natural ecological processes (Bolund and Hunhammar 1999; Elmqvist et al. 2015) for urban liveability and resilience. The second relates to the argument that functional benefits of UGS need to be deliberately designed, using the notion of design as 'an intentional change of landscape pattern for the purpose of sustainably providing ecosystem services while recognizably meeting societal needs and respecting societal values' (Nassauer and Opdam 2008). That is to say, benefits of greenery should be targeted rather than incidental. The overarching perspective adopted here is that all over the world, land use and land cover changes during urbanization drastically transform natural ecosystems and ecological processes occurring therein, as well as outside the boundaries of cities (McDonnell and MacGregor-Fors 2016). UGS then, as argued here, is a principal medium to restore natural ecological flows and functions which have not disappeared from urban landscapes, but have simply become highly altered or impaired.

Key examples of such changes are the highly altered flow of water, nutrients, and energy, as well as reduction of habitats for biodiversity that occur during urbanization. The range of ecological processes associated with the disturbances to these ecological functions is listed in Table 2.1. Perhaps one of the most striking ways in which ecological flows have undergone dramatic transformations in the course of urbanization is how humans have manipulated water supply, waste water treatment, stormwater drainage in urban areas, and the appropriation of water resources to support expanding urban demands (Forman 2014: 84). Indeed, in the urban history of urban settlements, such advances in urban water management have served human needs well, and have contributed remarkably to improving sanitary conditions in cities, which is considered as one of the most important medical advances for human health in our urban history (Larsen et al. 2016). However, there is now growing concern that largely because of resource depletion, pollution and climate variations, ensuring a safe and sustainable water supply for large urban regions of the world is increasingly become a global challenge (Padowski and Gorelick 2014; Vörösmarty et al. 2000). Inevitably, water scarcity also begins to threaten food security for large populations in both developed and developing countries (Hanjra and Qureshi 2010). In other words, while humans have through innovations, developed technologies and appropriated natural resources to satisfy our existential needs and well-being, and are seemingly detached from natural ecosystems, in reality, we are still fundamentally dependent on natural resources and ecological processes for our well-being in the long-term. This is well-illustrated in the concept of Daly Triangle (described in (Wu 2013), which essentially suggests that the pursuit of human well-being cannot be achieved without safeguarding natural ecological processes and functions as the earth's life-support systems. The integrity of natural ecosystems, or natural capital, is the ultimate means on which human well-being has to be built upon, whereas technological, economic and other human and societal pursuits that lead to accumulation of built and financial capital can only serve as intermediate means for this purpose.

Another useful conceptual model that illustrates the interdependence of human and ecological functions is the conceptual model proposed by Alberti (2008a). The model highlights the flawed notion of decoupling (and substitutability) between 'human functions' (human inventions, built capital and resource appropriation to satisfy human existential needs and well-being), and ecological functions. Human and ecological functions, as also embedded on the Daly Triangle, are interdependent and mutually reinforcing. As such, beyond a threshold or tipping point, the impairment of ecological functions with continuing urbanization based on a business-as-usual mode will lead to human functions being depressed as well (Fig. 2.2). Alberti further suggested that the state of this coupled human-environmental functions, as determined by urban forms and

Table 2.1 Ecological functions and processes which are influenced by urban landscapes

Socio-ecological functions	Impact of urbanization on ecological processes	Potential of urban green spaces to reduce impacts of urbanization
Ecological functions		
Primary production—*capacity of urban areas to produce photosynthate and accumulate biomass*	? Net primary production ? Carbon storage and sequestration	? Low potential for carbon sequestration (Pataki et al. 2011)
Hydrologic function—*capacity of urban areas to maintain natural hydrological cycle through interception, infiltration, evapotranspiration*	↑ Runoff ↑ Nutrients and pollutants discharge ↑ Erosion/sedimentation ↑ Increased water temperature ↑ Flashiness of runoff	High potential stormwater mitigation (Larsen et al. 2016; Pataki et al. 2011) High potential for water-quality improvement (Land et al. 2016; Larsen et al. 2016; Pataki et al. 2011) ? ? High potential for stormwater mitigation (Larsen et al. 2016; Pataki et al. 2011)
Nutrient cycling—*capacity of urban areas to retain, allow for biological uptake and accumulation, and recycling through decomposition or inorganic processes*	↑ Nutrient runoff and ↓ nutrient uptake ↑ Eutrophication ↑ Soil NO_3 and nitrification ? Trophic interactions ? Ecosystem metabolism	High potential for water-quality improvement (Land et al. 2016; Larsen et al. 2016; Pataki et al. 2011); ? ? ? ?
Habitat provision and biodiversity—*capacity of urban areas to support biodiversity, which is in turn the foundation of other ecological functions*	↓ Habitat diversity ↓ Species diversity ↓ Native species ↑ Tolerant/generalist species ↑ Invasives	Medium potential for habitat provision, as local and regional processes are important, and adequate habitat area is required for successful conservation (Beninde et al. 2015; Faeth et al. 2001; Strohbach et al. 2013)

(continued)

Table 2.1 (continued)

Socio-ecological functions	Impact of urbanization on ecological processes	Potential of urban green spaces to reduce impacts of urbanization
Disturbance regulation—*capacity of urban areas to withstand disturbances such as climate change, weather extremes, invasives, flooding, erosion, insect/pathogen outbreak*	↑ Weather variability	?
	↑ Flooding	?
	↑ Drought	?
	↑ Landslides	?
	↑ Heat stress	High potential for local cooling (Coutts et al. 2013; Davis et al. 2016; Pataki et al. 2011; Santamouris 2015)

Information presented are mainly extracted from review papers that synthesise current understanding. Ecological processes and urbanization impacts are extracted from (Alberti 2010): ↑↓ indicate direction of changes in functions, and ? indicates uncertainties in current knowledge. Potential of UGS to moderate changes in ecological processes: low, medium and high indicates potential contribution as used in (Pataki et al. 2011)

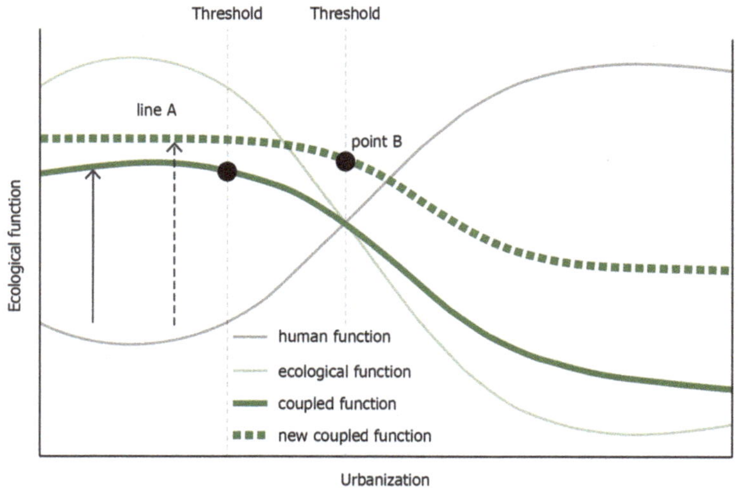

Fig. 2.2 As urbanization increases, ecological functions are reduced due to urbanization impacts on the environment, whereas human functions (built and financial capital, human well-being) increases. In reality, human and ecological functions are coupled and human well-being cannot be sustained beyond a threshold at which ecological functions are degraded. Focusing on ecological functions of UGS can push the coupled function to provide higher ecological and human functions (to *line A*), and extend the threshold of failure of ecological functions to *point B*. Numerous human-ecological functions are theoretically possible. Adapted from Alberti (2008a)

patterns, can also be instrumental in influencing urban resilience to external disturbances such as climate change (Alberti 2008a).

We highlight a role for UGS in this model: it is proposed here that UGS should serve as a medium to enhance ecological functions in cities. Enhancing ecological functions in cities could lead to two possible outcomes in this model. Firstly, UGS implementation can be directed to reduce resource consumption within cities, reduce resource appropriation from cities' hinterlands, restore ecological processes and functions in cities, and reduce waste generation (such as nutrients and pollutants). This reduces the rate of decline in ecological functions with urbanization, within and outside cities, and at the same time, through benefits conferred to urban dwellers, shifts the coupled human-ecological function to a more positive state (line A in Fig. 2.2). These cumulative effects could then increase the threshold beyond which human and ecological functions are impaired and become irreversibly transformed (to point B). In this way, the ability of urban systems to tolerate disturbances or stressors arising from impaired or disrupted ecological functions, a component of urban resilience, can also be enhanced. Viewed in this sense, greening of cities extends its role beyond creating a visually pleasing urban environment; greening should be viewed as an essential means to enhance both human functions and reduce environmental impacts.[4] Specific ways in which the ecological and human functions interact, and influence of socio-economic and environmental factors external to cities' environment would then mean that there will be a range of shapes and positions of line A, and where point B falls in this coupled function.

Is there evidence for such roles of UGS? This can gleaned from an increasing amount of studies on urban ecosystem services provided by UGS, which provide an emerging consensus of the types of ecological functions that are particularly valuable in urban areas. Several key examples are provided in Table 2.1, which summarizes the current evidence that UGS can mitigate changes to ecological processes and functions associated with urbanization. While not all types of ecological functions can be realistically or impactfully restored by UGS, current evidence suggests that localized heat mitigation, stormwater mitigation, stormwater quality improvement, and habitat provision are primary areas which UGC can make contributions to reduce our reliance on energy, conserve and protect water resources, and support biodiversity. Elmqvist et al. (2015) also recently demonstrated that enhancing ecosystem services in urban areas is not just ecologically and socially desirable, but is also an economically viable option. While this chapter only focuses on roles of UGS in enhancing ecological functions, it should also be noted that there is a large body of evidence highlighting that UGS has important roles in the socio-cultural aspects of urban living, particularly for individual and community well-being, social capital and sense of place. Aspects of these functions are discussed in Chaps. 5, 6, 8, and 13.

2.3 From Ecological Knowledge to Action: Approaches to Increase Ecological Functions of UGS

What approaches are useful to increase ecological functions in cities? As highlighted in earlier in this chapter, highly functional UGS requires knowledge to be used in deliberate design in professional practice. This is especially important in the context of an 'action gap' in translating knowledge from science into action in professional practice for the design of the urban environment (Wang et al. 2014; Opdam et al. 2013). Given that UGS is an integral and often sizeable part of physical make-up of cities, and our understanding of cities has been increasingly shaped by the burgeoning field of urban ecology and urban landscape ecology, it is logical to focus the approaches and narrow the action gap based on our understanding from these disciplines. Indeed, there is a growing call to use urban ecological knowledge to improve urban conditions, be it for urban liveability

[4]Understandably, UGS is important as an essential, but not sole means for this purpose. We have also suggested that focus on urban systems need to extend beyond the social and ecological components to be also directed to the built components of cities (buildings, roads and other infrastructures), given that the built components are usually dominant in cities (see Tan and Abdul Hamid 2014).

(McDonnell and MacGregor-Fors 2016), urban sustainability (Wu 2014), or urban resilience (McPhearson et al. 2016). Urban ecology and its allied disciplines are increasingly seen as a holistic foundation for the science of, and basis for pushing knowledge-to-action in urban systems (Childers et al. 2015; McPhearson et al. 2016).

It is also useful to note that urban ecology is in itself, a young and evolving discipline that will need to increasingly incorporate other disciplines, so that new methodologies, approaches, and concepts from other disciplines can be used to address urban challenges (McDonnell 2012). As an evolving discipline, the emphasis of urban ecology has also shifted over the past two decades. This shift has frequently been described as a shift from *ecology in cities* to *ecology of cities*. The former refers to studies on ecological conditions of green areas as analogues of wild or rural ecosystems occurring in urban areas and how organisms and biogeochemical processes are affected by urban conditions. The latter emphasizes cities as hybrid ecosystems with tightly coupled social and ecological components, and in which the understanding of human as an agent of, and the responders to changes occurring within the urban ecosystem is a principal means to manage cities as complex, adaptive ecosystems (Grimm et al. 2000; McPhearson et al. 2016;). More recently, Childers et al. (2015) described the need for urban ecological studies to build an *ecology for cities* emphasis, underscored by the need to use knowledge from urban ecology and its allied disciplines in urban planning, urban design, governance and engineering, etc., for a more action-oriented and transdisciplinary approach to meet urban challenges.

What do all the conceptual developments, and advances in tools and approaches mean for the aim of increasing the ecological functions in cities through UGS? How can gains in our knowledge of urban ecological systems be put to use in the design, implementation and management of UGS? These questions are not new. There has been an on-going effort to link ecological science to design (for example, see Hwang et al. 2016; Johnson and Hill 2002; Lovell and Johnston 2009; Nassauer and Opdam 2008; Pickett and Cadenasso 2008). This chapter draws upon insights from these works for specific application to UGS. In particular, a useful entry point for this effort is the synthesis of ecological principles from our understanding of the nature of cities through an ecosystem perspective. Ecological principles, as used here refer to basic assumptions about urban ecosystems, how they function and how these premises can be used to guide landscape planning and design. Principles are useful as they help to translate our conceptual understanding of complex systems into concise statements for the framing of important issues. Principles are also useful to provide general insights for a developmental pathway towards an ideal or desired condition, and are analogous to a compass for direction finding under different circumstances (Luederitz et al. 2013).

Various scholars have synthesized ecological principles that can be applied to the design of the urban environment. For instance, Flores et al. (1998) proposed five principles based on 'content', 'context', 'dynamics', 'heterogeneity' and 'hierarchies' for the management of green spaces of the New York City Metropolitan Area. Dale et al. (2000) under the Ecological Society of America's Committee on

Land Use proposed ecological principles based on themes of 'time', 'species', 'place', 'disturbance' and 'landscape' and through these, recommended guidelines as 'practical rules of thumb' for making decisions on land uses. Zipperer et al. (2000) compared the ecological principles of Flores et al. (1998) and Dale et al. (2000) and added the principle concerning 'connectivity' to highlight the need for functional connection between green patches, corridor and the landscape matrix in urbanizing landscapes. Subsequent efforts by Cadenasso and Pickett (2008), Wu (2008) and Pickett and Cadenasso (2013) added new perspectives on effects of urban forms and patterns, interacting social and ecological processes, and the role of urban design for experimentation.

Unfortunately, it is also clear from our review of the literature that confirmed generalizations on functioning of urban ecological systems are scant, as highlighted by McDonnell and Hahs (2013). In particular, specific rules based on ecological principles for direct application in design are also rare. Nevertheless, it is suggested here that principles provide 'the best bet' for linking science with design. The following section builds upon the earlier syntheses by other scholars, and translate the understanding of these principles into strategies focused on the design, implementation and management of UGS to enhance their ecological functions.

2.4 Ecological Principles and Strategies for UGS

This section describes the principles, applicable strategies and implications for UGS implementation and management. It is acknowledged here too that there needs to be a continuing effort to keep pace with the increasing knowledge from science and translate these into practice through other specific strategies and tactics. Table 2.2 lists the four principles and eight strategies focused on UGS.

Table 2.2 Principles and strategies for UGS implementation to enhance ecological functions

Principles	Strategies
Spatial patterns of UGS across different scales influence the ecological functions of UGS	(a) At the city or regional scale, conceive and implement a network of UGS that that can guide the development and management of UGS in the long-term (b) Consider scale in association with landscape pattern for UGS implementation
Heterogeneity of UGS determines its resilience to disturbances	(a) Maintain a diversity of UGS forms in cities to enhance landscape heterogeneity (b) Increase species diversity and functional diversity in UGS
Urban ecosystems are dynamic	(a) Design UGS to accommodate, not resist, changes brought about by natural growth processes
Ecological processes remain important in cities	(a) Design UGS to restore ecological flows (b) Layer ecological functions to achieve multi-functionality in UGS (c) Integrate ecological functions with built infrastructures

2.4.1 Principle 1—Spatial Patterns of UGS Across Different Scales Influence the Ecological Functions of UGS

This first principle relates to a key tenet of landscape ecology, that landscape patterns (configuration and composition) affect landscape processes and functions (Turner 1989). The distribution, size and shape of green patches govern the interactions between patches in the form of flow of materials and energy flow between UGS, and between UGS and its surrounding matrix. Alberti (2005) further extended this understanding to urban ecosystems, and cogently illustrated the evidence on the influence of urban patterns on urban hydrology, primary productivity, biodiversity, nutrient and materials cycle, etc. For instance, the pattern of UGS, both in terms of shape of individual green patches of UGS (e.g. edge to interior ratio), and extent of physical and functional proximity (e.g. connection of green patches by riparian corridors) can influence the ability of green patches to support biodiversity (Flores et al. 1998; Dramstad et al. 1996). Recent studies on cooling ability of UGS are also beginning to reveal the influence of distribution and configuration of UGS as cool islands (Kong et al. 2014; Chen et al. 2014). UGS patches have been shown to confer different levels of cooling depending on the shape, size, density of patches and vegetation composition of the green area. Patterns of UGS thus influence ecological functions. Unfortunately, the state of scientific knowledge does not allow confirmed generalizations to be drawn on cooling, and other operational relationships between patterns and a range of ecological functions important in urban areas. For instance, we still do not know from the limited literature, if a single large green space, or several small distributed green spaces are more effective for urban cooling, and how this relationship is influenced by built forms and climatic conditions of cities. There is nevertheless a range of guidelines developed from first principles, such as those synthesized by Dramstad et al. (1996) that could be also be tested in urban landscapes. The pattern-process-function principle also dictates that scale is important and that it is therefore desirable to have a multi-scaled view of UGS, from an overarching view at city scale, to regional and local considerations of patterns of UGS. Two strategies are suggested for application of this principle:

(a) At the city or regional scale, conceive and implement a network of UGS that that can guide the development and management of UGS in the long-term. Such a UGS network should emphasize functional connectivity between UGS patches, and not just focus on structural or physical connectivity. Functional connectivity refers to the level of spatial connectedness which allow for realized movement of organisms (Auffret et al. 2015). The underlying premise is that a functionally connected network of UGS is likely to support more ecological functions and tolerate disturbances than highly fragmented patches. For the support of biodiversity for instance, Abdul Hamid and Tan described in Chap. 12, a city-scale ecological network for biodiversity conservation in Singapore that takes into account current distribution and state of fragmentation of green spaces, habitat quality, and species dispersal requirements. The value of such a network is that it highlights parts of the network that are valuable for conservation, parts that are threatened with future developments and hence requiring specific landscape and urban design interventions, and parts which are currently isolated but which could be connected by greening of the urban matrix or restoration of degraded sites. A network identified at this scale thus provides a long-term guide for current and future land developments. On top of biodiversity enhancement, such a network can also be functionally differentiated with overlays of other uses such as recreation and urban hydrological management. This overview provides valuable information at a coarse scale that can guide other future urban developments.

(b) Consider effects of scale on ecological functions for UGS implementation.
The design of UGS to deliver enhanced ecological functions has to consider the relationships between scale of implementation and ecological processes driving those functions, as the relationship between pattern and process is scale dependent (Wu 2008). For instance, while the role of UGS in mitigating the Urban Heat Island effect is well-recognized, the overall benefits are highly dependent on relative amount of vegetated surfaces that confer cooling benefits, and built surfaces that retain and re-radiate heat. This relationship changes with scale. While a large tree next to a low-rise building may provide cooling benefits through shading or evaporative cooling at this microscale level, this benefit is reduced when the tree, or a row of trees are placed within a large development area with high-rise and high-density building towers. At the mesoscale, it should be expected that the effects of anthropogenic heat sources, wind-patterns and other aspects of regional climate also become important in the overall heat balance. In the area of blue-green infrastructure management (Chap. 10), scale considerations are particularly important. In Singapore, there are numerous installations of bioretention systems or rain gardens implemented as part of national efforts improve stormwater quality and reduce flood risk through water sensitive urban design. However, many of such UGS are installed over small areas within a development, in which treated stormwater in one sector is subsequently mixed with upstream, or downstream untreated stormwater collected in conventional drainage systems, leading to questionable overall net benefit of stormwater management (Fig. 2.3). Such installations fail to consider that urban stormwater management has to be tackled at watershed scale (Walsh et al. 2016), and not at the scale of a small green space or planted verge along a street. Designing without understanding scale effects on ecological functions can lead to counterproductive and wasteful use of resources.

Fig. 2.3 A rain garden designed to treat stormwater runoff from the service road of a residential estate in Singapore (marked in *white lines*). Note however that the treatment is highly localized; and runoff from upstream and downstream in the same road, as well as runoff from other paved areas in this precinct are not treated. Overall net benefits for stormwater management at the precinct scale can be questioned

2.4.2 Principle 2—Heterogeneity of UGS Determines Its Resilience to Disturbances

Landscape heterogeneity is the spatial patchiness of landscape patterns, both in configuration and composition of UGS patches. The influence of landscape heterogeneity on a range of ecosystem functions in natural ecosystems is well known. Landscape heterogeneity for instance, influences species presence and abundance, species composition, biotic and abiotic interactions, etc. (Chapin et al. 2011; Turner et al. 2013). A principal influence of heterogeneity highlighted by Flores et al. (1998) is that heterogeneity of green spaces helps to maintain greater species diversity, which is in turn a key feature that enables continued functioning of ecosystem during environmental changes. This strategy should also guide the planning and management of UGS.

(a) Maintain a diversity of UGS forms in cities to enhance landscape heterogeneity. UGS in cities can exist in different forms, from managed spaces such as parks, street planting verge, rooftop greenery, green walls, to natural or unmanaged spaces such as nature reserves, remnant primary and secondary forested areas, young regrowth woodlands, scrublands, etc. In relatively young and still urbanizing cities such as Singapore and Melbourne, continuing land developments on remnant native vegetation will slowly tilt the composition of UGS in the city to be dominated by managed forms of landscapes over unmanaged forms of landscapes (Tan, 2016: 190). This change in landscape type has implications which are often not recognized: for instance, the ability of urban landscapes to support biodiversity is reduced, as remnant native vegetation is critical to support urban biodiversity (Chong et al. 2014) and the extent of its conservation is known to be strong predictor of extinction rate of extant species of cities (Hahs et al. 2009). As Tan et al. (2016) also argued, there are also a range of ecological and socio-cultural values that are lost when secondary forests are developed as such values are often ignored in land use planning. As cities lose more of its native vegetation, and more managed forms of UGS appear in urban areas, there is a risk that UGS become dominated by urban landscapes of similar forms and similar floristic composition (Qian et al. 2016), leading to an emergence of 'landscape homogenization' with continued urbanization. Landscape homogenization, in limiting the ability of UGS to act as habitats for a wide range of biodiversity could then be a factor among other urban drivers that lead to the phenomenon of biotic homogenization, which show up through large similarities in species composition observed in cities (McKinney 2006). The strategy of enhancing urban landscape heterogeneity encourages conservation of valuable remnant native vegetation, which can be incorporated into a city or regional scale network of UGS described earlier, and also calls for increasing structural diversity of vegetation in the design of UGS (see also Chap. 7), and increasing plant diversity (point below).
(b) Increase species diversity and functional diversity in UGS
High species diversity is an insurance against changing environmental conditions (Alberti 2008b; Flores et al. 1998) and should be encouraged in UGS implementation. It has also been highlighted recently that increasing functional diversity, which is the diversity of plant functional traits, rather than species richness of plants alone, is key for increasing ecological functions (Beck 2013: 118). Plant selection for UGS should thus focus on enhancing species diversity and functional traits in aspects such as plant habit, height, leaf area, physiology and resource needs, etc.

One example of increasing landscape heterogeneity in urban areas is the use of tiered and cluster planting of street verges with high species variety, in contrast with conventional monoculture planting of trees at uniform spacing (Fig. 2.4). Another example is to recognise the value of spontaneous vegetation (Fig. 2.5) and urban woodlands (Chap. 13), and allow for their establishment in the built environment. These should have a recognized role in enhancing landscape heterogeneity, as well

Fig. 2.4 An example of cluster planting along a roadside planting verge in Singapore, with tree canopy, shrubbery and ground cover. Such a type of cluster planting creates a new aesthetics for streetscape, and provides refuge for small animals, and with the right plants, supports biodiversity (*Photo credit* P.Y. Tan)

Fig. 2.5 Contrast between the conventional turfed areas under trees on the *left*, and spontaneously generated area on the *right* when mowing was stopped in a landscaped area in Singapore. In addition to creating visual interest, the patch on the right supports a much higher level of biodiversity (*Photo credit* P.Y. Tan)

as for their role to provide a range of ecological functions (Robinson and Lundholm 2012).

2.4.3 Principle 3—Urban Ecosystems Are Dynamic

Natural ecosystems are constantly responding and adjusting to past and current changes in environmental conditions (Chapin et al. 2011). In urban ecosystems, in addition to environmental factors, social factors act as additional drivers that make urban ecosystems highly dynamic (Flores et al. 1998). For instance, the extent of UGS in a city, as highlighted earlier in the chapter, respond to policy changes. The composition of UGS, for instance, the vegetation in a park, also undergo natural successional processes, are subjected to environmental stresses of drought and temperature changes, and are influenced by societal factors such as level and quality of horticulture care and aesthetic preferences. The dynamism experienced simply reflect the basic fact that ecological processes continue to take place in managed urban ecosystems (see Principle 4), and UGS should be designed to work with, not against changes.

(a) Design UGS to accommodate, not resist, changes brought about by natural growth processes.
It is common in urban areas to see urban vegetation implemented that requires high level of maintenance. Extreme examples are lawns and playing fields that require high level of irrigation and fertilization to maintain them in conditions acceptable for their purpose. In such examples, the energy and carbon costs associated with maintenance operations are accepted as societal or economic decisions in return for the utilitarian values provided, such as for recreation. There are however, also many instances of UGS implementation in which maintenance is high but both utilitarian and ecological functions are low as a consequence of failure to accommodate the growing needs of plants. An example is the prevalent use of shrubbery on highly restricted spaces in roadsides that subsequently require regular pruning to prevent traffic obstruction, and yet which do not serve apparent ecological functions (Fig. 2.6). Another prevalent example is the creation of 'instant landscapes', in which a high planting density is used to create landscapes that look mature as required by the landscape industry. A usual consequence is that at very high plant density, thinning operations have to be carried regularly, and plants simply do not grow to their full potential because of space constraint and undue competition. Beck (2013: 92) suggested that constructed ecosystems should be allowed to 'self-design' by accommodating natural ecological processes and leveraging on the capacity of ecosystems to self-adapt. In so doing, it will be expected that energy input to maintain the system will be lower. A direct application of a self-designing strategy is to create planting schemes that allow successional processes to occur, and employ strategies that make use of plant

Fig. 2.6 Shrub planting in narrow planting verges in the centre divider of roads are common in cities: Singapore (*top-left*), Hong Kong (*top-right*), Bangkok (*bottom-left*), Tokyo (*bottom-right*). Apart from adding some greenness to the urban landscape, ecological functions are limited in such UGS. On the other hand, as these are often just abutting the carriageway, frequent trimming is needed, incurring high energy and carbon footprint in the process. The picture on *bottom-left* shows one worker trimming, and another collecting the prunings (*Photo credit* P.Y. Tan)

functional traits in the process, such as nitrogen fixing, species that support avifauna, and forest pioneer species to provide the suitable microclimate for other species to establish. In such a way, landscapes can be shaped dynamically over time (Hwang et al. 2016) and level of maintenance interventions can be reduced.

2.4.4 Principle 4—Ecological Processes Remain Important in Cities

Just as a natural ecosystem is defined by interactions between state factors comprising climate, organisms, topography, parent material, time, humans (Amundson and Jenny 1997), an urban ecosystem can also be defined by groups of state factors, which Tan and Abdul Hamid (2014) organized into natural environmental factors (urban climate, urban biogeochemistry, urban soils, urban biodiversity), built environment factors (buildings, infrastructure, telecommunications), and socio-economic factors (human communities, institutions, heritage, economy). Interactions between state factors are fundamentally mediated by the flows of energy and materials (water, nutrients), and in the case of urban ecosystems, information as well as goods and services. Although not visible to all, ecological processes remain important in urban ecosystems (Flores et al. 1998; Pickett et al. 2011). For instance, when a green field site is replaced with impervious surfaces during land development, the original natural hydrological processes of infiltration, runoff, storage and evapotranspiration have not disappeared, but have merely changed in relative importance. With an increased amount of impervious surfaces, there is reduced infiltration, storage and evapotranspiration, but increased runoff. Concomitant with such alterations in hydrological flows are changes in nutrient flows and energy fluxes due to reduction in latent heat and increase in sensible heat. Thus, ecological flows have not disappeared, but have merely been altered, often to the detriment of human and ecological health.

(a) Design UGS to restore ecological flows
UGS can be seen as a medium to restore ecological flows. The growing acceptance of green roofs, for instance, is partly driven by their use as a means to moderate stormwater discharge through increasing detention of stormwater in urban areas (Locatelli et al. 2014). The emergence of new concepts in urban water management, shown by the adoption of terms such as water sensitive urban design, low impact development, sustainable urban drainage system (Fletcher et al. 2014), are also premised on mimicking of natural hydrological flows through ecologically engineered systems such as bioretention systems, rain gardens, and constructed wetlands. Opportunities should also exist to use UGS to mediate the flow of nutrients in urban areas, although this aspect is far less developed compared to management of water cycle.

(b) Layer ecological functions to achieve multi-functionality in UGS
Just as ecological flows in nature are coupled, such as between energy and water, a useful goal of UGS design is to enable different types of ecological flows to be usefully connected. A clear example is the nexus between water and energy fluxes. While many urban developments incorporate bioretention systems, rain gardens, or constructed wetlands for stormwater management functions, few systems link the objective of management of stormwater to urban climate modifications by enhancing the coupling between water and energy fluxes. The evaporative cooling by plants is driven by the conversion of incoming solar radiation to latent heat during evapotranspiration, but this process is highly dependent on water availability to plants. As bioretention systems, rain gardens, and constructed wetlands are designed to preferentially channel stormwater, through appropriate design, the water can also be channelled to irrigate adjacent UGS and thereby promoting evaporative cooling (Coutts et al. 2013). At the same time, health of UGS can be maintained in periods of water scarcity. Forman's question (2014: 139) 'couldn't creative thinking [through design] and technology capitalize on the considerable rainwater and built surfaces to produce combined solutions for the urban heat, stormwater and water body problems?' in fact also points to the same quest for

multi-functional solutions incorporating UGS to be designed. Additionally, species composition in bioretention systems, rain gardens, and constructed wetlands can be used to support a range of urban biodiversity, such as macroinvertebrates, herpes and avifauna. It is also obvious that although the discussion in this chapter has focused on ecological functions, bioretention systems, rain gardens, and constructed wetlands can also be designed to support a range of socio-cultural functions. Viewed from the perspective of land scarcity in cities, layering of functions to derive multi-functional uses of UGS is also a means to optimize land use efficiency and reduce land for grey infrastructure needs (such as conventional drainage infrastructures).

(c) Integrate ecological functions with built infrastructures
Cities are urban ecosystems dominated by built and social components. Much as there is a high potential and multiple avenues for UGS through a range of ecological and socio-ecological functions to enhance urban conditions, it should be recognized that unless significant progress is also made in the way the built environment is constructed and human consumption patterns moderated, the ecological footprint of cities will continue to exert a strain on natural ecosystems to support human's needs. In other words, although UGS can function as a medium for ecological functions to be introduced to cities, the extent of ecological functions provided is ultimately limited by the amount of vegetated spaces versus the built surfaces in the city. As Tan and Abdul Hamid (2014) highlighted, the dominance of the built component should be seen as an opportunity for design intervention by understanding ecological flows in cities. An example of such intervention is to require buildings to compensate for loss of stormwater detention and infiltration compared to the pre-development state through ecologically engineered stormwater detention tanks and infiltration tanks. In the push for greening of buildings through green roofs and green walls, design should also focus on restoring the pre-development state of evaporative cooling through the use of vegetated surfaces to increase evapotranspiration, and reduce heat intake and storage on building walls. Through appropriate design, there is good potential for UGS and buildings to work in concert to restore hydrological, nutrient and energy flows in urban areas (Tan 2013: 185).

2.5 Conclusion

That UGS is an essential infrastructure in cities is well-accepted, and indeed, it currently forms to varying degrees a key part of the physical makeup of cities. However, it is useful to recognize that the value of UGS cannot be determined by its mere presence, as poor design often means that UGS is either not well-used by humans, or UGS incur higher carbon and energy footprint than the ecological benefits that it should provide. The burgeoning field of urban ecology provides useful conceptual understanding of the nature of cities as urban ecosystems, and principles for enhancing the ecological functions of cities. Similar perspectives can be applied to UGS through such an ecological lens. In this chapter, a role of UGS to enhance ecological functions through deliberate design was suggested and explained using a conceptual model. Four principles selected from the literature for translating urban ecological knowledge into strategies were described. Admittedly, this list of principles and strategies is not exhaustive, and neither is this attempt meant to be exhaustive for a short book chapter of this nature. It is an effort to highlight that urban ecological science can inform the design and implementation of UGS. Importantly too, a current major weakness in the field is that there are inadequate confirmed generalizations that relate to many aspects of UGS to their ecological functions, such as relationships between pattern (configuration, composition, distribution) and heat mitigation, biodiversity support, nutrient recycling, etc. However, the growing emphasis in urban greening worldwide, as many chapters in this book highlight, provides numerous opportunities for scientists, practitioners and policy makers to work in concert to improve the ecological functions of UGS for the betterment of urban conditions.

References

Alberti M (2005) The effects of urban patterns on ecosystem function. Int Regional Sci Rev 28:168–192

Alberti M (2008a) Futures of urban ecosystems. In: Advances in urban ecology: integrating humans and ecological processes in urban ecosystems. Springer, Boston, pp 225–250. doi:10.1007/978-0-387-75510-6_9

Alberti M (2008b) Population and community dynamics. In: Advances in urban ecology: integrating humans and ecological processes in urban ecosystems. Springer, Boston, pp 197–223. doi:10.1007/978-0-387-75510-6_8

Alberti M (2010) Maintaining ecological integrity and sustaining ecosystem function in urban areas. Curr Opin Environ Sustain 2:178–184

Amundson R, Jenny H (1997) On a state factor model of ecosystems. Bioscience 47:536–543

Auffret AG, Plue J, Cousins SAO (2015) The spatial and temporal components of functional connectivity in fragmented landscapes. Ambio 44(1):51–59. doi:10.1007/s13280-014-0588-6

Beck T (2013) Principles of ecological landscape design. Island Press, Washington, DC

Beninde J, Veith M, Hochkirch A (2015) Biodiversity in cities needs space: a meta-analysis of factors determining intra-urban biodiversity variation. Ecol Lett 18(6):581–592. doi:10.1111/ele.12427

Bolund P, Hunhammar S (1999) Ecosystem services in urban areas. Ecol Econ 29(2):293–301. doi:10.1016/s0921-8009(99)00013-0

Burkholder S (2012) The new ecology of vacancy: rethinking land use in shrinking cities. Sustainability 4(6):1154–1172. doi:10.3390/su4061154

Cadenasso ML, Pickett STA (2008) Urban principles for ecological landscape design and management: scientific fundamentals. Cities Environ 1(2):1–16

Chapin FS, Matson PA, Vitousek PM (2011) Landscape heterogeneity and ecosystem dynamics. In: Principles of terrestrial ecosystem ecology. Springer, New York, pp 369–397. doi:10.1007/978-1-4419-9504-9_13

Chen A, Yao XA, Sun R, Chen L (2014) Effect of urban green patterns on surface urban cool islands and its seasonal variations. Urban Forest Urban Greening 13(4):646–654. doi:10.1016/j.ufug.2014.07.006

Childers DL, Cadenasso ML, Morgan Grove J, Marshall V, McGrath B, Pickett STA (2015) An ecology for cities: a transformational nexus of design and ecology to advance climate change resilience and urban sustainability. Sustainability 7(4):3774–3791. doi:10.3390/su7043774

Chong KY, Teo S, Kurukulasuriya B, Chung YF, Rajathurai S, Tan HTW (2014) Not all green is as good: different effects of the natural and cultivated components of urban vegetation on bird and butterfly diversity. Biol Conserv 171:299–309. doi:10.1016/j.biocon.2014.01.037

Coutts AM, Tapper NJ, Beringer J, Loughnan M, Demuzere M (2013) Watering our cities: the capacity for water sensitive urban design to support urban cooling and improve human thermal comfort in the Australian context. Prog Phys Geogr 37(1):2–28

Dale VH, Brown S, Haueber RA, Hobbs NT, Huntly N, Naiman RJ, Risbsame WE, Turner MG, Valone TJ (2000) Ecological principles and guidelines for managing the use of land: an ESA report. Ecol Appl 10:639–670

Dallimer M, Tang Z, Bibby PR, Brindley P, Gaston KJ, Davies ZG (2011) Temporal changes in greenspace in a highly urbanized region. Biol Lett 7(5):763–766

Davis AY, Jung J, Pijanowski BC, Minor ES (2016) Combined vegetation volume and 'greenness' affect urban air temperature. Appl Geogr 71:106–114. doi:10.1016/j.apgeog.2016.04.010

De Sousa MRC, Montalto FA, Spatari S (2012) Using life cycle assessment to evaluate green and grey combined sewer overflow control strategies. J Ind Ecol 16(6):901–913. doi:10.1111/j.1530-9290.2012.00534.x

Dramstad WE, Olson JD, Forman RTT (1996) Landscape ecology principles in landscape architecture and land-use planning. Island Press, Washington, DC

Elmqvist T, Setälä H, Handel SN, van der Ploeg S, Aronson J, Blignaut JN, Gómez-Baggethun E, Nowak DJ, Kronenberg J, de Groot R (2015) Benefits of restoring ecosystem services in urban areas. Curr Opin Environ Sustain 14:101–108. doi:10.1016/j.cosust.2015.05.001

Faeth SH, Saari S, Bang C (2001) Urban biodiversity: patterns, processes and implications for conservation. In: eLS. Wiley, New York. doi:10.1002/9780470015902.a0023572

Fletcher TD, Shuster W, Hunt WF, Ashley R, Butler D, Arthur S, Trowsdale S, Barraud S, Semadeni-Davies A, Bertrand-Krajewski J-L, Mikkelsen PS, Rivard G, Uhl M, Dagenais D, Viklander M (2014) SUDS, LID, BMPs, WSUD and more—the evolution and application of terminology surrounding urban drainage. Urban Water J 1–18. doi:10.1080/1573062X.2014.916314

Flores A, Pickett STA, Zipperer WC, Pouyat RV, Pirani R (1998) Adopting a modern ecological view of the metropolitan landscape: the case of a greenspace system for the New York City region. Landscape Urban Plann 39(4):295–308. doi:10.1016/S0169-2046(97)00084-4

Flynn KM, Traver RG (2013) Green infrastructure life cycle assessment: a bio-infiltration case study. Ecol Eng 55:9–22. doi:10.1016/j.ecoleng.2013.01.004

Forman RTT (2014) Urban ecology—science of cities. Camridge University Press, Cambridge

Frazier AE, Bagchi-Sen S (2015) Developing open space networks in shrinking cities. Appl Geogr 59:1–9. doi:10.1016/j.apgeog.2015.02.010

Grimm NB, Grove JM, Pickett STA, Redman CL (2000) Integrated approaches to long-term studies of urban ecological systems. Bioscience 50(7):571–584

Hahs AK, McDonnell MJ, McCarthy MA, Vesk PA, Corlett RT, Norton BA, Clemants SE, Duncan RP, Thompson K, Schwartz MW, Williams NSG (2009) A global synthesis of plant extinction rates in urban areas. Ecol Lett 12(11):1165–1173. doi:10.1111/j.1461-0248.2009.01372.x

Hanjra MA, Qureshi ME (2010) Global water crisis and future food security in an era of climate change. Food Policy 35(5):365–377. doi:10.1016/j.foodpol.2010.05.006

Hwang YH, Feng YQ, Tan PY (2016) Managing deforestation in a tropical compact city (part B): urban ecological approaches to landscape design. Smart Sustain Built Environ 5(1):73–92. doi:10.1108/SASBE-08-2015-0023

Jim CY, Chan MWH (2016) Urban greenspace delivery in Hong Kong: Spatial-institutional limitations and solutions. Urban Forest Urban Greening 18:65–85. doi:10.1016/j.ufug.2016.03.015

Johnson B, Hill K (2002) Ecology and design: frameworks for learning. Island Press, Washington, DC

Kabisch N, Haase D (2013) Green spaces of European cities revisited for 1990–2006. Landscape Urban Plann 110:113–122. doi:10.1016/j.landurbplan.2012.10.017

Kong F, Yin H, James P, Hutyra LR, He HS (2014) Effects of spatial pattern of greenspace on urban cooling in a large metropolitan area of eastern China. Landscape Urban Plann 128:35–47. doi:10.1016/j.landurbplan.2014.04.018

Land M, Granéli W, Grimvall A, Hoffmann CC, Mitsch WJ, Tonderski KS, Verhoeven JTA (2016) How effective are created or restored freshwater wetlands for nitrogen and phosphorus removal? A systematic review. Environ Evid 5 (1). doi:10.1186/s13750-016-0060-0

Larsen TA, Hoffmann S, Lüthi C, Truffer B, Maurer M (2016) Emerging solutions to the water challenges of an urbanizing world. Science 352(6288):928–933. doi:10.1126/science.aad8641

Lim HS, Lu XX (2016) Sustainable urban stormwater management in the tropics: an evaluation of Singapore's ABC Waters Prorgam. J Hydrol 538:842–862

Locatelli L, Mark O, Mikkelsen PS, Arnbjerg-Nielsen K, Bergen Jensen M, Binning PJ (2014) Modelling of green roof hydrological performance for urban drainage applications. J Hydrol 519(Part D):3237–3248. doi:10.1016/j.jhydrol.2014.10.030

Lovell ST, Johnston DM (2009) Designing landscapes for performance based on merging principles in landscape ecology. Ecol Soc 14(1):44

Luederitz C, Lang DJ, Von Wehrden H (2013) A systematic review of guiding principles for sustainable urban neighborhood development. Landscape Urban Plann 118:40–52. doi:10.1016/j.landurbplan.2013.06.002

McDonnell M, Hahs A (2013) The future of urban biodiversity research: moving beyond the 'low-hanging fruit'. Urban Ecosyst 16(3):397–409. doi:10.1007/s11252-013-0315-2

McDonnell MJ (2012) The history of urban ecology: an ecologist's perspective. In: Niemala J, Breuste J, Elmqvist T, Guntenspergen G, James P, McIntyre NE (eds) Urban ecology: patterns, processes, and applications. Oxford University Press, Oxford

McDonnell MJ, MacGregor-Fors I (2016) The ecological future of cities. Science 352(6288):936–938. doi:10.1126/science.aaf3630

McKinney ML (2006) Urbanization as a major cause of biotic homogenization. Biol Conserv 127 (3):247–260. doi:10.1016/j.biocon.2005.09.005

McPhearson T, Pickett STA, Grimm NB, Niemelä J, Alberti M, Elmqvist T, Weber C, Haase D, Breuste J, Qureshi S (2016) Advancing urban ecology toward a science of cities. Bioscience 66 (3):198–212. doi:10.1093/biosci/biw002

Nassauer JI, Opdam P (2008) Design in science: extending the landscape ecology paradigm. Landscape Ecol 23(6):633–644. doi:10.1007/s10980-008-9226-7

Opdam P, Nassauer JI, Wang Z, Albert C, Bentrup G, Castella J-C, McAlpine C, Liu J, Sheppard S, Swaffield S (2013) Science for action at the local landscape scale. Landscape Ecol 28(8):1439–1445. doi:10.1007/s10980-013-9925-6

Padowski JC, Gorelick SM (2014) Global analysis of urban surface water supply vulnerability. Environ Res Lett 9(10). doi:10.1088/1748-9326/9/10/104004

Pan L, Chu LM (2016) Energy saving potential and life cycle environmental impacts of a vertical greenery system in Hong Kong: a case study. Build Environ 96:293–300. doi:10.1016/j.buildenv.2015.06.033

Pataki DE, Carreiro MM, Cherrier J, Grulke NE, Jennings V, Pincetl S, Pouyat RV, Whitlow TH, Zipperer WC (2011) Coupling biogeochemical cycles in urban environments: ecosystem services, green solutions, and misconceptions. Front Ecol Environ 9(1):27–36. doi:10.1890/090220

Pickett STA, Cadenasso ML (2008) Linking ecological and built components of urban mosaics: an open cycle of ecological design. J Ecol 96(1):8–12

Pickett STA, Cadenasso ML (2013) Urban ecology. In: Meyers RA (ed) Ecological systems: selected entries from the encyclopedia of sustainability science and technology. Springer Science + Business Media, New York, pp 273–300

Pickett STA, Cadenasso ML, Grove JM, Boone CG, Groffman PM, Irwin E, Kaushal SS, Marshall V, McGrath BP, Nilon CH, Pouyat RV, Szlavecz K, Troy A, Warren P (2011) Urban ecological systems: scientific foundations and a decade of progress. J Environ Manage 92 (3):331–362

Qian S, Qi M, Huang L, Zhao L, Lin D, Yang Y (2016) Biotic homogenization of China's urban greening: a meta-analysis on woody species. Urban Forest Urban Greening 18:25–33. doi:10.1016/j.ufug.2016.05.002

Quigley MF (2011) Potemkin gardens: biodiversity in small designed landscapes. In: Niemala J, Breuste JH, Elmqvist T, Guntenspergen G, James P, McIntyre NE (eds) Urban ecology: patterns, processes, and applications. Oxford University Press, New York, p 274

Robinson SL, Lundholm JT (2012) Ecosystem services provided by urban spontaneous vegetation. Urban Ecosyst 15(3):545–557. doi:10.1007/s11252-012-0225-8

Santamouris M (2015) Regulating the damaged thermostat of the cities—status, impacts and mitigation challenges. Energy Build 91:43–56. doi:10.1016/j.enbuild.2015.01.027

Spirn AW (1984) The granite garden—urban nature and human design. Basic Books, New York

Strohbach MW, Lerman SB, Warren PS (2013) Are small greening areas enhancing bird diversity? Insights from community-driven greening projects in Boston. Landscape Urban Plann 114:69–79. doi:10.1016/j.landurbplan.2013.02.007

Tan PY (2013) Singapore a vertical garden city. Straits Times Press, Singapore

Tan PY (2016) Greening Singapore: past successes, emerging challenges. In: Heng CK (ed) Fifty years of urban planning in Singapore. World Scientific, Singapore, pp 177–195

Tan PY, Abdul Hamid ARB (2014) Urban ecological research in Singapore and its relevance to the advancement of urban ecology and sustainability. Landscape Urban Plann 125:271–289. doi:10.1016/j.landurbplan.2014.01.019

Tan PY, Feng YQ, Hwang YH (2016) Deforestation in a tropical compact city (Part A): understanding its socio-ecological impacts. Smart Sustain Built Environ 5(1):47–72. doi:10.1108/SASBE-08-2015-0022

Tan PY, Wang J, Sia A (2013) Perspectives on five decades of the urban greening of Singapore. Cities 32:24–32

Turner MG (1989) Landscape ecology: the effect of pattern on process. Annu Rev Ecol Syst 20 (1):171–197. doi:10.1146/annurev.es.20.110189.001131

Turner MG, Donato DC, Romme WH (2013) Consequences of spatial heterogeneity for ecosystem services in changing forest landscapes: priorities for future research. Landscape Ecol 28 (6):1081–1097. doi:10.1007/s10980-012-9741-4

Vörösmarty CJ, Green P, Salisbury J, Lammers RB (2000) Global water resources: vulnerability from climate change and population growth. Science 289(5477):284–288. doi:10.1126/science.289.5477.284

Walsh CJ, Booth DB, Burns MJ, Fletcher TD, Hale RL, Hoang LN, Livingston G, Rippy MA, Roy AH, Scoggins M, Wallace A (2016) Principles for urban stormwater management to protect stream ecosystems. Freshwater Sci 35(1):398–411. doi:10.1086/685284

Wang Z, Tan PY, Zhang T, Nassauer JI (2014) Perspectives on narrowing the action gap between landscape science and metropolitan governance: practice in the US and China. Landscape Urban Plann 125:329–334. doi:10.1016/j.landurbplan.2014.01.024

Wu J (2008) Toward a landscape ecology of cities: beyond buildings, trees, and urban forests. In: Carreiro MM, Song Y-C, Wu J (eds) Ecology, planning, and management of urban forests: international perspectives. Springer, New York, pp 10–28. doi:10.1007/978-0-387-71425-7_2

Wu J (2013) Landscape sustainability science: ecosystem services and human well-being in changing landscapes. Landscape Ecol 28(6):999–1023. doi:10.1007/s10980-013-9894-9

Wu J (2014) Urban ecology and sustainability: the state-of-the-science and future directions. Landscape Urban Plann 125:209–221. doi:10.1016/j.landurbplan.2014.01.018

Zhao J, Chen S, Jiang B, Ren Y, Wang H, Vause J, Yu H (2013) Temporal trend of green space coverage in China and its relationship with urbanization over the last two decades. Sci Total Environ 442:455–465. doi:10.1016/j.scitotenv.2012.10.014

Zipperer WC, Wu J, Pouyat RV, Pickett STA (2000) The application of ecological principles to urban and urbanizing landscapes. Ecol Appl 10(3):685–688

Chapter 3
Imperatives for Greening Cities: A Historical Perspective

Yuanqiu Feng and Puay Yok Tan

Abstract Although co-ordinated and comprehensive efforts at citywide greening had only emerged in the last two centuries, green spaces and vegetation have been a vital part of urban history, across periods and locales. Urban dwellers have inserted and maintained urban greenery for a wide variety of reasons. Some of these imperatives are persistent throughout time, while others have emerged with new knowledge and societal developments. More often than not, multiple motivations are embedded in urban greening projects. Based on a review of existing literature spanning multiple disciplines, we provide an overview of the range of reasons why urban dwellers have embarked on greening projects in their environments. We identify nine themes that continue to be highly relevant today, and provide brief historical perspective on each. By extending the discussion of urban greening beyond its potential to meet contemporary challenges, we hope to provide a more long-ranging view of why greenery has been incorporated in cities.

Keywords Urban greening · Aesthetics · Recreation · Well-being · Community bonding · Social hierarchy · Economic value

3.1 Introduction

The mistaken view that city and nature are polarized objects, that nature is the opposite of the city, and that nature exists only outside cities, has left an indelible mark in the way humans have shaped the urban environment over the history of urban settlements. Such a view has contributed to conflicting characterization of

Y. Feng (✉) · P.Y. Tan
National University of Singapore, Singapore, Singapore
e-mail: akify@nus.edu.sg

P.Y. Tan
e-mail: akitpy@nus.edu.sg

P.Y. Tan and C.Y. Jim (eds.), *Greening Cities: Forms and Functions*,
Advances in 21st Century Human Settlements,
DOI 10.1007/978-981-10-4113-6_3

cities, which are portrayed as safe havens from a dangerous and uncontrollable nature, as corrupting and unhealthy environments drawing populations away from a moral and healthful countryside, and as blight upon a fragile nature. Increasingly however, it is recognised that while cities are human-dominated environments, shaped by numerous social forces that produce structures and spaces for living and working, they are rarely, if ever, devoid of natural elements. Thinkers, reformers, planners and residents constantly advocate and bring natural features and objects into urban environments, most prominently and visibly in the form of urban vegetation and green open spaces. From private home gardens and potted plants to public parks and green roofs and green walls, urban dwellers have found various motivations for introducing and maintaining vegetation in their surroundings.

The motivations for creating and maintaining vegetation in cities are diverse and interconnected. Some motivations could be rational and utilitarian, for example, urban farming to meet requirements for food and commodities, or urban greening to mitigate urban temperature. Some are social in nature: green spaces have been created as objects indicating social hierarchies and relations, and greening movements have been initiated to express various social values and beliefs. Other motivations are perhaps less explicable, traceable to a deeper, instinctive affinity for nature. Depending on the context and the agents involved, urban green spaces are also designed with more than one dominant purpose in mind.

We contend that in order for the greening of cities to be entrenched in urban development as not just being a desirable aim, but as an essential need, urban greening must be economically feasible, ecologically functional, and above all, satisfy diverse human needs. Such needs pertain not just to improving cities' physical conditions for human well-being, but also the spirituality and aspirations of diverse, multi-cultural groups that cities attract. In current discourse on urban greening, there is substantial emphasis on economic and ecological considerations. Economic considerations are important as urban greening can represent large investments of time and resources. Creating green spaces in urban environments can be costly given the value of urban lands, and that biotic and abiotic conditions in urban areas are often unfavourable to support plant life. Results are rarely instantaneous, and green spaces can take long periods to develop and mature. By studying the effects of existing and established urban green spaces, research in the last few decades have systematically identified and quantified their benefits, functions and values, often through cost-benefit analyses. There is also increasing attention the ecological potential of urban greening, for instance as highlighted in Chaps. 2 and 9.

However, urban greening need not adopt a wholly instrumental objective justified by net benefits over costs. Indeed, acknowledging that urban greening can be motivated by a plurality of motivations is to simply recognize that humans are intimately connected to nature in many ways, ranging from existential needs to spiritual connections. Such a dependence on nature has not disappeared, and urban development should always explore ways to forge meaningful connections between urban dwellers and their living environment through greenery. A useful starting point for this aim is to simply document the diverse reasons, from past to present, that have motivated urban dwellers to green their living environment. Looking at the past is also useful as cities exhibit path dependence that have implications on urban morphology and pattern, including urban landscapes (Hensley et al. 2014). Societal norms and decisions that defined urban forms in the past are often carried forward to the current and which influence future developmental patterns. As Boone et al. (2009) highlighted, legacy of past decisions has implications on present landscape. Examining urban history thus has the potential to 'reveal how actual decisions have been made in many varied contexts and the resulting short- and long-term implications of those decisions' (Redman 2011: 206). Being aware of past trends, drivers and motivations provide urban planners and designers with a richer basis from which the future of urban landscapes can be conceived.

This chapter explores the different reasons why urban dwellers have included and maintained various forms of green spaces in the history of urban settlements. We identify and elaborate on these motivations, in light of the roles that urban vegetation plays and the benefits they provide.

3.2 Motivations for Greening Cities

Through a review of the literature covering garden history, urban planning, philosophy and urban ecology, we identify and describe the evidence for the roles and functions of greening. The motivations for greening are then thematically organized into nine categories, namely: aesthetics and beautification, recreation and leisure, religion, spirituality and symbolism, social hierarchy and relations, social reform and community building, physical and mental well-being, food production and sustenance, ecological health and environmental sustainability, and economic value and competitive advantages. In each theme, we explain the relevance to human aspirations or well-being, and describe examples that illustrate it.

As the contents span across a wide range of periods and geographies, we briefly discuss here what we mean by 'urban' and 'urban greening'. The use of the term 'urban' is subjective and contextualized; the definition of 'urban' is specific to period and location, and to socio-political-economic realities. It is also subjected to the disciplinary lens from which the concept is discussed. Natural scientists and social scientists may define what is urban based on very different criteria. Urban can be defined based on measurable properties such as population density, energy consumption, quantity of built surface (McIntyre et al. 2008). It can also be more broadly defined based on less tangible aspects such as the presence of cultural signposts, the kinds of social relations and networks, types of labour and employment opportunities. In this chapter, we use the terms 'urban' and 'city' as a shorthand, or an approximation, for describing an area where population densities are higher and human activities more concentrated. By 'urban greening', we refer to concerted, organised or semi-organized efforts at creating and maintaining vegetated areas in urban or peri-urban environments.

3.2.1 Aesthetics and Beautification

The aesthetic function of urban vegetation is perhaps its most enduring and universally appreciated role. Plants are used to create pleasing visual compositions, add colour and seasonal variation, provide perfume and auditory effects, conceal the unsightly, among numerous other purposes. While aesthetic preferences and horticultural trends have varied dramatically across and within periods and cultures, designers, planners and residents consistently use the forms, colours, textures, and scents of plants to provide pleasure and delight in urban settings. Whether as a design material or an aesthetic object, it is widely acknowledged that urban vegetation offers sensory relief in the hardscaped city, where the dominance of buildings and infrastructure can result in oppressive and monotonous environments with adverse consequence on human health (Verbrugge and Taylor 1980). From the elaborate and intensively managed gardens of the wealthy and powerful to the pedestrian wayside tree, aesthetic considerations feature strongly in various forms and manifestations of urban greenery.

Despite the long history of using plants to beautify the human-dominated environment, major discussions on the aesthetics of nature have emerged only fairly recently in Western discourse. Eighteenth century founders of modern aesthetic theory considered nature the most exemplary objects and scenes of disinterested aesthetic appreciation. Arguing that aesthetic responses to nature are intellectually and morally purer than responses to the artificial, Immanuel Kant writes, '…the interest in the beautiful of art (…) gives no evidence at all of a mind attached to the morally good, or even inclined that way. But, on the other hand, I do maintain that to take an immediate interest in the beauty of nature (…) is always a mark of a good soul' (Kant 1911: 298). Nature thus offered the consummate aesthetic experience, unfettered by the individual's utilitarian interests. The aesthetic writings of Immanuel Kant, Edmund Burke and William Gilpin distinguished between three aesthetic dimensions of nature: the *beautiful*, the *sublime* and the *picturesque*.

Of interest to the discussion on urban green space is the notion of the *picturesque*. While the *beautiful* referred to the cultivated landscapes and gardens of Europe and the *sublime* described the awe-inspiring wilderness, the *picturesque* referred to a form of painterly beauty, 'which please from some quality capable of being illustrated in painting' (Gilpin 1792: 3). Defining qualities were asymmetry, roughness, sudden variation, and irregularity. Elaborating on the concept, Udevale Price explains that is not 'the smooth young beech nor the fresh and tender ash, but the rugged old oak or knotty wych elm that are picturesque' (Price 1810: 57). These notions formed the basis for the development of the English landscape garden, a vast departure from the then-dominant formal French style gardens. Highly influential, the naturalistic English-style landscape garden would shape the design of urban green spaces in the 18th and 19th century in Europe and North America.

Among other reform-minded objectives that we will discuss in the following sections, public officials and visionaries designed these parks to provide pleasure and relief for urban residents living in the ever-densifying city. They were conceived as a landscape to counterpoise the realities of the city, and faithfully replicated the idealized images of nature and pastoral life popularly depicted in landscape paintings of the era. Some of the best-known examples of 18th and 19th public parks such as Alphand's public parks in Paris, Olmsted's Central Park in New York and the Englischer Garten in Munich were executed based on the landscaping principles of the picturesque. Achieving the ideal aesthetic vision was a compelling reason for large investments into urban greening projects. In pursuit of the idealized 'natura' aesthetic, immense engineering efforts were often required. Constructing *Bois de Boulogne*, for example, involved five years of hydraulic works and earthworks in order to transform a flat site into an undulating landscape of hills and waterbodies (Moncan 2009). Similarly, the construction of Central Park moved over 500,000 ft^3 (1416 m^3) of soil and used more gunpowder than was fired at the battle of Gettysburg in order to remove boulders on site (Slavicek 2009).

The inseparability of beauty, human behaviour and utility was most explicitly drawn by the early 20th century City Beautiful Movement in North America (Wilson 1989). Driven by middle and upper class standards of propriety (social order, cleanliness and neatness, etc.) and aesthetic preferences for the monumental city designs in Europe as embodied by Haussmann's reconstruction of Paris, proponents believed that the beautification of cities would yield corresponding improvements in public morality and social order (Pregill and Volkman 1999). Although the movement was brief in its ascendency, important urban greening standards and approaches were pioneered by affiliated designers. For instance, Kessler pioneered a park and boulevard system for Kansas City by designating and developing the most visually outstanding natural areas as parks, and linking them with a planted boulevard system. In Washington D.C., Olmsted incorporated parks and connective systems in the McMillan plan and provided residents with naturalistic green spaces within a density built-up city. Burnham's work in Cleveland and Chicago also involved the insertion of large green areas within the dense city grid and connecting outlying forest preserves to the city with boulevard systems. In Philadelphia, the Fairmont Parkway (renamed as Benjamin Franklin Parkway) is considered one of the most notable City Beautiful Movement projects (Fig. 3.1). The City Beautiful Movement, while criticized for being overly concerned with aesthetics at the expense of meeting the needs of urban residents and failing to meet its social reform objectives, left a lasting impact on 20th century urban planning with its aesthetic ideals. From Manila to Canberra to Brasilia, sprawling public greens were incorporated within important districts for monumental effect under its influence.

The primacy of aesthetic considerations in urban greening has faded somewhat in contemporary design and planning discourse as other imperatives such as ecology, health and food sustainability rise to the fore. Nevertheless, we remain fascinated about how and why natural elements elicit aesthetic responses, and numerous philosophical, psychological and evolutionary theories have since been proposed. Scholars in design fields also continue to emphasize the importance of beauty and aesthetic quality even while designing with other agenda in mind, so as to ensure that the resultant project would be accepted by the community which may

Fig. 3.1 Benjamin Franklin Parkway (renamed from Fairmont Parkway) in Philadelphia, a product of the City Beautiful movement completed at the turn of the twentieth century (*Photo credit* P.Y. Tan)

not so readily appreciate the other services (ecological or otherwise) provided by the green space. Persistent and universal, the aesthetic value of urban greenery cannot be downplayed. In the words of Girot (2016: 13), 'it is precisely our ability to cultivate a strong poetic response to human needs and beliefs that will help us to find better expressions of nature'.

3.2.2 Recreation and Leisure

Urban green spaces are tied to their recreational and leisurely functions, from sports and games to more passive and solitary pastimes. Since antiquity, elaborate gardens, pleasure grounds and hunting preserves have adjoined the private estates of the privileged class, and served as exclusive spaces of leisure and entertainment for the wealthy and powerful. Today, parks and tree-lined promenades are favoured destinations for strolls, exercise, play and sporting activities among members of the public. Dedicated recreation grounds such as football fields and golf courses comprise significant proportions of green space in cities, and recreation remains an important consideration in the planning and design of urban green spaces.

The greening of linear spaces, in particular, has historically created important recreational and leisurely opportunities along urban infrastructure. Early examples dating from the 16th century can be found along the fortress walls of medieval cities such as Lucca in Italy (Fig. 3.2), where rows of poplars planted to fortify walls and earthworks added an attractive recreational function to the defensive structure (Crandell 2013). Tree-lined boulevards such as the Cours la Reine in 17th century Paris and Olmsted's parkways in early 20th century North America are other prominent surviving examples of greening efforts along transportation infrastructure to pleasurable riding experiences through the city. Linear green spaces remain an important typology in dense urban environments, being less land-intensive while

Fig. 3.2 The planted medieval city walls of Lucca, as seen from within (*top*) and outside (*bottom*) the old city (*Photo credit* Y. Feng)

Fig. 3.3 The daylighted Cheonggyecheon stream in Seoul has become a highly popular place for leisure in a dense city. It is a linear blue-green space (*top*) and a space for cultural events such a lantern festival (*bottom*) (*Photo credit* P.Y. Tan)

maximizing contact with the surrounding city. Some of the most notable contemporary examples of urban greening are retired urban infrastructure that have been planted and retrofitted, and presently serve important public recreation roles—Seoul's Cheonggyecheon (Fig. 3.3) and New York's Highline being two of the most oft-cited projects.

Urban green spaces dedicated to specific athletic and sporting activities have also long been ubiquitous in cities, and comprise some of its most intensively developed and maintained urban green spaces. Hunting parks preserved specifically for the recreation of the aristocratic classes have existed since Roman times. These were elaborate, heavily-managed spaces that sometimes involved importing exotic game such as elephants, giraffes and lions (Wilkinson 1988). Since the 16th century the popularity of various lawn games such as lawn bowling, pall-mall, croquet, cricket, tennis, polo, football and soccer saw the creation of numerous closely mown lawn spaces throughout cities influenced by western culture. The Pall Mall in London, as well as the Unter den Linden in Berlin, are examples of these dedicated recreation spaces for privileged classes that have since been converted into important streets after years of urban growth. The popularity of games such as golf, football and soccer today mean that significant tracts of urban land are set aside to meet public demand. Golf courses, in particular, are especially land and resource-intensive green spaces. Yet, an estimated 350 new courses are added every year to the global approximate of 25,000 (Wheeler and Nauright 2006). In Singapore, golf courses occupy more than 14 km^2 of land (Neo and Savage 2002)—a significant amount when public parks and nature reserves in the city amount to just 29 km^2 (Parks 2015).

The need to provide recreation opportunities for residents in the densely built urban environment continues to persuade cities to set aside land and transform existing spaces and structures for these purposes. Urban greening, to varying degrees of intensity and sophistication, is often an integral part of these efforts. In numerous projects, vegetation has proven to be an important transformative element that allows for functional layering and repurposing in the urban fabric.

3.2.3 Religion, Spirituality and Symbolism

Archetypes such as the ‘Tree of Life’ or the ‘Sacred Grove’ are widespread among world religions and mythologies. Green spaces and individual plants—especially trees—can possess great religious and spiritual significance. Various religions and cultures have vested symbolic meaning in plants, and urban green spaces provide places of contemplation and reflection in the bustle of the city. The creation and preservation of green spaces in cities for religious, spiritual and symbolic reasons are, unsurprisingly, a widespread practice.

The belief that divine forces may dwell in or be connected with natural features are pervasive in early civilisations and cultures. Ancient Greeks, for example, deliberately planted sacred groves around some temples and sacred precincts such as the Temple of Zeus at the city of Priene, defining the boundaries of the holy grounds (Barnett 2007). These groves were commonly dedicated to deities associated with wild nature, such as Artemis and Pan and can be interpreted as a threshold between the cultivated human world and the untamed forces of nature (Polignac 1995). The preservation and creation of sacred groves also feature prominently in Japan’s ethnic Shinto religion. A fine example is the 70-ha Meiji Jingu in the heart of Tokyo. Its forest, which consists of 120,000 planted trees, was an intensive urban greening effort by Japanese civic groups in the 1920s

Fig. 3.4 Meiji Shrine, Tokyo built in 1920 with a large planted forest more than a hundred-years old that now also serves as a refuge for local biodiversity and people (*Photo credit* P.Y. Tan)

(Meiji-Jingu 2016) (Fig. 3.4). Now a prominent attraction in the city, it was created to host the deified spirit of the Emperor Meiji and the Empress Shoken, following the Shinto belief that the Emperor was God, and that the *kami* (divine essence) resided in natural features such as trees, rocks and streams (Matsui 1996).

In Islamic traditions, royalty and nobility have created highly symbolic gardens in their palaces and private compounds. Designed in the image of paradise as described in the Qur'an, Moorish and Mughal gardens reflect Islamic cosmological order and provide a facsimile of paradise on earth. These gardens reminded devotees of the immanence of God, and the lushly planted vegetation were understood as a 'shadow of their heavenly archetypes, their beauty a radiation of God's glory on earth' (Clark 2011). Architectural elements enclosed these gardens and courtyards, creating a private and inward-looking space that shields devotees from the busy external world and provides a space for contemplation and prayer (Clark 2004). In the Benedictine Monastery of St. Gall in Switzerland, cloister gardens were similarly insular and enclosed. Scholars suggest that the cloister gardens helped to separate the monastic life of the friars from the secular world of the labourers, and eliminate distractions from their chores and prayers (Pregill and Volkman 1999).

This spiritual and meditative characteristic is also prominent in Chinese and Japanese gardens. Elements of the Chinese garden possess an inherent symbolic purpose, meant to help its occupants engage harmoniously with nature, in accordance with Taoist beliefs that human life ought to be aligned with nature. The small, labyrinth-like gardens of the Chinese literati gardens in Suzhou remain the finest examples. Through aspects such as the selection of plant materials, the placement of rocks, the design of pathways, the composition and shifting of views and the naming of the garden, gardeners weave dense layers of meanings and allusions into the design, creating a setting for self-cultivation and spiritual refinement. While similarities can be observed with Japanese gardens, the motifs and aesthetic principles since the Heian period clearly reflected a Buddhist philosophical underpinning. In contrast to Chinese gardens that still had strong social functions, shoguns and samurais dedicated their gardens to solitary reflection and meditation, favouring simplicity over ostentation (Pregill and Volkman 1999).

We feel a strong affinity with trees and attach various symbolic meanings to them. Urban green spaces and urban forests are sometimes created to serve as memorials for fallen war heroes and founding fathers. Scholars highlight the symbolic value of urban trees as representation of people, citing the way parts of the body are mapped to trees and vice versa, the use of tree-based metaphors for families, and the common practice of memorializing a departed person by planting a tree (Dwyer et al. 1991). The symbolic power of tree planting is particularly poignant in the aftermath of disasters. Memorials such as the 9/11 memorial in New York City features a 'survivor tree' surrounded by a planted urban forest. The 'survivor tree', a Callery pear tree (*Pyrus calleryana*) that was recovered from the rubble, became a 'living reminder of resilience, survival and rebirth' (911memorial.org 2016). A programme annually gives seedlings from the tree to communities that had recently gone through difficult times. Similarly, after the destruction wrought by Hurricane Katrina, organizations worked with local residents on a number of reforestation activities. In an ethnographic study, residents interviewed indicated that trees were symbols of longevity, hope and growth, among others; the act of planting trees was a symbol of their commitment to the future (Tidball 2014).

3.2.4 Social Hierarchies and Relations

Historically, the display and aesthetic appreciation of gardens and horticultural objects have been significant to the social and cultural life of cities. The expression of social status, cultural sophistication and civic propriety have also been a powerful impetus for creating and maintaining green spaces in the city. As social creations, urban green spaces reflect societal relationships and patterns through their functions, configurations, scales, distributions, and accessibility.

Until the last few centuries, a minority with social power and influence had planted and maintained the vast majority of vegetation in urban environments. Royalty, aristocracy, religious bodies, colonialists, and so on, often initiated public greening projects with some intention to symbolize and express power over the city. In the age of political absolutism in 17th century France, Louis XIV dismantled city walls of Paris to eliminate risks of political grip over the city, and thereafter planted a promenade of elms upon its remains. In reference to the project, Lawrence (1995: 31) describes the green promenade as 'window dressing on the destruction of the independence of the city'. Also in France, the tree-lined boulevards initiated by Napoleon III in the 19th century provided a processional space for displays of military might and offered police access to rebellion-prone neighbourhoods that needed to be kept under surveillance (Lawrence 1993; Mumford 1961), in addition to its role as a public amenity. Colonial-era landscape interventions by the British in their colonized cities likewise used public gardens and landscaping features that distinguished the quarters of the masters from their subjects. In Calcutta, extensive gardens were created around government dwellings to protect the privacy of their British occupants. Where the governor-general lived, residences landscaped in the style of picturesque English gardens were created, linked to the city of Calcutta by a formal tree-lined promenade (Pregill and Volkman 1999). Greenery in this case provided both a semblance of home for British subjects and asserted the political control of a foreign territory by overlaying its landscape upon the colonized city.

Various urban green space typologies have also traditionally been exclusive to people of specific social classes, serving as markers of social status. Lavish private gardens and grounds adjoined the estates of privileged classes—symbolizing power and status, testifying to wealth and good taste. To borrow Bordieu's (1986) concept, private green spaces sometimes served as a form of cultural capital. Since the Middle Ages, the phenomenon of creating gardens and villas by European political families reflected the importance of gardens to socio-political motivations. For instance, in 16th century Genoa, Magnani (2008: 55) argues that the gardens of these families implied power and status 'through size, decorative repertoire, and complexity'.

The political intrigue and genesis of Vaux-le-Vicomte and Versailles provides another interesting case for discussion. A ground-breaking project commissioned by Nicolas Fouquet, then the Superintendent of Finances in France, Vaux-le-Vicomte is widely considered the predecessor to the extravagant Versailles Palace and the pioneering project of the French formal garden. Already feeling threatened by the influential and refined Fouquet, Louis XIV had the unsuspecting Fouquet imprisoned under charges of corruption soon after an unparalleled fete thrown in the newly completed Vaux-le-Vicomte, in which the king was the honoured guest. Before the creative breakthroughs displayed in Fouquet's new estate and grounds, the king's own palaces seem to pale in comparison. Schama (1991) wrote that 'the real trouble with Fouquet' had less to do with his accumulation of wealth and built property, which were common among ministers, but rather with 'his stupendously good taste', embodied by the architecture and gardens of Vaux-le-Vicomte. Tellingly, the king seized numerous possessions from Fouquet's chateaux, from tapestries to orange trees, and later engaged the same creative team behind Vaux-le-Vicomte: Andre le Notre, Charles le Brun and Louis le Vaux in the same year to collaborate on the palace and gardens of Versailles—a project that would overshadow Vaux-le-Vicomte in scale and extravagance.

In the east, traditional Chinese gardens played similar roles in social the lives of urbanites. In his essay on urban gardens in Ming Dynasty Jiangnan, Hammond (2008) highlights the importance of gardens as a tool for the old literati class to assert superiority over a burgeoning merchant class in a period where intense economic growth broadened literacy and offered greater social mobility. As spaces for the production of literary works and socialization of upper-class gentlemen, gardens were an important part of the identity of the cultured elite. The creation and display of gardens was a way of promoting family stature and differentiating *guifu jia* ('families of wealth and honour') from the 'the cultural pretensions of the *nouveau riche*' (Hammond 2008: 47–48).

Apart from acting as a potent instrument in the competition for social prestige, private urban greens play an obvious role in promoting sociability and cultivating good social relations, having long been used as a space to host social events and entertain guests. In dynastic China, traditional Chinese gardens served as a 'conversation piece' and 'a generator of social relationships' through the culture of exchanging horticultural gifts (Metailie 2008: 35), while a neatly mowed lawn in a front yard in suburban America conveys something about the respectability and neighbourliness of the residing family (Wigley 1999). Whether they are being used to reinforce and express social hierarchies or improve social relations between urban dwellers, the creation and maintenance green spaces by private parties for social positioning is age-old. The association between socio-economic status and urban greenery extends to modern urban living, a phenomenon described by urban ecologists as 'luxury effect' (Hope et al. 2006) and the 'ecology of prestige' (Grove et al. 2014).

3.2.5 Social Reform and Community Building

More ambitiously on a public scale, urban greening has also been seen as a means of reforming society and improving social welfare in the city. These wide-reaching reformist objectives for urban greening are perhaps more recent, developed when social and environmental problems that emerged with the Industrial Revolution spurred a search for solutions to urban ills. Thinkers of the time perceived that the evils of the industrial city, ranging from poor environmental quality to disease to moral degradation, had much to do with the severed connection with nature. To remedy this, the city had to be reconciled with nature. Planners and thinkers approached this goal in two major ways: first, by inserting natural elements into the existing urban fabric through various urban greening projects, and second, by rethinking the urban-rural divide through utopian visions that proposed alternative ways of planning new towns within the larger regional setting (Hirt 2011). By reconnecting urban dwellers with nature, city planners expected to see improvements in social well-being and public morality, in addition to better health outcomes.

Socializing and civilizing a poor, uneducated, newly-urban working class was a prominent issue in the early years of Victoria's reign, motivated by both genuine concern for the welfare of the poor as well as fear for the safety of the rich (Taylor 1995). To this end, numerous parks were built and donated by the wealthy and powerful. Between mid-century and century's end in Victorian London, the total number of public parks swelled from a dozen to more than 200 (Sexby 1898, cited in Dreher 1997). Taylor (1995) argues that these parks were modelled after landscape visions of a nature that was 'organized, civilized,—suburbanized, even', reflecting the ideal Victorian society—rational, civilized and orderly. Visits to these urban parks were intended to be not just physically invigorating, but also morally uplifting and educational (through the display of eclectic cultural artefacts and scientific organization of plants). Parks were constructed as spaces of public virtue in the city, where visitors are expected to dress well and behave respectably, countering the negative perception of cities and encourage civic pride among urban dwellers (Dreher 1997). Along similar lines, Cranz (1982) identifies a 'Reform Park' model popularized by social workers in early 20th century United States, focused on providing public spaces for children, immigrants and the working class. In contrast to the large pleasure grounds of the late 19th century that were located too far from the city centre to be easily accessible for the poor working class, 'Reform Parks' were small green spaces created in the inner city that aimed to reduce class conflict, Americanize immigrants and educate children, among other social goals (Cranz and Boland 2004). Around the same period, the School Garden Movement in the United States initiated gardening activities in schools, vacant urban lots and homes, among other open areas, to cultivate good social behaviour and civic values in children, immigrants, delinquents and the infirm (Lawson 2004). At the height of its role as a tool for civic education, one of these school gardens, the Worcester Good Citizen's Factory, was designed as a microcosm of the city, with streets, boulevards, amenities and children roleplayed as city governments and officials (Lawson 2004).

Towards similar urban reform objectives, other planners and thinkers offered more radical utopian visions that re-imagined how cities can be reorganized to avoid the ills of existing models. Compared to the public park projects earlier discussed, these utopian plans sought to deliver more far-reaching and definitive solutions to intractable urban problems. Ebenezer Howard's Garden City, for example, was meant to address problems such as rural-urban migration, slum expansion and overcrowding, growing social inequity, and 'the exclusion of the benefits of city life from the residents of rural areas, and vice versa' (Batchelor 1969). As implied by the name, the Garden City comprises extensive greenery, allowing the country to extend into the town, promoting urban self-sufficiency and a favourable living environment that combines the economic advantages of the city with the benefits of living close to nature. Other reformist utopian projects of the early modernist era also prioritised the maximization of urban green space—Le Corbusier's Radiant City, for example, housed its residents in high-rise towers to preserve 95% of urban land for parks. His Contemporary City, in the same spirit, emphasised vertical grow to 'increase open space and diminish the distances to be covered' (Le 1929: 163). Frank Lloyd Wright, who is famously known for his love of nature, proposed his automobile-centred Broadacre City that offered every household at least an acre of space. Effectively, the 'city' is dispersed within a natural landscape, connected only by highways.

Reform-oriented motivations for urban greening remain strong. Today, urban revitalization and transformation projects often involve urban greening as a significant component. Researchers and planners suggest that urban green spaces should be designed to promote social inclusion in an age of migration and globalization by providing a space for social learning where people from various social categories may meet and interact. A study examining the role of urban parks in promoting multicultural integration among youths in Zurich, for example, suggested that green spaces provided an easily and freely accessible platform where Swiss and immigrant youth may meet and interact, more so than other types of spaces in the city (Seeland et al. 2009). In a similar vein, Kweon et al. (1998) argued that urban greening can help to create more socially supportive environments for elderly persons living in inner city neighbourhoods. Other studies also suggest that well-maintained urban vegetation is significantly correlated with lowered rates of crime and anti-social behaviour (Troy et al. 2012; Wolfe and Mennis 2012). Among grassroots organizations worldwide, urban greening initiatives have been started in order to help improve the well-being of neighbourhoods. For their proponents, community gardens and planting activities promote community cohesion through cooperation, participation and involvement. Community greening activities are also believed to foster a sense of ownership, create pride in the neighbourhood and empower residents to take on greater challenges.

3.2.6 Physical and Mental Well-Being

The healthful effects of nature on urban residents is a persistent key theme in the history of urban greening. Studies in recent decades provide scientific evidence for what many societies have long believed: the positive effects of living close to nature (albeit a form that is modified to be pleasing to humans) on physical and mental health (van den Berg et al. 2015). From Greek Asclepeian sanctuaries to Swiss sanatoriums, restorative spaces have traditionally been located close to natural settings. Natural environments, it was reasoned, offered better air, light and water quality than the densely built environment, and had a purifying effect on both the body and the psyche. Current research demonstrates a number of ways in which green spaces can benefit human health: restoration from stress and mental fatigue; buffering impacts of stressful life events; increasing and prolonging physical activity; decreasing feelings of aggression; reduce physical pain and promote recovery from surgery; improve air quality; reduce symptoms of behavioural problems, and so on. Potentially improved public health outcomes are therefore an attractive reason for urban greening. Health facilities, in particular, are paying increasing attention to greening and landscaping in their compounds, to create a more conducive environment for healing and recovery.

Before the advent of modern medicine, disease was commonly attributed to either supernatural forces or miasma ('bad air') throughout various medical traditions. The miasma theory implied a direction relationship between health and environment. The Hippocratic Corpus, for example, held that foul odours would cause diseases, while perfumes and fresh air could prevent it. Later in the medieval ages, herb gardens were often the only defence against the scourge of disease, and they were commonly established in hospitals and religious houses. On top of their provisioning functions, these gardens were sometimes used as a form of horticultural therapy for patients. At leper hospitals in Oxfordshire and Devon, for example, inmates were required to tend to these gardens as part of their daily regiment, so as to provide exercise and prevent depression (Rawcliffe 2008). Additionally, views of natural elements and proximity to green spaces were thought to hasten recovery. On the healing capacity of nature, a 15th century physician writes, '…if you dwell as frequently as possible among plants which have a pleasant smell… By this odour they restore and invigorate you on all sides, as if by the breath and spirit of the life of the world'. (Ficino 1499, cited in Rawcliffe 2008). For the unwell and infirm, gardens not only supplied the remedies for their ailments, but also provided the ideal setting for recovery.

The notion of parks as the 'lungs' of a city is now well-worn. In the context of the miasma theory that remained in popular circulation in the 19th century, the significance of parks to public health as suggested by this metaphor can be even better appreciated. It resonated with residents in industrial-age London, where the crowded, unsanitary and polluted conditions bred diseases such as tuberculosis, cholera and typhoid. In the aftermath of the first major cholera outbreak in England, a law requiring urban centres to establish a public walk or park was proposed by the 1833 Select Committee on Public Walks (Schuyler 1986, cited in Ward Thompson 2011). Parks provided open space and a spot of fresh air in the otherwise unhealthy city and were seen as a solution to poor public health outcomes in London. In the United States, Olmsted echoes similar beliefs with a particular emphasis on the positive effects of nature on mental health, observing that the artificial environment negatively affects the 'mental and nervous system and ultimately (a person's) entire constitutional organization' (Olmsted 1886, cited in Ward Thompson 2011). On the restorative effect of nature, he wrote, '…the occasional contemplation of natural scenes of an impressive character, particularly if this contemplation occurs in connection with relief from ordinary cares, change of air and change of habits, is favourable to the health and vigor of man and especially to the health and vigor of their intellect beyond any other conditions which can be offered to them…' (Olmsted 1865).

In the last three decades, the work of researchers such as Roger Ulrich and the Kaplans provide scientific evidence that green spaces do have a quantifiable impact on physical and mental health. Ulrich's (1984) study titled 'View through a window may influence recovery from a surgery' demonstrated that a view of greenery had positive effects on post-surgery recovery. Compared to patients with a window view of a brick wall, patients with a view of greenery required less pain medication, received fewer negative evaluative comments from nurses, and were discharged earlier. Echoing the arguments made earlier by Olmsted, the Attention Restoration Theory developed by Stephen and Rachel Kaplan asserted that concentration can be restored by spending time in natural surroundings (Kaplan and Kaplan 1989). Based on various observations in experiments, the Kaplans developed the theory after observing how mentally fatigued persons performed better at given tasks after spending time in nature. Further substantiating these theories, recent research has provided evidence of physiological changes in blood pressure, muscle tension, salivary cortisol levels etc. after spending time in nature. A growing body of research also suggests that urban green spaces help to prevent illnesses related to sedentary modern lifestyles by increasing time spent outdoors and encouraging physical activity and socialization. To account for the strong preference that humans seem to have for natural environments, scholars have offered a number of evolutionary and biological explanations. Wilson and Kellert's (1993) 'Biophilia Hypothesis', which postulates that humans are biologically wired to feel deeply affiliated with nature and other lifeforms as a result of evolutionary processes, is particularly influential. Biophilic design and biophilic urbanism, which applies insights and concepts from the biophilia hypothesis to the built environment, advocates actively incorporating direct, indirect and symbolic experiences of nature into the city to improve human well-being (Kellert et al. 2008).

3.2.7 Food Production and Sustenance

Providing food and producing agricultural commodities is a critical function that urban green spaces can provide, especially in periods of hardship and disaster. Although the large-scale production of food and agricultural commodities is typically confined to the rural landscape, there are strong incentives to allocate resources for agricultural production within urban boundaries. For the urban poor and residents in war-torn cities, urban agriculture can sometimes be the chief source of food. Health-conscious and environmentally conscious urbanites may perceive home-grown food to be a healthier and less polluting alternative due to reduced use of chemical fertilizers and pesticides. Urban agriculture also helps to mitigate food insecurity for dense megacities and cities without sufficient rural hinterlands that are currently reliant on long supply chains to feed their populations.

Depictions of antiquity gardens show them filled with useful and edible plants. A famous fresco from the tomb of Nebamun, a mid-ranking official in Ancient Thebes, for instance, illustrates a garden full of fig trees, date palms, fish and ducks. Since early civilization, the gardens of urban residents have been commonly planted with useful agricultural species: fruit tree, herbs, edible roots, vegetables, and so on. Of the various forms of urban agriculture, urban gardening—in home gardens, allotment gardens and community gardens (Drescher et al. 2006)—is ubiquitously practised by individuals.

Periods of food shortage heighten the value and prevalence of these gardens. Allotment gardens in Europe originated in the 19th century, as lands offered to members of the poor working class to supplement income in a period when the industrial urbanism resulted in the privatization of common lands. During the two world wars, the numbers of these allotment gardens surged in both Europe and the United States. Governments rallied citizens to join in the food production effort with slogans such as 'Dig for Victory' and 'Hoe for Liberty' in Britain and the United States, respectively. Local authorities transformed open spaces into productive land, from vacant lot to parks and playfields. In the case of Britain, quantities grew from 600,000 to 1,500,000 gardens in WWI, and then from 800,000 to 1,400,000 in WWII (Barthel et al. 2014). American civilian gardeners grew over $520 million worth of food in 1918 to supplement food requirements in WWI, and produced over 42% of the nation's vegetable needs at their most productive in WWII (Lawson 2004). In WWII, German allotment gardens rose from 450,000 to 800,000 (Gröning 1996, cited in Barthel 2014), while French home gardens produced enough vegetable to supply 40% of vegetables consumed in the United States (Basset 1979, cited in Barthel 2014).

Periods of economic downturn also prompted a similar focus on agricultural production on urban lands. During the Great Depression of the 1920s and 1930s, the Federal Emergency Relief Administration put in place relief gardening programmes to assist the unemployed, providing gardeners with a wage for cultivating land and distributing their produce to the needy. In New York City, a gardening campaign cultivated over 5000 vacant lots, producing over $2.8 million of food by 1934 (Tucker 1993). More recently in Cuba, urban gardening for agricultural production similarly became an important source of food with the loss of trade following the collapse of the socialist bloc in 1989 (Altieri 1999). As imports decreased by 75%, the food crisis forced urban dwellers in Havana, many of whom were first-time gardeners, to begin cultivating (Altieri et al. 1999). In 1996, urban agriculture in Havana produced 8500 tons of agricultural produce, 7.5 million eggs, and 3650 tons of meat (Campanioni 1996, cited in Altieri et al. 1999). Aside from their importance in sudden economic crises, urban gardens are also critical for the urban poor in developing cities. Informal settlers in urban and peri-urban regions of Asia, Africa and Latin America continue to rely on self-cultivation to provide for household needs and income generation in the informal economy.

The continued increase of the global population, much of it projected to be in urban regions, urges cities around the world to look closer at urban agriculture as a means of supplying the city's nourishment needs. Climate change also brings about greater uncertainty for traditional agricultural production, emphasizing the importance of developing more controllable alternatives of agricultural production for example, in plant factories and vertical farming systems (Fig. 3.5). In contrast to the necessity-driven urban farming in cities facing food shortage, urban agriculture in today's developed regions are more often driven by social movements and changing public opinions about the city's relationship with food. Proponents argue that growing food in urban areas can potentially offer fresher produce, decrease 'food miles' accumulated from long-distance transportation, reduce agricultural pollution, allow nature to regenerate in abused farmlands, reduce agricultural wastage, and produce safer and higher quality food for a lower price. Technologies and techniques are, however, still at a nascent stage. Intensive agricultural production on a significant scale in the cities of the Global North would require a favourable confluence of technological, socio-political and economic factors. Please also refer to Chap. 8 on additional benefits as well as constraints on urban agriculture.

Fig. 3.5 An experimental plant factory in Chiba University, Japan. Plant factories, although requiring high capital investment, can achieve much higher productivity in leafy vegetables production compared to conventional farming (*Photo credit* P.Y. Tan)

3.2.8 Ecological Health and Environmental Sustainability

Preserving ecological health and ensuring environmental sustainability are now pressing concerns in cities. As evidence elaborating the repercussions of major global challenges such as mass species extinctions, loss of ecosystem services, environmental pollution and climate change mounts, more questions are also being asked on how cities can be better designed and planned based on ecosystem awareness. Often, proposed solutions involve extensive greening of the urban landscape. Experts and professionals increasingly recognize the numerous ways in which urban greenery contributes to the ecological health of the city: carbon sequestration, removal of excess nutrients and pollutants, air quality improvement, water detention, soil protection, evapotranspiration, biodiversity enhancement, climate amelioration, among others. As such, scores of new urban greening initiatives and projects are justified based on their perceived ecological benefits. Compared to other urban greening imperatives highlighted in this chapter, urban greening for the purpose of improving the biological and biophysical health of cities is a more recent development.

While the works of 19th century designers such as Olmsted and Cleveland have implicitly embedded ecological concerns in their works, the Scottish biologist Patrick Geddes was perhaps the first apply ecological thinking to urban planning in an explicit manner through his 1909 ‘Valley Section’, which connected social processes to its larger geographical context. However, interest in ecological urban planning lulled until the 1960s due to changes in social attitudes and public opinion (Botequilha Leitão and Ahern 2002). In the late 1960s and 1970s, Ian McHarg’s (1969) seminal work ‘Design with Nature’ and increasing public awareness of the shocking environmental impacts of cities revived planning interest in ecological concerns. Concurrently, the burgeoning fields of landscape ecology and urban ecology have further increased interest in applying ecological concepts to address various environmental concerns and mitigate the city’s ecological impacts.

The city of Berlin, now a renowned example of an ecologically planned city, provides useful examples of urban greening to protect threatened habitats and biodiversity. Locked within Soviet-controlled territories of the German Democratic Republic, the Berlin wall marooned West Berlin within the Berlin wall for the duration of the Cold War, without access to the countryside. As a result, urban green spaces grew in importance for residents, and ecologists had to focus their research on urban sites. Ecologists in Berlin were highly involved in city land-use planning and sought to integrate their insights in city planning (Lachmund 2013). One of their most important endeavours was the ‘biotope’ mapping of West Berlin, which framed the urban landscape as a patchwork of 57 pre-defined habitat types. Represented in a manner that was easily legible to city administrators and urban planners, the mapping re-framed the city as a hybrid natural and social ecosystem (Lachmund 2013). The Species Protection Programme was then introduced to guide landscape management and development. In line with the guidelines of this programme, urban spaces were systematically greened: roofs were topped with gardens, lawns were transformed into meadows, decaying urban sites were left for spontaneous vegetation takeover, and an urban greenbelt linking large habitats was proposed.

Few other cities have executed ecological planning of similar comprehensiveness and scale with a focus on biodiversity conservation. However, many cities now consciously implement urban greening to achieve other environmental goals that more directly affect human well-being, such as the management of urban water. Known variously as 'low-impact development', 'sustainable urban drainage systems' or 'water-sensitive urban design' (Fletcher et al. 2014) throughout the world, these approaches emphasize effectively managing the urban water cycle in city planning (please see Chap. 10 for more elaboration of the concept and terms). Environmental concerns addressed by these schemes include the treatment of polluted urban runoff, flood control, improving infiltration rates and groundwater recharge, water conservation, wastewater treatment, and the protection of natural waterbodies. To achieve these objectives, designers commonly use softscape and green elements to replace urban hardscape and concrete infrastructure. For example, systems of raingardens, vegetated swales and naturalized waterways replace conventional concrete drainage systems to slow down flow rates, allowing sedimentation, increasing infiltration rates and improving water quality through plant uptake of pollutants and excess nutrients. Constructed wetlands, which mimic the ecosystem functions of natural wetlands, also provide an alternative to standard industrial wastewater treatment plants. Australian cities under water stress such as Melbourne, in particular, have embraced these planning principles and technologies as a strategy to maximize water resources. The Australian Water Sensitive Cities Blueprint highlights the importance of urban greenery to urban water management, reasoning, 'as stormwater runoff is generated across distributed areas, distributed green infrastructure presents the best opportunity for delivering multiple benefit outcomes while managing stormwater impacts' (CRCWSC 2013).

Global climate change and its potentially disastrous consequences loom over cities worldwide, and calls for the development of climate mitigation and adaption strategies. Of the numerous challenges posed by climate change, projected sea level rises and increasing incidences of extreme weather events are regarded as particularly serious threats to major cities, three-quarters of which are located along coastlines (UNEP 2005). With less vegetation cover to moderate the effects of global warming, urban surfaces are likely to experience exacerbated climate scenarios (Wilby 2003). Increasing the percentage of vegetation cover in cities therefore appears to be a direct and cost-effective way to address the problem. Urban green spaces and vegetation can help to sequester carbon, ameliorate the urban heat island effect and manage the effects of increasing precipitation intensity. In one study on the potential role of urban green space to climate change adaptation, researchers modelled the urban region of Greater Manchester (Gill et al. 2007). It was suggested that by adding another 10% of vegetation in the high-density residential areas, maximum surface temperatures could be decreased by 2.5 °C, based on 2080 temperature projections on a high emissions scenario. Conversely, losing 10% of vegetation cover could lead to a further increase of up to 3.5 °C by 2080 in the same emissions scenario. Strengthening the case for urban greening, the same study also modelled a further hypothetical scenario where all existing roofs in high-density areas were converted into roof gardens. In this scenario, maximum surface temperature could be lowered by 7.6 °C, and runoff volume could be reduced up to 47.2%.

Urban greening can also help to protect vulnerable urban shorelines and cities from the destructive effects of storm surges. The devastation wreaked by Hurricane Sandy on New York City in 2012 clearly demonstrated how vulnerable the city could be in extreme storm events, which are likely to become more severe and common. The New York City Panel for Climate Change estimated that 1-in-100 year floods were likely to happen five times as frequently, and sea levels are projected to rise by 2.5 ft by 2050 (NYCPCC 2013). Towards the end of 2012, the Special Initiative for Rebuilding and Resiliency was set up in the wake of Hurricane Sandy as a task force to develop recommendations and strategies to safeguard the city in future storm events. In June 2013, a report entitled 'A Stronger, More Resilient New York' (City of New York 2013) was published to present their findings and recommendations. Under the coastal protection recommendations, the report supported the enhancement of wetlands and coastal forests, as well as the creation of planted dune systems along certain shoreline stretches to attenuate waves. The report also encouraged the creation of more Greenstreets as it was assessed that many of these streets had performed well during the disaster, absorbing water equivalent up to 31 times its area—a value that is approximately 30 times more than a conventional street in the city.

Ecosystem services provided by urban green spaces promise to reduce the energy consumption of cities and conserve depleting resources. As environmental problems are increasingly foregrounded in public discourse, virtually every major proposal for creating new green spaces in the city today are accompanied by justifications of the project's ecological benefits and its contribution to environmental sustainability. Yet, there is still a need for much knowledge from ecological and biophysical studies to be operationalized in practice (please refer to Chap. 2). Increasing the efforts for collaboration between scientists, practitioners and policy makers is still needed for ecological benefits of greening to be fully realized.

3.2.9 Economic Value and Competitive Advantages

Access to green spaces is highly prized in most modern cities. After the completion of a prominent urban greening project, properties in the surrounding areas can usually look forward to a rise in value. Awareness of this advantage has encouraged city mayors and developers to invest in urban greening, with the knowledge that spill over economic benefits generated would offset initial costs and bear returns over the long run. On a larger scale, a greener urban environment can shape an identity for the city and create a more attractive environment to live and work in, providing a key competitive advantage as cities jostle for highly mobile global talent.

With few exceptions, research shows that well-landscaped properties with a higher amount of tree cover tend to fetch higher housing values and rental rates (Morales 1980), increase perceptions of productivity, and may even induce greater consumer spending (Wolf 2009). Hedonic analyses since the 1970s have shown that in neighbourhoods where greenery is scarce, tree cover and landscaped grounds commanded a market premium. A 1976 study on the contribution of trees to property value in a suburban town in Connecticut concluded that trees added an additional 6% to property value (Morales et al. 1976), while a more recent study in Quebec City in 2002 suggested that the presence of mature trees increased housing values by up to 15% in neighbourhoods with higher socio-economic status (Thériault et al. 2002). In a commercial setting, aesthetically-pleasing landscaping added an additional 7% to office rental prices in Cleveland, as did good building shade provided by trees (Laverne and Winson-Geideman 2003). Despite the higher costs associated with maintaining extensive landscaping and greenery in a property, developers increasingly recognize that investments in greening properties can translate to faster sales and larger profits.

Proximity to urban green spaces also exerts a well-documented effect on property prices. Buyers are willing to pay more for properties located close to parks and other urban green spaces. Olmsted and the New York Parks Commission presented some of the earliest empirical evidence on the economic impact of urban parks when they conducted a study on the impact of Central Park on the property tax base of three proximate wards. According to the reported figures, the assessed property tax base value in these three adjacent wards was placed at approximately $26 million in 1856, increasing to $236 million in 1873 after the completion of the park (Fox 1990, cited in Crompton 2001). Even after accounting for the overall doubling in property values experienced throughout New York City in the study period, the impact of the park remains impressive (Crompton 2001). A similar appreciation in adjacent property value more recently in the same city can be observed after the Highline was completed. While residential properties in the vicinity were valued at 8% lower than the average in Manhattan in 2003, prices for properties within a 5-minute walk from the Highline increased by 103% by 2011 to exceed borough-wide values (NYCEDC 2011). Many other studies have drawn similar conclusions about the effect of large greening projects on property values. In a review of approximately 30 studies on the topic, Crompton (2001) observes that the effect of parks on real estate value was frequently such that the resultant increase in property tax on the surrounding neighbourhood was sufficient to compensate for the loss of tax revenue when designating land as a public park. A 1978 regression analysis estimating the relationship between greenbelt proximity and property prices in Boulder, Colorado showed that property prices decreased by $4.20 for every foot away from the greenbelt, all else being comparable (Correll et al. 1978). Designating urban land for parks—particularly when they are intended for passive recreational purposes—can have a significant economic impact on its surrounding neighbourhood.

Urban theorists and economists widely recognize that the economic success of cities are centrally tied to the city's capacity to attract and retain human capital (Florida 2004). Where talented and highly educated people congregated, cities developed faster and prospered. Appealing to this class of individuals is therefore an important long-term strategy for sustained economic growth. Singapore, which has given itself the monikers of 'Garden City' and 'City in a Garden' since independence, has employed citywide urban greening as a strategy for attracting businesses and talent. Early greening efforts in the city were championed and prioritized by the founding Prime Minister, Lee Kuan Yew, who made the analogy that a neat and well maintained garden implied the presence of a dedicated gardener, just as a clean and green city implied the presence of a capable and efficient government (Neo et al. 2012). Greenery was believed to help Singapore's image in the bid to attract foreign investors, and Lee argued that being clean and green was essential, for 'if [Singapore] had First World standards, then businessmen and tourists would make [Singapore] a base for their businesses and tours of the region' (Neo et al. 2012). The state implemented extensive greening programmes and policies, ensuring that all but the most highly urbanized areas are vegetated. Across the island, vegetation flank public infrastructure such as roads and waterways. In 2015, street trees alone numbered approximately 1.5 million.

As the city-state prospered, urban greening projects retained its prominent role in urban development. The nature of newer projects, however, takes on a distinctively different tone and approach to reflect shifting global demands. To maintain the city's competitive advantage, innovation, vibrancy and sustainability had to be included in Singapore's long-standing image of efficiency and competency. The Marina Bay development, conceptualized as 'a 24/7, thriving and energetic place where people live, work and play' and 'a place for people from all walks of life to explore, exchange and entertain', features a horticultural extravaganza on prime-land. This planning decision was made based on the assessment that the new park would increase property values and serve as a major tourist attraction, while creating aesthetic excitement on the flat reclaimed landscape. The Gardens by the Bay, immediately identifiable by a series of futuristic structures, is visually spectacular and technologically sophisticated. Reflecting the city's desire to become an innovative international hub, horticultural specimens were sourced from all over the world to be housed in two cutting-edge 'coolhouses', which are plant conservatories designed as the climatic inverse of temperate hothouses. In a milieu where sustainable design is the order of the day, energy-conserving features and marketing strategies mediate and circumvent the resource-intensive nature of such an endeavour, while showcasing technological capabilities to an international audience. These internationally visible Gardens project a powerful image of Singapore, and were seen as being tied 'directly and intrinsically to the future identity and success' of the nation from its conception (Koh 2012).

3.3 Conclusion

For a wide range of reasons, we make space for greenery in our cities. In this chapter, we have identified a number of common reasons for urban greening in its long history, some more recently articulated than others. Intuitively, visionaries and early proponents of urban green spaces were cognizant of their potential benefits and research continues to validate their insights. Imperatives for urban greening emerge from an innate human desire for the presence of nature, the current state of knowledge, as well as the events and changing structures of our societies. More incentives are likely to surface over time, closely reflecting the new priorities and challenges of contemporary cities. For example, in response to the homogenizing forces of globalization, designers increasingly value urban green spaces for their potential to craft or highlight place identity. As cities grow increasingly faceless and similar, landscape approaches seem to offer attractive solutions to counteract the fading of place identity. While constructing architectural marvels that turn cities into 'gigantic sculpture gardens' (Tuan 2008: 29) can appear contrived, landscape and vegetation can seem more authentic, being unambiguously unique to the climate and biogeography of a place (Murphy 2008). The list of potential benefits continues to lengthen, and the case for investing in urban green spaces will likely become stronger. It is clear too that there are other reasons that we have omitted in our description, and other methods of rationalizing the organization of these reasons. We believe that such discourse should be encouraged, as it serves to strengthen our understanding of the complexities and plurality of meanings associated with greening the city.

Few however, would dispute the value of having green spaces in cities today. Rather, the more contentious debates are related to their purpose and accessibility, given the limits on space and resources in the city. Which greening imperatives should be prioritized over others and how should that change the way they are planned and designed? Who should benefit from these green spaces and how should they be distributed and maintained?

Providing equitable access to public green spaces and resisting the effects of gentrification in neighbourhoods around new park projects remain a challenge. Although public parks have been opened to the public in Europe and North America since the 19th century, inequities in access to urban green spaces continue to persist in contemporary cities and differentiate between social classes, albeit in less stark terms. Studies have shown, for example, that urban green spaces still tend to be disproportionately more accessible for communities that are affluent. With increased understanding of the importance of green spaces to critical issues such as human health and climate change adaptation, the unequal distribution of green spaces along various axes of difference such as race, income, gender and (dis)ability is seen as a serious environmental justice issue (Byrne et al. 2009). Yet, attempts to solve the problem by simply creating more green spaces in neighbourhoods with poorer park access may not have the intended effect. When the neighbourhood becomes a more expensive place to live in due to its increased attractiveness, intended beneficiaries can be pushed out in a process of gentrification. Recognizing this paradox, scholars advocate for solutions that are more explicitly tailored to the community's needs as opposed to conventionally appealing park designs (Wolch et al. 2014).

Such issues raised by scholars serve to remind urban planners that greening of cities approached from a purely technocratic view of improving urban climate, hydrology and ecology, with inadequate attention on meeting other diverse human needs of culture, heritage, community mediated by connection with nature, is short-sighted. As our knowledge of cities as socio-ecological systems where social and ecological processes interact in complex feedback loops improve, multi-functional approaches to urban greening with the intent to reveal and alter these intertwining socio-ecological processes, creating more positive feedback loops will become compelling. An urban green space effectively designed to be multifunctional may concurrently provide aesthetic landscape improvements, deliver ecological services, offer recreational opportunities and generate economic benefits. Seeing clearly the often-intertwining functions, values and rationales for greening is contingent on understanding not just current urban challenges, but also looking backwards to decipher why greenery has always been part of our urban history.

References

911memorial.org (2016) The Survivor Tree. https://www.911memorial.org/survivor-tree

Altieri MA, Companioni N, Cañizares K, Murphy C, Rosset P, Bourque M, Nicholls CI (1999) The greening of the 'barrios': urban agriculture for food security in Cuba. Agric Hum Values 16(2):131–140. doi:10.1023/a:1007545304561

Basset T (1979) Reaping on the margins: a century of community gardening in America. Landscape 25(2):1–8

Barnett R (2007) Sacred groves: sacrifice and the order of nature in ancient greek landscapes. Landscape J 26(2):252–269. doi:10.3368/lj.26.2.252

Batchelor P (1969) The origin of the garden city concept of urban form. J Soc Architectural Historians 28(3):184–200. doi:10.2307/988557

Barthel S, Parker J, Folke C, Colding J (2014) Urban gardens: pockets of social-ecological memory. In: Tidball GK, Krasny EM (eds) Greening in the red zone: disaster, resilience and community greening. Springer, Dordrecht, pp 145–158. doi:10.1007/978-90-481-9947-1_11

Barthel S, Parker J, Folke C, Colding J (2014) Urban gardens: pockets of social-ecological memory. In: Greening in the Red Zone. Springer, pp 145–158

Boone CG, Buckley GL, Grove JM, Sister C (2009) Parks and people: an environmental justice inquiry in Baltimore, Maryland. Ann Assoc Am Geogr 99(4):767–787

Bourdieu P (1986) The forms of capital. In: Richardson J (ed) Handbook of theory and research for the sociology of education. Greenwood, New York, pp 241–258

Botequilha Leitão A, Ahern J (2002) Applying landscape ecological concepts and metrics in sustainable landscape planning. Landscape Urban Plann 59(2):65–93. doi:10.1016/S0169-2046(02)00005-1

Byrne J, Wolch J, Zhang J (2009) Planning for environmental justice in an urban national park. J Environ Plann Manage 52(3):365–392. doi:10.1080/09640560802703256

Campanioni N (1996) El Huerto Intensivo en la Agricultura Urbana de Cuba. Seminario Taller Regional 'La Agricultura Urbana y el Desarrollo Rural Sostenible:39–48
City of New York (2013) A stronger, more resilient New York. City of New York, New York
Clark E (2004) The art of the Islamic garden. Crowood, Marlborough
Clark E (2011) The symbolism of the islamic garden. http://islamic-arts.org/2011/the-symbolism-of-the-islamic-garden/
Correll MR, Lillydahl JH, Singell LD (1978) The effects of greenbelts on residential property values: some findings on the political economy of open space. Land Econ 54(2):207–217. doi:10.2307/3146234
Crandell G (2013) Tree gardens: architecture and the forest. Princeton Architectural Press, New York
Cranz G (1982) The politics of park design: a history of urban parks in America. MIT Press, Cambridge
Cranz G, Boland M (2004) Defining the sustainable park: a fifth model for urban parks. Landscape J 23(2):102–120. doi:10.3368/lj.23.2.102
CRCWSC (2013) Blueprint 2013: stormwater management in a water sensitive city. Cooperative Research Center for Water Sensitive Cities, Victoria, Australia
Crompton JL (2001) The impact of parks on property values: a review of the empirical evidence 33 (1)
Dreher NH (1997) The virtuous and the verminous: turn-of-the-century moral panics in London's public parks. Albion 29(02):246–267. doi:10.2307/4051812
Drescher AW, Holmer RJ, Iaquinta DL (2006) Urban homegardens and allotment gardens for sustainable livelihoods: management strategies and institutional environments. In: Kumar BM, Nair PKR (eds) Tropical homegardens: a time-tested example of sustainable agroforestry. Springer, Dordrecht, pp 317–338. doi:10.1007/978-1-4020-4948-4_18
Dwyer JF, Schroeder HW, Gobster PH (1991) The significance of urban trees and forests: toward a deeper understanding of values. J Arboric 17(10):276–284
Fletcher TD, Shuster W, Hunt WF, Ashley R, Butler D, Arthur S, Trowsdale S, Barraud S, Semadeni-Davies A, Bertrand-Krajewski JL, Mikkelsen PS, Rivard G, Uhl M, Dagenais D, Viklander M (2014) SUDS, LID, BMPs, WSUD and more—the evolution and application of terminology surrounding urban drainage. Urban Water J. doi:10.1080/1573062X.2014.916314
Fox T (1990) Urban Open Space: an investment that pays. The Neighborhood Open Space Coalition. New York
Florida R (2004) America's looming creativity crisis. Harvard Bus Rev 82(10):122–136
Gill SE, Handley JF, Ennos AR, Pauleit S (2007) Adapting cities for climate change: the role of the green infrastructure. Built Environ 33(1):115–133. doi:10.2148/benv.33.1.115
Gilpin W (1792) Three essays: on picturesque beauty; on picturesque travel; and On sketching landscape: to which is added a poem, on landscape painting. Printed for R. Blamire, London
Girot C (2016) The course of landscape architecture. Thames & Hudson Ltd., London
Grove JM, Locke DH, O'Neil-Dunne JPM (2014) An ecology of prestige in New York City: examining the relationships among population density, socio-economic status, group identity, and residential canopy cover. Environ Manage 54(3):402–419. doi:10.1007/s00267-014-0310-2
Gröning G (1996, September 26–29). Branching out: linking communities through gardening. Paper presented at the 1996 Annual Conference of the American Gardening Association (ACGA). Montreal, Canada
Hammond K (2008) Urban gardens in Ming Jiangnan: insights from the essays of Wang Shizhen. In: Conan M, Chen W, Dumbarton O (eds) Gardens, city life and culture: a world tour. Dumbarton Oaks Research Library and Collection; Distributed by Harvard University Press, Washington, D.C.; [Cambridge, Mass.], pp 47–48
Hensley M, Mateo-Babiano D, Minnery J (2014) Healthy places, active transport and path dependence: a review of the literature. Health Promot J Austral 25(3):196–201. doi:10.1071/HE14042

Hirt S (2011) Integrating city and nature: urban planning debates in Sofia, Bulgaria. In: Brantz D, Dümpelmann S (eds) Greening the city urban landscapes in the twentieth century. University of Virginia Press, Charlottesville, pp 17–36

Hope D, Gries C, Casagrande D, Redman CL, Grimm NB, Martin C (2006) Drivers of spatial variation in plant diversity across the Central Arizona-Phoenix ecosystem. Soc Nat Resour 19 (2):101–116

Kant I (1911) Critique of judgement (trans: Walker JC). Oxford University Press, Oxford; New York (Original work published 1790). Retrieved from http://bradleymurray.ca

Kaplan R, Kaplan S (1989) The experience of nature: a psychological perspective. Cambridge University Press, Cambridge

Kellert SR, Heerwagen J, Mador M (2008) Biophilic design: the theory, science, and practice of bringing buildings to life. Wiley, Hoboken

Kellert SR, Wilson EO (1993) The biophilia hypothesis. Island Press, Washington, DC

Koh BS (2012) Perpetual spring, Singapore's Gardens by the Bay. Marshall Cavendish Corp., Marshall Cavendish Editions, Singapore

Kweon B-S, Sullivan WC, Wiley AR (1998) Green common spaces and the social integration of inner-city older adults. Environ Behav 30(6):832–858. doi:10.1177/001391659803000605

Lachmund J (2013) Greening Berlin: the co-production of science, politics, and urban nature. MIT Press, Cambridge

Laverne RJ, Winson-Geideman K (2003) The influence of trees and landscaping on rental rates at office buildings. J Arboric 29(5):281–290

Lawrence HW (1993) The neoclassical origins of modern urban forests. Forest Conserv Hist 37 (1):26–36. doi:10.2307/3983816

Lawrence HW (1995) Changing forms and persistent values: historical perspectives on the urban forest. In: Bradley GA (ed) Urban forest landscapes: integrating multidisciplinary perspectives. University of Washington Press, Seattle

Lawson L (2004) The planner in the garden: a historical view into the relationship between planning and community gardens. J Plann Hist 3(2):151–176. doi:10.1177/1538513204264752

Le C (1929) The city of tomorrow and its planning. Architectural Press, London

Marsilio Ficino (1989) Three Books on Life. In: Kaske CV and Clark JR (eds) Medieval and Renaissance Texts and Studies No. LVII (Binghampton), p 291

Magnani L (2008) Genoese gardens: between pleasure and politics. In: Conan M, Chen W, Dumbarton O (eds) Gardens, city life and culture: a world tour. Dumbarton Oaks Research Library and Collection; Distributed by Harvard University Press, Washington, D.C.; [Cambridge, Mass.], p 55

Matsui T (1996) Meiji Shrine: an early old-growth forest creation in Tokyo. Ecol Restor 14(1):46–52. doi:10.3368/er.14.1.46

McHarg IL (1969) Design with nature. Published for the American Museum of Natural History [by] the Natural History Press, Garden City, NY

McIntyre NE, Knowles-Yánez K, Hope D (2008) Urban ecology as an interdisciplinary field: differences in the use of 'urban' between the social and natural sciences. In: Marzluff JM, Shulenberger E, Endlicher W et al (eds) Urban ecology: an international perspective on the interaction between humans and nature. Springer, Boston, pp 49–65. doi:10.1007/978-0-387-73412-5_4

Meiji-Jingu (2016) Meiji Jingu official website: nature at Meiji Jingu. http://www.meijijingu.or.jp/english/

Metailie G (2008) Gardens of Luoyang: the refinements of a city culture. In: Conan M, Chen W, Dumbarton O (eds) Gardens, city life and culture: a world tour. Dumbarton Oaks Research Library and Collection; Distributed by Harvard University Press, Washington, D.C.; [Cambridge, Mass.], pp 15–31

Moncan P (2009) Les jardins du baron Haussmann. Les Éditions du Mécène, [Paris]

Morales DJ (1980) The contribution of trees to residential property value. J Arboric 6(11):305–308

Morales DJ, Boyce BN, Favretti RJ (1976) The contribution of trees to residential property value: Manchester, Connecticut. Valuation 23(2):26–43

Mumford L (1961) The city in history: its origins, its transformations, and its prospects. Harcourt, Brace & World, New York

Murphy P (2008) Economy and affect: people-place relationships and the metropolis. In: Ruan X, Hogben P (eds) Topophilia and topophobia: reflections on twentieth-century human habitat. Routledge, London

Neo H, Savage VR (2002) Shades of green, fields of gold: representations, discourse and the politics of golf in Singapore. Landscape Res 27(4):397–411. doi:10.1080/01426390220000023952

Neo BS, Gwee J, Mak C (2012) Growing a city in a garden. In: Gwee J (ed) Case studies in public governance: building institutions in Singapore. Routledge, Oxon, pp 11–64

NYCEDC (2011) August 2011 economic snapshot. New York City Economic Development Corporation, New York

NYCPCC (2013) Climate risk information 2013: observations, climate change projections, and maps. New York City Panal for Climate Change, New York

Olmsted FL (1865) The Yosemite valley and the Mariposa Big Tree Grove

Olmsted FL (1886) Notes on the plan of franklin park and related matters. Boston: Printed as a supplement to the City of Boston Eleventh Annual Report of the Board of Commissioners of the Department of Parks for the Year 1885.

Parks N (2015) Living with nature: annual report 2014/15. National Parks Board's Publication, Singapore

Polignac Fd (1995) Cults, territory, and the origins of the Greek city-state. University of Chicago Press, Chicago

Pregill P, Volkman N (1999) Landscapes in history: design and planning in the eastern and western traditions. John Wiley, New York

Price U (1810) Essays on the picturesque, as compared with the sublime and the beautiful; and, on the use of studying pictures, for the purpose of improving real landscape. Printed for J. Mawman, London

Rawcliffe C (2008) 'Delectable sightes and fragrant smelles': gardens and health in late Medieval and early Modern England. Garden Hist 36(1):3–21

Redman CL (2011) Social-ecological transformations in urban landscapes—a historical perspective. In: Niemelä J, Breuste J, Elmqvist T, Guntenspergen G, James P, McIntyre NE (eds) Urban ecology—patterns, processes and applications. Oxford University Press, Oxford, pp 206–212

Schuyler D (1986) The New Urban Landscape: the redefinition of city form in nineteenth-century America. John Hopkins University Press, Baltimore

Schama S (1991) Palaces and pleasures; Vaux-le-Vicomte: a perfect Chateau, Envied by a King. The New York Times, 20 Oct 1991

Sexby JJ (1898) Gardens and Open Spaces of London: Their history and associations. The municipal parks, London pp 625−634

Seeland K, Dübendorfer S, Hansmann R (2009) Making friends in Zurich's urban forests and parks: the role of public green space for social inclusion of youths from different cultures. Forest Policy Econ 11(1):10–17. doi:10.1016/j.forpol.2008.07.005

Slavicek LC (2009) New York City's Central Park. Chelsea House Publishers, New York

Taylor HA (1995) Urban public parks, 1840-1900: design and meaning. Garden Hist 23(2):201–221. doi:10.2307/1587078

Thériault M, Kestens Y, Des Rosiers F (2002) The impact of mature trees on house values and on residential location choices in Quebec City. In: Rizzoli AE, Jakeman AJ (eds) First biennial meeting of the international environmental modeling and software society 478–483

Tidball KG (2014) Seeing the forest for the trees: hybridity and social-ecological symbols, rituals and resilience in postdisaster contexts. Ecol Soc 19(4). doi:10.5751/ES-06903-190425

Troy A, Morgan Grove J, O'Neil-Dunne J (2012) The relationship between tree canopy and crime rates across an urban–rural gradient in the greater Baltimore region. Landscape Urban Plann 106(3):262–270. doi:10.1016/j.landurbplan.2012.03.010

Tuan YF (2008) Time, space and architecture: some philosophical musings. In: Ruan X, Hogben P (eds) Topophilia and topophobia: reflections on twentieth-century human habitat. Routledge, London, pp 31–43

Tucker DM (1993) Kitchen gardening in America: a history. Iowa State University Press, Ames

Ulrich R (1984) View through a window may influence recovery from surgery. Science 224 (4647):420–421. doi:10.1126/science.6143402

UNEP (2005) Coastal area pollution: the role of cities. United Nations Environment Program, Nairobi, Kenya

van den Berg M, Wendel-Vos W, van Poppel M, Kemper H, van Mechelen W, Maas J (2015) Health benefits of green spaces in the living environment: a systematic review of epidemiological studies. Urban Forest Urban Greening 14(4):806–816. doi:10.1016/j.ufug.2015.07.008

Verbrugge LM, Taylor RB (1980) Consequences of population density and size. Urban Aff Rev 16 (2):135–160. doi:10.1177/107808748001600202

Ward Thompson C (2011) Linking landscape and health: The recurring theme. Landscape Urban Plann 99(3–4):187–195. doi:10.1016/j.landurbplan.2010.10.006

Wheeler K, Nauright J (2006) A global perspective on the environmental impact of golf. Sport Soc 9(3):427–443. doi:10.1080/17430430600673449

Wigley M (1999) The Electric Lawn. In: Teyssot G (ed) The American lawn. Princeton Architectural Press with Canadian Centre for Architecture, Montréal, New York, pp 154–195

Wilby RL (2003) Past and projected trends in London's urban heat island. Weather 58(7):251–260. doi:10.1256/wea.183.02

Wilkinson PF (1988) The historical roots of urban open space planning. Leisure Stud 7(2):125–143. doi:10.1080/02614368800390121

Wilson WH (1989) The City Beautiful movement. Creating the North American landscape. The John Hopkins University Press, Baltimore

Wolch JR, Byrne J, Newell JP (2014) Urban green space, public health, and environmental justice: the challenge of making cities 'just green enough'. Landscape Urban Plann 125:234–244. doi:10.1016/j.landurbplan.2014.01.017

Wolf KL (2009) Strip malls, city trees, and community values. Arboric Urban Forest 35(1):33–40

Wolfe MK, Mennis J (2012) Does vegetation encourage or suppress urban crime? Evidence from Philadelphia, PA. Landscape Urban Plann 108(2–4):112–122. doi:10.1016/j.landurbplan.2012.08.006

Part II
The Functions of Urban Green Spaces

Chapter 4
Urban Greening and Microclimate Modification

Evyatar Erell

Abstract Vegetation is promoted widely all over the world as a means of creating a better quality of life in cities. Plants are credited with lowering air temperature and development of green spaces is considered one of the main strategies for mitigating the urban heat island. This chapter examines the mechanisms by which plants can modify the urban microclimate, with an emphasis on air temperature and outdoor thermal comfort. It outlines a scheme for classifying urban vegetation according to its location in the city and its intended role, which may be useful for planners and landscape architects. The chapter concludes with a methodology for integrating vegetation in the urban planning process to best achieve the desired microclimatic effects.

Keywords Air temperature · Mean radiant temperature · Evapotranspiration · Thermal comfort · Urban heat island (UHI) · Park cool island (PCI)

4.1 Introduction

Plants are familiar to everyone and are regarded almost universally as desirable. Among other benefits, vegetation is credited with contributing to the microclimate of urban areas, providing shade and reducing air temperature. As cities all over the world seek to mitigate the effects of urban heat islands and to adapt to potentially harmful effects of global climate change, green infrastructure is increasingly seen as a key component of sustainable urban design.

Microclimate modification by means of integrating vegetation in built-up areas typically has two broad aims: to improve the environment for people outdoors and to improve the environmental performance of buildings. Each of these aims will be introduced in brief, but this chapter focuses on the former. Readers interested in the effect of vegetation on buildings are encouraged to read Chap. 11.

E. Erell (✉)
Ben-Gurion University of the Negev, Beersheba, Israel
e-mail: erell@bgu.ac.il

P.Y. Tan and C.Y. Jim (eds.), *Greening Cities: Forms and Functions*,
Advances in 21st Century Human Settlements,
DOI 10.1007/978-981-10-4113-6_4

People in modern societies spend approximately nine-tenths of their lives indoors: whether working, playing, eating or sleeping—we rely on controlled conditions indoors to provide an appropriate environment for most of our daily activities. We mostly venture outdoors while travelling from one building to another—even when the weather conditions are favourable to outdoor activity. Yet most people, when asked, report that they enjoy spending time outdoors. We often go outdoors on vacation, but we could also spend much more time outdoors during our daily routine, if conditions were comfortable. Time spent outdoors is less time spent indoors, less energy required to air condition buildings—and is also healthier.

Buildings are physical enclosures that define part of space set apart from the surroundings to varying degrees. Building occupants increasingly breathe air that is heated or cooled by mechanical systems, require artificial light and consume substantial amounts of energy. Energy consumption may be reduced by environmentally conscious design, but the building is affected by ambient conditions, by the presence of neighbouring buildings and by the characteristics of its surroundings. Judicious use of vegetation may reduce the environmental load on a building, but in dry countries such as Israel or Australia, plants may require irrigation, and their contribution in dense urban locations might be limited. How can vegetation contribute to more efficient operation of buildings?

Before describing in detail the processes through which plants interact with their surroundings, it may be useful to consider the context in which their effects will be assessed. This requires a clarification of two terms: 'microclimate' and 'urban'. 'Climate' may be defined as 'the average course or condition of the weather at a place, usually over a period of years, as exhibited by temperature, wind velocity, and precipitation' (Merriam-Webster). A certain climate may be attributed to a large area or to a specific location. Accordingly, a 'microclimate' is the climate of a very small space, which in spite of its small physical extent experiences conditions that differ from those of the surrounding area in a meaningful way (Erell et al. 2011). A microclimate may affect an area that may range in size from as little as a few centimetres to several kilometres. A microclimate is considered distinct from the general climate of the region as the result of the interaction between the atmosphere and the surface. This interaction is particularly complex in cities due to the great variability they exhibit in their land use and land cover (LULC). This variability has led to the development of a classification scheme that identifies 'Local Climate Zones' (LCZs) that differ from each other in systematic ways, and may be used to describe the whole continuum of built-up areas (Stewart and Oke 2012). Thus, the use of the general term 'urban' in the context of microclimate implies a certain simplification.

Plants affect their surroundings in complex ways, and the mechanisms by which they do so are often over-simplified and misunderstood. Furthermore, because their effects are so multi-faceted, it is easy to emphasize one aspect and neglect others. It is beyond the scope of this chapter to address all of these effects in a comprehensive manner. It will therefore focus on the effect of vegetation on air temperature, a topic that has received great attention in recent years. After reviewing the evidence for reduction of air temperature, the chapter will discuss the effect of plants on pedestrian thermal comfort, with an emphasis on warm climates. The chapter concludes by proposing a methodology to promote integration of vegetation in urban planning.

4.2 Measuring Urban Air Temperature

Numerous studies have investigated the effect of vegetation on air temperature. Reducing air temperature in built up areas, sometimes referred to as 'urban heat island mitigation', is in fact often described as one of the primary benefits of plants. Regrettably, the magnitude of this effect is often over-stated. Far too much focus has been accorded to air temperature: it is only one factor affecting human thermal comfort, and only one factor among many that affect building energy consumption. However, the fact that it appears to be easy to measure and that people seem to understand it readily has made air temperature a headline item.

Despite the wealth of studies on this topic, the evidence for the extent of temperature modification is in fact somewhat mixed. This is partly due to the type of study: empirical studies based on field measurements, remote sensing or simulation studies using computer modelling. Each of these methods has inherent drawbacks, as well as advantages.

Although outdoor air temperature is possibly the most widely recorded meteorological datum, measuring it accurately in the urban environment is in fact quite difficult. This is because temperature sensors are affected not only by convective exchange with the air, but also by radiant exchange with their surroundings. To counter error resulting from radiant exchange, instrument screens are used. However, even the venerable Stevenson screen used in most standard weather stations is incapable of proving adequate shielding from intense sunlight in warm conditions with little wind. Temperature readings from such stations in these conditions may be higher than the 'true' air temperature by 2 K or more (Erell et al. 2005). The magnitude of the error depends not only on the environmental conditions, but also on the physical dimensions of the sensor: larger sensors are more prone to error through insufficient radiant protection, while sensors smaller than about 0.1 mm in diameter need no shielding in most conditions. Some sensor-shield combinations may result in errors of 5 K or more!

To obtain a simultaneous record of temperature across a large area such as a city, many observational studies employ remote sensing, mostly satellite-based imaging. The surface temperature recorded must then be processed to estimate air temperature. Such images invariably show cooler vegetation contrasted with warmer road or building surfaces. The surface temperature of vegetation is indeed often lower than that of concrete or asphalt, but the processing of remote sensing data to derive canopy-level air temperature in complex urban environments is difficult and may be inaccurate. Urban areas pose a particularly difficult challenge, because the satellite 'sees' a two-dimensional surface (roads, roofs and the ground) while cities have complex three-dimensional surfaces that often display marked thermal anisotropy.

Urban geometry thus affects the microclimate but cannot be fully described by remote sensing (Voogt and Oke 2003). Satellite images also suffer from relatively low spatial resolution or limited temporal coverage.

4.3 Mechanisms for Air Temperature Reduction

How do plants lower air temperature? In the absence of advection, air temperature is governed mainly by the energy balance at the surface. A warm surface will give off more heat by convection to the air in contact with it than a cooler surface exposed to the same environmental conditions. Plants may lower surface temperature by reducing exposure to sunlight or by evaporative cooling.

The magnitude of evaporative cooling depends on the balance between the rate of transpiration by the trees and the rate at which water vapour is transported away from the canopy, by wind or diffusion. The transpiration rate is limited by water availability (Rahman et al. 2011), which is frequently restricted in streets where nearly the entire surface is paved. Additionally, transpiration depends on the environmental conditions: warm sunny conditions promote photosynthesis, so that the stomata of leaves are open, but stomata may be closed under excessive heat and high radiant loads to reduce water loss. The environmental stress is often exacerbated in trees surrounded by pavement (Kjelgren and Montague 1998), leading to reduced transpiration and thus less cooling. Finally, the transpiration rate is related to the species and the size of the canopy.

Two measures of evapotranspiration are relevant in this context. The total annual water loss is a guide to irrigation requirements where there is insufficient precipitation to sustain the trees or where there is an extended dry season. The maximum hourly evaporation rate is an indication of the cooling potential of the tree by this mechanism, which may vary in response to weather and water availability.

Field measurements of evapotranspiration rates in urban trees are rare. Studies of trees in forests may provide some useful estimates of the range of water loss rates, though they are necessarily representative of the species in question and the conditions in the locality, especially soil moisture and climate. A series of studies in Mediterranean conditions (Schiller et al. 2002, 2003, 2007) found that the annual water loss ranged from only 4 L/tree/day (*Quercus calliprinos*, Palestine Oak, a broadleafed evergreen) to as much as 56 L/tree/day (*Quercus ithaburensis,* Tabor Oak). The maximum hourly and daily evapotranspiration rates in each species occurred at different times of the year. The large range of values reported may explain some of the discrepancy among studies of the evaporative cooling effect of urban trees.

In the absence of measured evapotranspiration data, evapotranspiration for a particular crop or species of tree (ET_C) may be estimated by multiplying a crop coefficient (K_C) by the reference evapotranspiration (E_{TO}), which is the evapotranspiration from a tall, cool-season grass in a well-watered field (Garbesi 1992):

$$ET_C = K_C E_{TO}$$

The crop coefficients for a large variety of trees (as well as shrubs) may also be found in the same publication (Garbesi 1992), and range from a low of 0.04 (*Pinus canariensis*, Canary Island Pine) to as much as 1.32 (*Podocarpus macrophyllus*, Yew Pine).

For a single tree, the volumetric rate of evapotranspiration (V_{ET}) is obtained by multiplying the crown area (A_C) by the evapotranspiration rate of the tree (ET_C):

$$V_{ET} = ET_C A_C = K_C ET_O A_C$$

Computer simulation of the evaporative cooling effect of typical street trees has yielded estimates of between 0.7 K (Saxena 2001) and 1.6 K (Gromke et al. 2015). This range of results reflects not only different simulation techniques but also the underlying assumptions regarding the tree characteristics, water availability and environmental conditions.

The contribution of evapotranspiration to the cooling effect of trees may be evaluated in the field by comparing the vapour content of the air beneath the canopy with the air in adjacent areas with no trees. Shashua-Bar and Hoffman (2000) noted that in a Tel Aviv avenue the monitored increase in moisture was negligible, suggesting a limited role for evaporation. This also suggests that although vegetation may be more abundant in humid regions than in dry climates, the potential for cooling by evapotranspiration is most likely limited because the air is already close to saturation.

Vegetation may also reduce surface temperature by intercepting some of the incident sunlight: in other words—by casting a shadow. The sunlight thus absorbed is partly transformed into chemical energy through the process of photosynthesis. Some of the energy is transformed into latent heat, as water is converted to vapour by evapotranspiration. Thus, although plants have an albedo of only 0.15–0.25, and absorb almost as much solar energy as fresh asphalt, their surface remains much cooler, so less sensible heat is given off to the air.

The importance of shading as a cooling mechanism implies that cooling by trees may be more substantial in climates with intense solar radiation, such as deserts or warm Mediterranean regions, than in overcast tropical areas.

4.4 Classification of Urban Vegetation

There is a great diversity of studies on the effect of vegetation on air temperature, which have been carried out in different climates, different times of the day and of the year, with different kinds of vegetation in diverse urban contexts. Bowler et al. (2010) suggested that although

> Most studies investigated the air temperature within parks and beneath trees and are broadly supportive that green sites can be cooler than non-green sites… the current evidence base does not allow specific recommendations to be made on **how** best to incorporate greening into an urban area.

Comparing results and interpreting them requires a classification scheme. The primary feature adopted here is the location of the vegetation in the urban setting, which may be building-integrated (roof or facade); on the building lot; along the street; or in parks and public gardens of various sizes and shapes. Other classification schemes may be equally valid, but this method is deemed most likely to generate conclusions that may be applied usefully in urban planning and landscape architecture. A secondary classification is based upon the type of vegetation, making a distinction between surface-cover vegetation (especially grass), and trees. Trees may create a shaded volume of air accessible to pedestrians beneath them, while surface-cover plants do not (although they may still shade the soil surface itself). Finally, all green areas may have localized effects on air temperature, but their effects may also extend to adjacent areas, the extent of which has been the subject of several investigations.

4.4.1 Parks

Most of the studies of the effects of vegetation on the urban microclimate compare conditions in parks and gardens of various sizes with conditions in adjacent built-up areas. As noted above, it is generally acknowledged that air temperature in urban parks is lower than adjacent streets, a condition reflected in the term 'Park Cool Island' (PCI). However, parks are not necessarily cooler than built-up areas at all times, and parks may display different temperature patterns throughout the day. This is because parks vary not only in size but also in the type of vegetation and surface cover.

During daytime, the temperature measured at a surface is affected by the presence or absence of shade, albedo, water availability and thermal properties of the underlying soil. These properties govern the receipt of solar radiation, its absorption and the role of evaporative cooling. At night, the thermal properties of surfaces and the radiative geometry are the major controls on cooling. Urban parks vary substantially with respect to the above factors, and may be classified according to the arrangement of vegetation (Spronken-Smith and Oke 1998): grass; grass with tree border; savannah (grass with isolated trees); garden; forest; and multi-use (Fig. 4.1).

Park cool islands may develop either during the daytime or at night (Table 4.1). However, a given urban park will display a regular diurnal pattern, indicating that the formation of PCIs may be the result of a number of mutually exclusive factors. Daytime PCIs form as a result of the combined effects of soil moisture and shading (Spronken-Smith and Oke 1998): Trees shade the surface, while grass is typically cooler than most solid surfaces during the daytime if it is well-irrigated. The relative coolness of irrigated parks therefore peaks in the afternoon (forest type), or early

Fig. 4.1 Classification of parks (after Spronken-Smith and Oke 1998; *Photo credit* E. Erell)

Table 4.1 Characteristics of park cool islands (Erell et al. 2011, following Spronken-Smith and Oke 1998)

	Daytime PCI	Night-time PCI
Type of park	Irrigated park with substantial tree cover	Dry parks with sparse tree cover
Mechanisms involved	Evaporation and shading: Trees shade the surface, while grass is typically cooler than paved surfaces—if it is well-irrigated	Long-wave radiant cooling: Sky view factor close to unity
Temporal pattern: time of maximum intensity	Afternoon (forest type) or early evening (garden, savanna and multi-use types)	Several hours after sunset
Comments		Warmer during the day than neighbouring urban areas

evening (garden, savannah and multi-use types). However, trees also inhibit nocturnal long-wave radiative cooling by blocking off part of the sky, while excess moisture increases the thermal storage of the soil and slows down surface cooling.

Night-time PCIs, on the other hand, typically form in relatively dry urban parks with a sparse tree cover. They are driven by long-wave radiative cooling, especially if there are few trees and no adjacent buildings. Areas where the sky view is reduced by obstructing features, such as perimeter trees or buildings, may be slightly warmer than exposed areas, at a distance of up to about 2.2–3.5 times the height of the object (Spronken-Smith and Oke 1998). In such parks, daytime temperatures may sometimes be higher than in neighbouring urban areas.

4.4.2 Streets

The consensus of most studies is that the presence of trees reduces peak daytime air temperature in streets. The magnitude of this reduction is, however, very frequently overstated, due to insufficiently-shielded sensors. Thus, although the improvement in pedestrian thermal comfort is quite real, it is mostly due to shading rather than to a reduction in air temperature. High-quality studies show a maximum reduction of 1–3 K during the warmest hours of the day, and in some cases the reported reduction in streets was negligible.

Quantifying the effect of street trees on air temperature at the individual tree or even street level is difficult because the cooling is localized and highly variable, both spatially and temporally (Coutts et al. 2016). The magnitude of daytime tree cooling depends on the amount of shading, street geometry and the local meteorological conditions. Trees have a greater effect in broad, shallow streets than in narrow, deep ones, where the urban morphology often overwhelms their effects. Furthermore, although trees may lower air temperature during daytime, they may limit nocturnal radiant cooling, so that dense avenues may feel warm and muggy at night in comparison to open spaces that are exposed to the sky and enjoy cooling breezes.

Although many studies attribute the cooling effect of street trees to evapotranspiration, this mechanism is most likely responsible for only a small portion of the temperature change. This is because photosynthesis takes place mostly in leaves that are exposed to sunlight—at the top of the tree canopy. Water vapour released by evapotranspiration tends to diffuse upwards, where vapour concentration is lower, or it may be advected by wind. The air beneath the canopy experiences little evaporative cooling. In fact, Shashua-Bar and Hoffman (2000) attribute some 70–80% of the cooling measured in a Tel Aviv avenue to shading by the tree canopies, rather than evapotranspiration.

4.4.3 Building Lots

Private gardens may have a substantial effect on the microclimate of low-rise neighbourhoods, and especially of detached single-family homes. They may also have an effect on the communal space of multi-story apartment buildings, but their direct contribution to the energy performance of the building declines as the number of storeys increases and the respective components of the building envelope are further removed from the vegetation.

A second consideration is the overall urban form, which has a great impact on the energy balance of the area in question. In fact, the massing of the buildings and the relation of the green space to the buildings may have a greater effect on temperature than the properties of the vegetation. Middel et al. (2014) report that in the case of a hot dry climate (Phoenix, Arizona), urban form has a larger impact on daytime temperatures than landscaping. Furthermore, small patches of grass planted in isolation do not result in a significant daytime cooling benefit compared to compact urban forms. Nevertheless, the study reports, as do numerous others, that mesic landscaping will tend to promote slightly cooler conditions than oasis plans which comprise relatively small, isolated patches of vegetation, while xeric landscaping that conserves water is likely to be warmest.

A study of the effect of several landscaping strategies on the microclimate of small courtyards by Shashua-Bar et al. (2009) may provide a useful indication of the maximum potential for reduction in air temperature by vegetation in typical urban gardens. The study, carried out in a hot-dry desert environment, reported a maximum reduction in air temperature of about 2 °C in a courtyard surrounded on three sides by buildings, shaded by trees and covered by grass (both irrigated liberally), compared to an identical adjacent courtyard with no vegetation. It is likely that the reduction in air temperature will be smaller in more humid conditions, where evaporative cooling is less effective; in more exposed sites (where the cooling effect is more likely to be dissipated by wind); if irrigation is more restricted; or if vegetation does not incorporate both trees and surface cover vegetation (which in combination provide the maximum cooling). It is of course quite possible that the cumulative effect of numerous gardens across an entire neighbourhood will generate greater cooling, or that the temperature reduction in large urban-scale gardens will be larger. But in most realistic landscaping plans the actual temperature reduction is likely to be modest. Any reduction of temperature will almost certainly require substantial inputs of water and will depend to a great degree on exposure to wind and the density of the buildings (Bonan 2000).

4.4.4 Building-Integrated Vegetation

The contribution of building-integrated vegetation to the urban microclimate has been the subject of numerous studies—yet there does not seem to be a consensus on the potential of green roofs and walls to reduce air temperature. Differences in reported findings are partly due to the different scenarios examined, and most certainly to the local climate and the base case used as a reference. Unfortunately, there have been no field studies that have been able to demonstrate a substantial reduction of air temperature that may be attributed to green roofs directly. This may be because the physical extent of most roofs is too small to generate a measurable effect, which may only be measured if a large proportion of the roofs in a dense urban area is covered by vegetation.

However, there are several factors that suggest that the potential for microclimate modification through green roofs is in fact fairly limited:

(a) Most green roofs have a shallow layer of soil, typically no more than 10–15 cm thick. This depth may only support grass or surface cover vegetation such as sedum, favoured in cool and dry climates because it is hardy and tolerates conditions of extreme drought and low temperature—but not trees or even bushes. Several studies of ground-based landscaping have shown that while grass lawns may reduce air temperature during the hot hours of the day, the effect is fairly small. There is no reason to expect that the effect of roof-supported cover vegetation will be greater.

(b) While surface-cover plants may have a surface temperature that is much lower than that of so-called 'black roofs', this is primarily due to evapotranspiration. This means that in the absence of regular precipitation, the cooling benefit of green roofs may be minimal, because the water storage capacity of the substrate is small. Although plants such as sedum may survive extended dry spells, their effect on air temperature in such conditions is minimal.

(c) Since roofs are exposed to airflow in the part of the atmospheric boundary layer described as the 'mixing layer', any cooling effect generated by evapotranspiration is more likely to be dissipated throughout the urban boundary layer than transported through the canopy layer to ground level.

In the absence of large-scale experimental studies of the microclimatic effects of green roofs, it may be useful to examine the results of thermal simulation. Li et al. (2014), using the Weather Research and Forecasting (WRF) Model to evaluate potential UHI mitigation strategies for the Washington-Baltimore urban area, found that replacing 50% of existing roofs with green roofs could reduce peak near-surface air temperatures during a heat wave by 0.26 K. This value is very sensitive to the assumed moisture content of the soil supporting the green roofs: A dry soil, such as might be expected after several dry days, results in a slight warming effect relative to the base case, while an increase in moisture to 0.45 m^3/m^3 (water-logged conditions) leads to a further reduction of air temperature by about 0.55 K.

4.4.5 Spatial Extent of Cooling

The spatial extent of cooling by street trees is fairly limited. Firstly, though this is often forgotten, the cooling effect is felt primarily downwind, and the upwind transport of cool air is limited to a relatively short distance (Upmanis and Chen 1999). Secondly, the diffusion of cool air is affected by the orientation and aspect ratio of adjacent streets: Cool air may be transported by wind along streets that intersect the street acting as the source, but if it is blocked by tall and dense buildings, it will be mixed with the above-roof air and have a much weaker effect on non-adjoining streets. Although as a rule of thumb the cooling effect of vegetation in an urban park may be felt up to a distance equal to about one park width away, the effect of a narrow linear pattern of street vegetation may be limited to a distance of only 100 m in most instances.

4.5 Human Thermal Comfort Outdoors

4.5.1 Assessing Thermal Comfort

Traditional comfort studies have sought to compute an energy balance for the human body, accounting for the environmental factors, clothing, and metabolic heat, as well as the potential cooling effect of sweating. The underlying assumption of these studies was that thermal comfort necessarily required thermal equilibrium, so that a person would most likely be comfortable when the body is neither gaining nor losing heat. The heat balance equations proposed in complete form by Fanger (1970) were, however, appropriate for indoor conditions, where net radiant exchange is typically fairly small, air speed is low and asymmetry in the thermal environment is limited. These conditions rarely apply to the outdoor environment, so several indices have been developed that seek to describe the effect of large and asymmetric radiant fluxes or strong winds.

Perhaps the two foremost indices of outdoor thermal comfort are the Physiological Equivalent Temperature (PET) (Hoppe 1999) and the Universal Thermal Climate Index (UTCI) (Jendritzky et al. 2012). Both indices are based on a complete energy balance between the human body and the surroundings, accounting for the effect of clothing and metabolic activity, but they are expressed as a 'temperature'. The PET is defined as '*the air temperature at which, in a typical indoor setting (without wind and solar radiation), the heat budget of the human body is balanced with the same core and skin temperature as under the complex outdoor conditions to be assessed.*' This approach creates a simple analogy between the compound effect of the prevailing outdoor conditions and a familiar indoor setting. Table 4.2 illustrates the effect of different combinations of environmental conditions on the value of PET.

Like PET, the UTCI was derived conceptually as an equivalent temperature (ET). Thus, for any combination of outdoor air temperature, wind, radiation, and humidity, the UTCI is defined as '*the isothermal air temperature of the reference condition that would elicit the same dynamic response (strain) of the physiological model*' (Jendritzky et al. 2012). UTCI incorporates an advanced adaptive clothing model that accounts for behavioural adaptation, different local insulation values for different body parts, and a reduction of the thermal and evaporative resistance of clothing caused by wind and movement of the wearer. UTCI values may be translated to categories of thermal stress on a 10-point scale ranging from 'extreme cold' to 'extreme heat'.

The calculation of both PET and UTCI requires inputs that may be obtained fairly easily, such as dry bulb air temperature, but also the mean radiant temperature (T_{mrt}), which is particularly difficult to assess in practice as it varies substantially

Table 4.2 Values of the physiological equivalent temperature for different environmental conditions

	T_a (°C)	T_{mrt} (°C)	v (m/s)	VP (hPa)	PET(°C)
Typical room	21	21	0.1	12	21
Winter, sunny	−5	40	0.5	2	10
Winter, shade	−5	−5	5.0	2	−13
Summer, sunny	30	60	1.0	21	43
Summer, shade	30	30	1.0	21	29

T_a is air temperature, T_{mrt} is the mean radiant temperature, v is wind speed and *VP* is the vapour pressure of the air

across very small distances.[1] The importance of this particular metric cannot be overestimated and it is of particular relevance in analysing the effect of vegetation on thermal sensation: The effect of plants on T_{mrt} is often far greater than their effect on air temperature or humidity (Shashua-Bar et al. 2011).

A different approach is adopted by a less well-known model, the Index of Thermal Stress (ITS), proposed by Givoni (1963) and developed by Pearlmutter et al. (2007, 2014). Suitable for warm conditions only, it is defined as '*the rate of sweat (in terms of its equivalent latent heat, in watts) required for the body to maintain thermal equilibrium with its surroundings through evaporative cooling*'. Calculation of this index requires input of all the radiant fluxes (which could also be used to estimate Tmrt), and is thus dependent on an accurate assessment of the surface temperature of all elements of the surroundings as well as the solar flux.

4.5.2 *Effects of Vegetation on Human Thermal Sensation*

The use of the ITS to estimate the effects of vegetation on human thermal sensation in hot dry climate was demonstrated in a study carried out at Sde Boqer, Israel (Shashua-Bar et al. 2009, 2011). Two adjacent courtyards with different landscaping treatments were monitored in detail, recording a wide range of environmental parameters. The ITS was calculated for six different combinations of surface cover (paved or grass) and shade (trees, shade mesh or no shade).

Figure 4.2 shows that replacing bare soil or pavement with grass is sufficient to reduce thermal stress on a standing person by as much as 200 W, resulting in a shift of thermal sensation from 'very hot' to 'warm'. The improvement is almost entirely

[1]Estimates of T_{mrt} may be obtained from field measurements employing the globe thermometer (Thorsson et al. 2007). Computer simulation of T_{mrt} is provided by programs such as ENVI-Met (Bruse and Fleer 1998), which computes the temperature of the ground, building and vegetation and requires extremely detailed inputs; or simpler tools such as RayMan (Matzarakis et al. 2007) or SOLWEIG (Lindberg et al. 2008), which do not require input of the thermal properties other than albedo of the surfaces and which thus rely on an approximation of the surface temperatures.

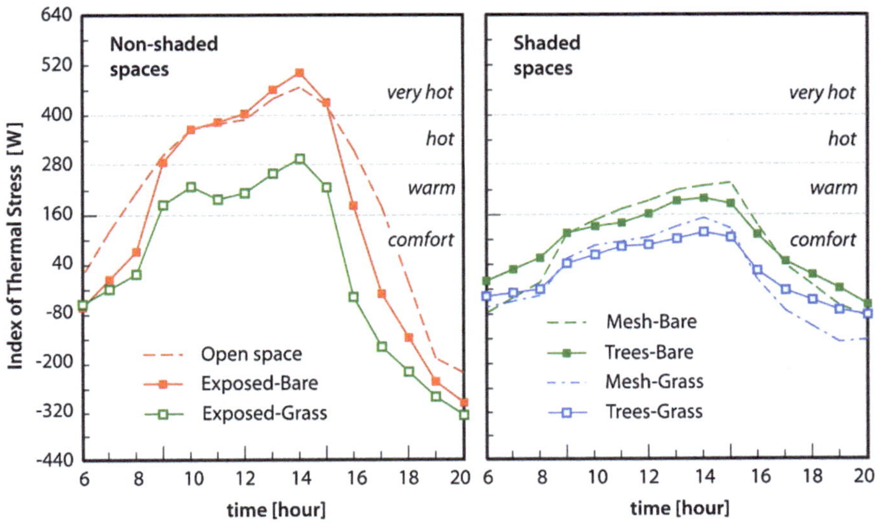

Fig. 4.2 Reduction in thermal stress by different landscaping strategies (Shashua-Bar et al. 2011)

due to the reduction in long-wave radiation emitted by the surface, as grass was much cooler than either the soil or concrete pavers. This significant improvement was obtained even though air temperature near the surface was reduced by less than 1 °C.

An even greater decrease in heat stress was achieved by adding shade, either a tree canopy or fabric mesh that provided a comparable reduction of solar radiation (about 70%). In typical summer conditions at the Negev desert, shade by itself is almost sufficient to offer thermal comfort. The addition of grass in combination with shade can further suppress thermal stress, sufficient to provide comfort. It should be emphasized that although the vegetation, which was irrigated liberally during the experiment, lowered air temperature by up to 2.5 °C during the hottest hours of the day—this in itself provided only minor relief from the heat. Rather, the major contribution to comfort was the reduction in radiant flux (both short-wave and long-wave). Figure 4.2 shows that the shade provided by an artificial mesh, which led to no reduction in air temperature, gave only slightly less benefit to thermal sensation compared to trees—either in combination with grass or alone.

As demonstrated in this experiment, surface-cover vegetation may have an important contribution to thermal comfort. This effect is often attributed to the reduction of air temperature, which is typically rather small. In fact, the effect of surface-cover vegetation on radiant exchange is often more important. This is because uniquely among all of the surfaces to which a person may be exposed, vegetation combines a very low albedo with a very low surface temperature. This means that a lawn reflects as little as 20% of incident sunlight, about the same as aged asphalt, but emits considerably less infrared radiation than other surfaces.

Grass is by far the most common type of surface-cover vegetation, and urban lawns are in fact the most widespread 'crop' in the United States (Milesia et al. 2005). However, grass is also a voracious consumer of water, and may require anything from 0.5–3.0 m^3/m^2/year of irrigation water in hot dry climates. Other surface-cover plants, notably various succulents, may contribute to a reduction of thermal stress almost as much as grass, while requiring only a fraction of the water (Snir et al. 2016). This is because these plants limit evapotranspiration during the daytime to conserve water, but their daytime temperature remains only a little higher than grass.

Figure 4.3 illustrates the effect of planted surfaces on the thermal stress of a person standing in full sunlight, in conditions typical of Sde Boqer in the summer. The calculation is based on measured meteorological conditions and surface temperatures, and thermal stress is estimated using the ITS. It should be emphasized that the changes in thermal stress are independent of possible effects of the surface cover on air temperature, because a small patch of vegetation would have little effect on temperature.

It is instructive to note that although artificial turf resembles natural grass in appearance and texture, it does not carry out photosynthesis and is not cooled by evapotranspiration. It becomes extremely warm, with surface temperatures over 70 °C observed during the experiment in Sde Boqer. It is thus a very poor alternative to natural grass in thermal terms, at least in hot climates.

Air movement is essential to outdoor human thermal comfort in warm-humid conditions (Sharma and Ali 1986; Mueller et al. 2014), and calculation of widely used indices of thermal stress such as the PET (Hoppe 1999) and UTCI (Jendritzky et al. 2012) requires wind data. The benefits of trees, expressed as reduction of air temperature during the daytime and especially their contribution to lowering radiant

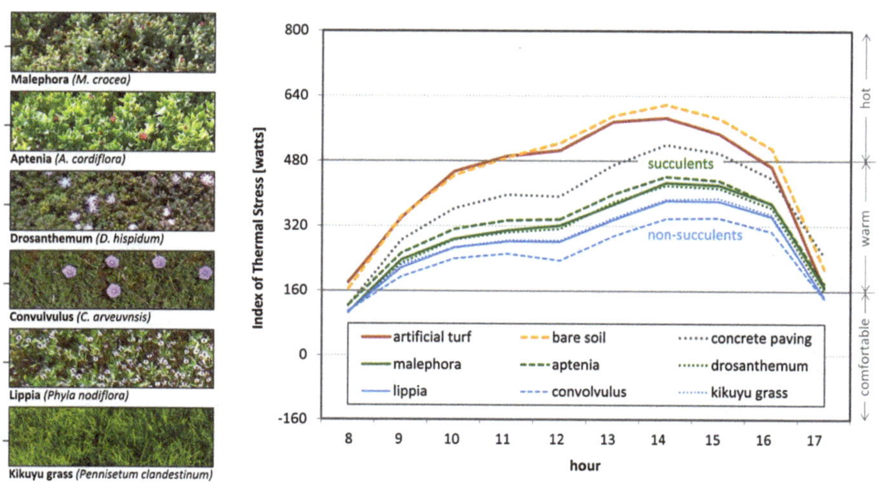

Fig. 4.3 Calculated values of the Index of Thermal Stress (ITS) over the daytime hours on a summer day, for a pedestrian standing in a hypothetical open space with different ground surface types. Plant types are illustrated on the *left* (Snir et al. 2016)

loads, have obscured their effect on airflow. The reduction in wind speed by vegetation has been studied in a cold-climate context, where trees and especially hedges are employed as wind breaks (Wang and Takle 1996), but there is little systematic research of the impacts of different species and different planting schemes in warm climates.

Finally, it should be noted that human thermal comfort is affected not only by the 'objective' environmental conditions—which may be measured, and the body's own behavioural and physiological response to these conditions (including change in clothing)—which may also be measured, but also by complex social and psychological influences that are harder to assess. The role of vegetation is considered beneficial in most cases beyond its direct effect on the environmental factors, which may account for the widely held belief that vegetation 'cools' the environment—a sentiment that is often expressed even after the sum of the effect of plants on actual heat exchange processes is accounted for.

4.6 Integrating Vegetation in Urban Areas

To be successful, the integration of vegetation in urban design must be carried out with great care and attention to detail. As illustrated in Sect. 4.5, the contribution of vegetation may be difficult to assess even within a specific context. Practices shown to be effective in one location may not deliver similar results elsewhere. Not only may the primary effect be difficult to duplicate, but the side-effects may result in an overall degradation of the environment through unintended consequences. This section will not provide specific recommendations. Rather, a methodology will be proposed to reduce the likelihood of such an occurrence, and to promote guidelines for good practices that may have wide applications.

The planning process should comprise the following steps:

4.6.1 Goal Setting

The expected benefits of the vegetation should be defined, so that appropriate planting strategies may be selected. The objectives must be specific and appropriate to the context. For example: maximizing pedestrian thermal comfort in warm summer afternoons may be achieved through radiant control and maximizing airflow, rather than the more general 'heat island mitigation'. The list of outcomes should be comprehensive, and include a list of positive goals as well as attention to potential undesirable effects that should be avoided.

4.6.2 Greening Location

The location of the vegetated areas should be selected in accordance with their roles—in conjunction with the availability of appropriate land and the potential for integration with other elements of the landscape, whether they are natural or man-made.

The first step is to identify the location of existing natural vegetation, often near streams. Preservation of natural environments within the city may be easier than establishing new ones in artificial parks. The second step is to identify existing or potential links among green areas in the city. A network of vegetation comprising parks and linear elements along streets is far more effective in supporting desirable microclimates than isolated parks, grand as they may be. Once the outlines of the network become apparent, it may be possible to identify the 'missing links'—locations where additional vegetation would not only enhance the local environment but could also contribute to the integrity and comprehensiveness of the network as a whole.

A network of vegetation has an intrinsic value, and may also support ecological corridors that allow movement of animals across the urban landscape (see Chap. 12). Networks have an additional value if they also support human activity. This means that parks not only modify microclimate, but serve as 'destinations' at different functional and physical scales, ranging from small neighbourhood gardens to vast metropolitan parks, where people may spend leisure time. If these destinations are then connected by linear elements such as boulevards, which are also enhanced by vegetation, the network may provide a functional alternative to the road network. If the parks are disconnected, pedestrians may be forced to rely on automobiles or public transport to access and to enjoy the amenities.

The procedure outlined here focuses on vegetation in parks and streets, which are almost invariably public, rather than building-integrated plants or vegetation on individual lots, which are typically owned privately, or are not accessible to the public. The desirable cross-section of streets is of particular importance in this context, both because of the challenges they may impose on the planner and because streets form the backbone of cities and are therefore—potentially—the primary pedestrian spaces.

The relationship between vegetation and water is also important. This will require prior analysis of the urban watershed, to determine the location and magnitude of water flows after storm events; possible points for intervention, and particularly where vegetation may be integrated with surface runoff control elements; and the contribution to quality of life that may be obtained from such intervention.

4.6.3 Size of Planted Area

The size of the planted areas will in the first place depend on the availability of suitable sites in the existing (or planned) urban fabric, and the intended function of the vegetation. For example, a linear element comprising shade trees along a sidewalk, supported only by seasonal flows of storm water; or a small neighbourhood garden shaded by trees but also including a small landscaped area supported by water from a larger source area, fed to a bio-filter and irrigated as necessary throughout the year.

As explained above, the potential for urban-scale modification of the urban air temperature through planting is probably limited. Furthermore, temperature reduction has a relatively minor effect on thermal comfort. Therefore, the role of the vegetation is local: to enhance thermal comfort in designated areas which people will stay or pass through. The dimensions of such spaces will thus be determined not by considerations of urban climate, but primarily by their functional requirements. Landscaping, especially trees but also surface-cover vegetation, may then be employed to make them more comfortable.

4.6.4 *Type of Vegetation*

The type of vegetation will be appropriate to its proposed function within the urban landscape (Fig. 4.4). As noted above, the potential of trees to ameliorate microclimate is generally much greater than surface cover plants, and grass in particular. As a general rule, indigenous plants are preferable to exotic species, as most landscape architects will testify. However, the contribution of local species to microclimate may not always be comparable to non-native species. This is particularly true in desert climates, because local species have evolved to be frugal in their water consumption. Hence, their potential for evaporative cooling is very limited. To reduce exposure to the excessive solar radiation, their canopies may be small in area, and thus cast only a very small shadow. It is in fact very difficult to find tree species that are well-suited to the desert, simply because deserts are not home to many trees.

Fig. 4.4 Selection of trees in parking lots. Palm trees may be transplanted easily, even fully-grown, and require less maintenance—but contribute very little to the microclimate of the parking lot in an urban mall in Be'er Sheva (*right*); Red maples provide shade to the small parking lot in Beaverton (*left*), but require much more maintenance *Photo credit* E. Erell)

The key factor in most urban landscaping designs will be the selection of trees. In addition to the suitability of the species to the local climate and soil, the designer may consider the extent and location of the area shaded by the trees and the degree to which the trees reduce wind speed. A dense continuous tree canopy may provide welcome shade in a warm tropical climate, but might also block breezes that could mitigate the heat stress. A compromise may be found—but this requires not only familiarity with the local tree species and the dimensions of their canopies when fully grown, but also an intimate understanding of the local wind regime.

4.6.5 Integration with Storm Water Management

Plants are a major component of urban water management practices. Blue-green infrastructure is dealt with at some length in a separate chapter, so it may just be worthwhile to emphasize here that the design of water management elements should not focus solely on the most efficient means of reducing total runoff, moderating peak flows and minimizing water contamination—to the exclusion of the microclimatic benefits of plants. The type of water infrastructure element and its location should also take into account the added amenity provided by such elements, including benefits to the microclimate from the presence of vegetation. This will require coordination of plans at different spatial levels, and indeed among different legal entities: vegetation on individual building lots, including green roofs, has a major role to play in urban water management, as do nature strips accessible to the public but maintained by the owners of the adjacent properties. Both may also have a key contribution to the urban microclimate.

Vegetation incorporated in storm water management elements is generally expected to require no irrigation. However, in Mediterranean climates, which have a prolonged hot, dry season with no precipitation, most types of vegetation—with the important exception of some tree species—require irrigation to prosper, or even to survive. Thus, the planning of a water-sensitive urban environment that incorporates vegetation must not only consider the potential benefits but also the cost in terms of water for irrigation. Studies carried out in Israel have demonstrated that trees are preferable to surface-cover vegetation in terms of their cooling efficiency, defined as the ratio between the sensible cooling effect and the latent heat of evaporation of water required to generate the desired effect (Shashua-Bar et al. 2009). The outcome may be measured as the reduction in thermal load on a pedestrian in the street or on the roof of a building.

4.6.6 Maintenance

Almost all plants in urban areas require maintenance (with the possible exception of large 'nature reserves' established within a metropolitan area). Maintenance costs often explain the lack of investment in vegetation in public areas such as parking lots. The construction cost of a large asphalt parking lot is undoubtedly smaller if it has no trees. But the contributions of trees to thermal comfort in hot climates should outweigh such considerations.

4.7 Conclusion

The capacity of vegetation to lower urban air temperature is sometimes overstated, but its contribution to the urban microclimate is nevertheless real. A simplified approach regarding the integration of vegetation in urban landscaping could follow the general tenet that 'the more the better'. While this may sometimes be true, urban vegetation almost always comes at a cost—of precious land, scarce water or expensive maintenance. Thus a more nuanced approach is preferable, which seeks to integrate vegetation where its benefits are best enjoyed and where costs are minimal, accounting for the complex and sometimes unexpected side-effects of plants. Amelioration of microclimate at the local and urban scale is a worthy objective, which is most likely to be achieved if landscape architects and urban planners jointly promote such goals.

References

Bonan GB (2000) The microclimates of a suburban Colorado (USA) landscape and implications for planning and design. Landscape Urban Plann 49:97–114

Bowler D, Buyung-Ali L, Knight T, Pullin A (2010) Urban greening to cool towns and cities: a systematic review of the empirical evidence. Landscape Urban Plann 97:147–155

Bruse M, Fleer H (1998) Simulating surface-plant-air interactions inside urban environments with a three dimensional numerical model. Environ Model Softw 13:373–384

Coutts A, White E, Tapper N, Beringer J, Livesley S (2016) Temperature and human thermal comfort effects of street trees across three contrasting street canyon environments. Theoret Appl Climatol 124:55–68

Erell E, Leal V, Maldonado E (2005) Measurement of air temperature in the presence of a large radiant flux: an assessment of passively ventilated thermometer screens. Bound-Layer Meteorol 114:205–231

Erell E, Pearlmutter D, Williamson T (2011) Urban microclimate: designing the spaces between buildings. Earthscan/James & James Science Publishers, London

Fanger PO (1970) Thermal comfort: analysis and applications in environmental engineering. McGraw-Hill, New York

Garbesi K (1992) Estimating water use by various landscaping scenarios. In: Akbari H, Davis S, Dorsano S, Huang J, Winnett S (eds) Cooling our communities: a guidebook on tree planting and light coloured surfacing. Lawrence Berkeley National Laboratories and the US Environmental Protection Agency, Washington, DC, pp 157–172

Givoni B (1963) Estimation of the effect of climate on man—development of a new thermal index. PhD Thesis, The Technion—Israel Institute of Technology

Gromke C, Blocken B, Janssen W, Merema B, van Hooff T, Timmermans H (2015) CFD analysis of transpirational cooling by vegetation: case study for specific meteorological conditions during a heat wave in Arnhem, Netherlands. Build Environ 83:11–26

Hoppe P (1999) The physiological equivalent temperature—a universal index for the biometeorological assessment of the thermal environment. Int J Biometeorol 43:71–75

Jendritzky G, de Dear R, Havenith G (2012) UTCI—why another thermal index? Int J Biometeorol 56:421–428

Kjelgren R, Montague T (1998) Urban tree transpiration over turf and asphalt surfaces. Atmos Environ 32:35–41

Li D, Bou-Zeid E, Oppenheimer M (2014) The effectiveness of cool and green roofs as urban heat island mitigation strategies. Environ Res Lett 9:055002

Lindberg F, Holmer B, Thorsson S (2008) SOLWEIG 1.0—modelling spatial variations of 3D radiant fluxes and mean radiant temperature in complex urban settings. Int J Biometeorol 52:697–713

Matzarakis A, Rutz F, Mayer H (2007) Modelling radiation fluxes in simple and complex environments—application of the RayMan model. Int J Biometeorol 51:323–334

Middel A, Häb K, Brazel AJ, Martin CA, Guhathakurta S (2014) Impact of urban form and design on mid-afternoon microclimate in Phoenix Local Climate Zones. Landscape Urban Plann 122:16–28

Milesia C, Elvidge C, Dietz J, Tuttle B, Nemani R and Running S (2005) A strategy for mapping and modelling the ecological effects of US lawns. International Society for Photogrammetry and Remote Sensing Archives XXXVI–8/W27.

Mueller N, Kuttler W, Barlag A (2014) Counteracting urban climate change: adaptation measures and their effect on thermal comfort. Theoret Appl Climatol 115:243–257

Pearlmutter D, Berliner P, Shaviv E (2007) Integrated modeling of pedestrian energy exchange and thermal comfort in urban street canyons. Build Environ 42:2396–2409

Pearlmutter D, Jiao D, Garb Y (2014) The relationship between bioclimatic thermal stress and subjective thermal sensation in pedestrian spaces. Int J Biometeorol 58:2111–2127

Rahman MA, Smith JG, Stringer P, Ennos AR (2011) Effect of rooting conditions on the growth and cooling ability of *Pyrus calleryana*. Urban Forest Urban Greening 10:185–192

Saxena K (2001) Microclimate modification: calculating the effect of trees on air temperature. MSc Thesis, Arizona State University

Schiller G, Ungar ED, Cohen Y (2002) Estimating the water use of a sclerophyllous species under an East-Mediterranean climate: I. response of transpiration of *Phillyrea latifolia* L. to site factors. For Ecol Manage 170:117–126

Schiller G, Unger ED, Moshe Y, Cohen S, Cohen Y (2003) Estimating water use by sclerophyllous species under east Mediterranean climate: II. the transpiration of *Quercus calliprinos* Webb. In response to silvicultural treatments. For Ecol Manage 179:483–495

Schiller G, Cohen S, Ungar ED, Moshe Y, Herr N (2007) Estimating water use of sclerophyllous species under East-Mediterranean climate: III. Tabor oak forest sap flow distribution and transpiration. For Ecol Manage 238:147–155

Sharma MR, Ali S (1986) Tropical summer index—a study of thermal comfort of Indian subjects. Build Environ 21:11–24

Shashua-Bar L, Hoffman M (2000) Vegetation as a climatic component in the design of an urban street: an empirical model for predicting the cooling effect of urban green areas with trees. Energy Build 31:221–235

Shashua-Bar L, Pearlmutter D, Erell E (2009) The cooling efficiency of urban landscape strategies in a hot dry climate. Landscape Urban Plann 92:179–186

Shashua-Bar L, Pearlmutter D, Erell E (2011) The influence of trees and grass on outdoor thermal comfort in a hot-arid environment. Int J Climatol 31:1498–1506

Snir K, Pearlmutter D, Erell E (2016) The moderating effect of water-efficient ground cover vegetation on pedestrian thermal stress. Landscape Urban Plann 152:1–12

Spronken-Smith RA, Oke TR (1998) The thermal regime of urban parks in two cities with different summer climates. Int J Remote Sens 19:2085–2104

Stewart I, Oke TR (2012) 'Local Climate Zones' for urban temperature studies. Bull Am Meteorol Soc 1879–1900

Thorsson S, Lindberg F, Eliasson I, Holmer B (2007) Different methods for estimating the mean radiant temperature in an outdoor urban setting. Int J Climatol 27:1983–1993

Upmanis H, Chen D (1999) Influence of geographical factors and meterological influences on nocturnal urban-park temperature differences – a case study of summer 1995 in Goteborg, Sweden. Climate Res 13:125–139

Voogt J, Oke TR (2003) Remote sensing of urban climates. Remote Sens Environ 86:370–384

Wang H, Takle E (1996) On shelter efficiency of shelter belts in oblique wind. Agric For Meteorol 81:95–117

Chapter 5
Urban Greening and Its Role in Fostering Human Well-Being

Christine A. Vogt, Cybil Kho and Angelia Sia

Abstract Population growth coupled with urbanisation has led to a decline in natural ecosystems throughout the world. Particularly in cities, urban developments continue to displace natural ecosystems and lead to cities being dominated by concrete and steel. However, with increasing recognition of the benefits of human interaction with nature, planning and design professionals are now making more deliberate attempts to introduce greenery into the built environment. Indeed, the fields of urban planning, public health, and park planning provide a rich account of the role that urban greening plays in human well-being. History of urban planning and greenspace began in Europe and America in the early 1800s. Early park settings were intended to benefit urban dwellers and factory workers who lacked exposure to clean air and greenery, whereas today planners develop green recreation areas for passive and active leisure pursuits. An interesting programme that started in the United States is 'Park Prescriptions', which is 'designed in collaboration with healthcare providers and community partners to utilise parks, trails, and open space for the purpose of improving individual and community health'. Horticultural therapy is another compelling initiative which promotes greater inclusion of greenery and active movement within healthcare settings, both inside buildings and the nearby environment. In such initiatives, park planners and managers work closely with urban planners and neighborhood developments to enhance access and leisure services in order to maximise associated physical and mental health and social benefits.

Keywords Parks · Landscapes · Health · History · Benefits

C.A. Vogt (✉)
Arizona State University, 411 N. Central Ave. Suite 550, Phoenix, AZ 85004, USA
e-mail: chrisv@asu.edu

C. Kho · A. Sia
National Parks Board, Singapore, Singapore
e-mail: kho_cybil@nparks.gov.sg

A. Sia
e-mail: angelia_sia@nparks.gov.sg

P.Y. Tan and C.Y. Jim (eds.), *Greening Cities: Forms and Functions*,
Advances in 21st Century Human Settlements,
DOI 10.1007/978-981-10-4113-6_5

5.1 Introduction

Population growth coupled with urbanisation has led to a decline in natural ecosystems throughout the world. Such changes in land composition are seen most starkly in cities, where dense urban developments with concrete and other building materials continue to displace natural ecosystems. In many urban developments, urban planners and landscape architects also attempt to reintroduce greenery but these often occupy a small footprint of the built-up spaces. In rural areas, where population growth is less, nature has a better chance of retaining its natural states, particularly where parks and green corridors have been purposefully preserved. Urban planners, urban citizens, developers, and policymakers in cities are beginning to collectively see that urban greening and parks are tied directly to quality of life and indirectly to economic value. The global cities of Paris, London, Tokyo, Singapore, New York City, and Chicago have long histories of creating parks and green areas within their growing cities. Parks and gardens afford urban dwellers a place that is different than their apartment or house where a yard is lacking or small. The landscapes were originally conceived as a playing or gazing ground for the elite and their grandiose houses. Urban parks originated as social places where adults and children could spend their leisure time and mingle as a family or with friends. The exposure to sunlight and fresh air along with being able to view growing trees, grass and flowers provide renewal to the otherwise stressed city worker or wealthy landowner. These reasons to be outdoors and in a leisure green space still resonate today.

This chapter captures the importance and contribution of parks and urban greening to human well-being. We will look into a few hundreds years of history. The British have always loved their parks and gardens, and during its colonisation history, introduced numerous parks in the areas which they colonised. More recent history illustrates that we continue to revere urban greenery. These views, that parks and greenery are essential for healthy living, have been substantiated with an increasing body of scientific evidence from social and environment disciplines highlighting a myriad of benefits. These benefits range from visual stimulation through looking at nature, yielding attention and focus benefits, to healing of patients in hospitals by engaging in horticultural activities. We propose that parks and green spaces are the basis of quality living in urban areas.

In this chapter we first provide a history of parks contributing to well-being. This section provides a historical overview of the theme of health in green space planning in early urban park movements. The second section features recent concepts of preventative health care in conjunction with the placement of parks and greenery close to where we live, work, and transit. Recent efforts around the world have touched on daily doses of nature and visits to parks as a way to overcome inactivity. The third section of the chapter provides an overview of the health and social benefits derived from park exposure. Modernisation has resulted in lifestyles that are increasingly sedentary and urban dwellers spend most, if not all, of their time indoors. Physical benefits from being outdoors and active in outdoor recreation, as well as the psychological benefits from observing nature and resting from work and household activities are described. The social benefits of being outdoors, whether alone or with others are highlighted. Along with benefits, an overview of the types of green spaces necessary to appeal to various users is profiled. The final section of the chapter considers how the evidence about parks contributing positive health and social outcomes can be leveraged in future plans, designs, and implementation efforts, particularly in urban settings.

5.2 History of the Modern Park's Contributions to Human Well-Being

A series of factors empowered the urban park movement starting in the 1830s in both Europe and America. Without precedence, these early modern societies were facing a set of unique challenges of an unfamiliar urban environment newly brought about by the Industrial Revolution in the 1800s. Rapid industrial growth and an influx of migrants resulted in urbanisation and crowded dwellings in industrial towns. In the city, dismal living conditions were especially pronounced for the working class who, residing in overcrowded districts, had few outlets for enjoyment (Jordan 1994; Taylor 1995). In Victorian society, the authorities recognised the pressing need to respond to the social issue of urbanisation, as highlighted in the Report from the Select Committee on Public Walks (Parliament House of Commons 1833) on the need for the provision of open space.

In response, one of the first public parks, Derby Arboretum, created by philanthropist Joseph Strutt was opened in England in 1840. Contrary to the idea of 'parks' as an exclusive space for the private affairs of royalty and nobility, Strutt desired an available space for the enjoyment of the working class. He saw the establishment of Derby Arboretum as providing the pioneer step of mitigating the government's concern of the health of the urban population through provision of open spaces for people's use (Butterton 1993). The park movement found itself spreading to America, with the design of Derby Arboretum inspiring the design of New York City's Central Park, as New York City also found itself facing similar social issues as a result of unprecedented wave of urbanisation due to a high influx of migrants living in harsh conditions in tenement housing (Bial 2002).

Adopting a deterministic view of cause and effect, park proponents thought that changing the urban environment through ameliorating the living conditions of the working class could engender a better society (Jordan 1994; Taylor 1995, 1999). They attributed the surge of vices in the city to alienation of urban residents from nature. The wealthy lived in the urban area but they were also likely to own estates and working landscapes such as farms and forests outside cities. These estates preserved nature around the urban core and gave jobs to many in the working class society. Woods and meadows were viewed as restorative to the mind, provided places to stroll for socialisation and fresh air, and allowed for hunting, trapping, and fishing. Over time, parks were developed in the urban core and outside the city to improve society's well-being through reproducing nature in an urban environment or restoring original or fragmented nature. Contending that commercial activity was wearying for the working class, immersion in the park would result in visitors feeling recharged and ready to engage in another round of work (Olmsted 1881).

With the primary objective of encouraging nature in fast-growing cities, pioneer American park designers deemed it best to have a large park located on the edge of a city where people could be recharged through indulging in contemplative activities while immersed in nature. Olmsted (1881) thought of the pastoral landscape as a way of counteracting what he considered to be the 'severe and excessive exercise of the mind'. The creation of a *rus in urbe* environment—one which provided the illusion of the countryside by a park within a city—was deemed essential. In contrast, activities such as energetic children's play or athletics were not encouraged. Olmsted did not favour recreation that would lead to overstimulation. Instead, he believed that gentle exercise relieved the brain. This first model (Table 5.1) of the modern urban park, commonly known as the Pleasure Ground (1850–1900) (Cranz 1982; Cranz and Boland 2004), was premised on the ideal pastoral landscape with buildings subordinate to the overall landscape. Creating generic spaces within this park was also key in the design of the ideal restorative environment; its

Table 5.1 A comparison of park models over the centuries

	Pleasure ground (1850–1900)	Reform park (1890- early 20th century)	Open-space planning (early 20th century)	'New-Age' park (present)
Description	Strong emphasis on creating an illusion of the countryside by a park within a city, its aesthetics, and tranquilising recreation	Less emphasis on the importance of being a picturesque environment, instead balancing 'utility' with 'beauty'	Parks are imagined as part of larger network connecting different green space and public lands	Adult playgrounds, intergenerational parks, restorative gardens
Public goal	Provision of open spaces to prevent alienation of lower-wage workers living in overcrowded urban conditions due to unprecedented urbanisation	Greater involvement of park advocates and designers to directly intervene and directly link the role of parks to change the health status of urban residents through more active recreation	Offer extensive recreational opportunities through provision of larger continuum of space for urban dwellers	Parks of various nature, catering to various social issues such as an ageing population, higher prevalence of chronic illness

Adapted from Cranz and Boland (2004)

design would therefore consist of uniform landscape and plantings. One of the ideas of the first modern urban park was realised with the Olmsted and Vaux plan for Central Park in New York City in 1858.

Later in the 1890s, park designers adopted more rationalistic interpretations of nature into their work. Exemplifying this was the Reform Park, an urban park model principled upon achieving both 'beauty' and 'utility'. While early park advocates made general claims about the therapeutic nature of parks, there were no direct interventions and measurements of the actual health changes of urban residents. Once the parks were designed and developed, the health improvements were left to chance. To achieve greater use of parks for health outcomes, a new group of reformers sought to link the goals of the sanitary reform movement to the goals of the parks and recreation movement more deliberately. Hence, rather than purely providing an immersive nature experience, the park was observed to slowly evolve to include opportunities for a greater variety of activities such as athletics and safer places for children to play (Cranz and Boland 2004; Young 1995). While these reformers subscribed to the health-giving character of parks, the newer parks were however, not built with an eye towards the picturesque or for the purpose of tranquilising recreation. Rather, they were specifically built for active recreation. Furthermore, generic spaces—a homogeneous landscape with uniform selection of plant species—that distinctly characterised the Pleasure Ground were replaced by ornamental horticulture. The latter concept of colourful plantings meant that certain locations within the park would gain prominence over others, thus contrasting with Olmstead and Vaux's concept of achieving an immersive experience in nature through geometric regularity (Young 1995). As it was difficult to fund elaborate parks based on Olmsted and Vaux's concept of an ideal park, smaller parks became popular due to the financial constraints faced by many cities.

The third model, Open Space Planning, which adopted a new philosophy of imagining parks, gained traction in the early 20th century. In contrast to the idea of parks as specialised places for recreation purposes, parks are instead imagined as part of a bigger network of public open spaces. Even smaller land parcels were viewed as potentially valuable in this network, meaning that areas within the city, streets, or an abandoned railway site could be part of this integrated green network together with parks. The same period also welcomed the concept of Greenway Planning, a continuous system of green and urban spaces linked by recreational and beautified corridors. Greenways provide alternative corridors which offer attractive visual form of greenery and solace and extensive recreational opportunities for urban dwellers in a larger continuum of space. The President's Commission on American Outdoors (1987) visualised the role of the Greenway network as such:

> A living network of greenways… to provide people with access to open spaces close to where they live, and to link together the rural and urban spaces in the American landscape… threading through cities and countrysides like a giant circulation system (President's Commission 1987: 102).

Since then, greenways have evolved from their role as park linkages to being a park space where active forms of recreation take place. According to the East Bay Greenway Health Impact Assessment (Heller and Bhatia 2007), trail users were found to have more than a 50% chance of meeting the Center for Disease Control's (CDC) recommendations for exercise. The concept of greenways as a linear park is especially useful in cities with high population density. Singapore is one example where the greenway concept has been successful in enhancing the recreational experience. Faced with acute land scarcity from various competing land use within its land area of 718.3 km^2, the need to optimise limited land space and yet cater to the recreational need of its urban populace is indeed crucial. Greenways are referred to as the Park Connector Network (PCN) in Singapore. Initially serving as linkages between major parks and nature sites, and also acting as ecological corridors in Singapore's Master Plan 2003 (Urban Redevelopment Agency 2008), the plan has since expanded to make parks more accessible to the general populace. The PCN will be complemented by seven loops, totalling a length of 360 km by 2020, and a 150 km round-island route that creates higher accessibility to nature sites by allowing people to walk, jog and cycle close to the coastline and greenery (Abdullah 2015). The PCN also provides the frame and the tributaries to provide a non-motorised or greener form of transportation in a city (Tan 2006).

Parks are a reflection of society's goals and its underlying issues and challenges. According to an United Nations report (2013) 'World Population Aging 2013', the aged population of the more developed regions tripled between 1950 and 2013, from 94 million to 287 million. The aging population is expected to increase further in coming decades, reaching 417 million in 2050. Coupled with the growing challenges of higher prevalence of chronic diseases resulting from a sedentary lifestyle, the fourth and current style of parks has been moulded into a vastly different model from the ones of the 20th century. The New Age Park include 'adult playgrounds'—public open-air exercise equipment for grown-ups which include frictionless cross-trainers, benches for sit-ups and leg exercisers—and they are a popular feature in many urban parks worldwide. Also, intergenerational parks based on the principle that seniors perform better while surrounded by people of all age groups, are also increasingly popular in urban parks in many developed countries.

The health values of parks and green spaces have been entrenched in the fundamentals of park planning and design since the pioneering park movement during the 19th century. The role that parks and trails play in encouraging physical activity or contemplative recreation have been well recorded in history. Moreover, current research suggests that physical activity is necessary for everyone's well-being, and physical activity in outdoor settings is more effective than equivalent activity performed indoors (Pasanen et al. 2014). This will be highlighted in the next section of this chapter on human well-being.

5.3 Green Space as a Health Intervention Measure

> At his office in Washington, D.C., Robert Zarr, a pediatrician, writes prescriptions for parks. He pulls out a prescription pad and scribbles instructions — which park his obese or diabetic or anxious or depressed patient should visit, on which days, and for how long — just as though he were prescribing medication.
>
> (Hamlin 2015 October)

The Park Prescriptions movement is an example of a salutogenic, or benefits-based, approach to stimulating interest in using outdoor recreation resources as a tool to improve public health. Originating in the United States, this recent movement seeks to reintroduce people to parks in a bid to increase outdoor physical activity to prevent or treat health problems resulting from a sedentary lifestyle and poor diet through strengthening the connection between the healthcare system and public spaces (Institute at the Golden Gate 2010) (http://www.parkrx.org). Proponents of this movement are not merely advocating physical activity, but are urging people to indulge in the outdoors. They base their claim on the findings of a growing body of research which have suggested that exposure to nature and outdoor exercise has significant health benefits ranging from physiological and psychological stress reduction, to psychological restoration.

This salutogenic approach, coined by Israeli American Sociologist Antonovsky (1996), refers to the focus on provisions in the living environment that make some people more resilient when faced with stress in their daily lives. In contrast to traditional healthcare which takes a pathogenic approach that seeks better medical interventions to cure chronic diseases, the salutogenic orientation aims to identify causes of health and implement these predictors of good health in our environment. As nature has been found to positively influence health, salutogenic design often engages the senses to connect people with nature, through both active and passive experiences.

Yet, this belief in the therapeutic effect of exposure to nature can be found as far back as ancient Egypt, where court physicians prescribed walks in palace gardens for royalty who were mentally disturbed (Toyoda 2012; Shoemaker 2004). Around 500 BC, the Persians created gardens that combined beauty, fragrance, music (flowing water) and cool temperatures. This belief was later supported by Dr. Benjamin Rush in 1812, when he reported that patients who worked in gardens had better recovery rates from mental illness compared to those who did not have the same gardening experience. Consequently, veterans of World War II were assigned to on-site gardens in Veterans Administration hospitals for rehabilitation therapies (Sydney et al. 2014).

The concept of using nature to improve human health and well-being was further developed through key research during the 1970s and 1980s. Psychology professors Rachel and Stephen Kaplan proposed the Attention Restoration Theory, which suggests that engrossed attention in performing a task can lead to mental fatigue. Recommending natural settings as a remedy to improve and restore attention, they attribute this restorative power of the natural environment to four characteristics. They include: (1) the feeling of being away, (2) fascination value of natural elements, (3) extent that the natural environment is replicated in a smaller and manageable one, and (4) the special resonance people have with nature as compared to an urban environment (compatibility) (Kaplan and Kaplan 1989). Wilson (1986) attributes the benefits to 'biophilia', which proposes that humans are more comfortable in nature because that is where they have evolved. Ulrich's research (1984) also demonstrated that patients with views of trees had shorter hospital stays and needed less medication. These studies resulted in an increasing interest and body of scientific evidence on the health benefits of parks and urban greenery.

The practice of horticultural therapy is one example of using nature to restore people's health and well-being. The American Horticultural Therapy Association (2012) defines the practice as 'the engagement of a client in horticultural activities facilitated by a trained therapist to achieve specific and documented treatment goals'. Benefits of this form of therapy have been time-proven, with its techniques employed to assist participants to learn new skills or regain those that were lost. Horticultural therapy helps improve memory, cognitive abilities, task initiation, language skills, and socialisation. A study by Western Michigan University (Wagenfeld and Atchison 2014) revealed that 60 out of 80 occupational therapists used gardening as a therapeutic intervention. The researchers found that occupational therapists thought of gardening as a therapy intervention to be meaningful and purposeful (94%, n = 56), motivating (80%, n = 48), fun (62%, n = 37), and client-centered (32%, n = 19). Such studies support the larger role which flowers (and greenery) play in patient's recovery and rehabilitation, rather than a mere form of emotional cheer.

Beyond Europe and America, there is also growing appreciation for the therapeutic use of greenery in other parts of the world. Asian countries, especially those with aged or ageing societies, such as in Japan, Korea, Taiwan, and China, have shown great reception towards employing horticultural therapy for rehabilitation. In Japan, the Awaji Landscape Planning & Horticulture Academy, an institution authorised by the Hyogo prefectural government, trains healthcare professionals to provide healing through structured programmes which include the usage of plants and exposure to the outdoors, and through gardening activities (Toyoda 2012). Before each horticultural therapy session, the client's general mental state is measured using a scale developed by the academy. The scale consists of ten check items each with four levels of rating, and evaluates dimensions such as 'affections', 'mental functions' and 'communication ability'. Likewise, in South Korea, this practice is also becoming increasingly popular. Horticultural therapy is offered in about 1700 facilities such as social welfare organisations, job rehabilitation facilities, hospitals, public health centres, and schools, supported by a pool of about 2000 qualified horticultural therapists (Park et al. 2012).

In an environmental scan on the receptiveness of the healthcare industry in adopting greenery in therapies in Singapore by a research team from the National Parks Board (NParks), the practice of rehabilitating patients with plants and gardens was found to have already been established in several local medical institutions by occupational therapists. One such medical institution is the Institute of Mental Health (IMH), which has developed a Friendship Garden for its patients, staff, and volunteers (Cheong 2015). About 30 long-stay patients take turns to visit the garden daily, joined by IMH staff and nurses. Thomas (2010), an occupational therapist who oversees some of the gardening activities, shared,

> Tending plants delights the patients to see that their plants grow well. Gardening allows patients to soil their hands while enjoying a constructive activity. The patients also enjoy each other's company as they receive instruction from nurses, teaching assistants and occupational therapists.

Therapeutic horticulture is being adopted in many other healthcare institutions in Singapore. One of them is the Saint Andrew's Nursing Home, where its Dementia Care Ward leads directly to an enclosed garden where patients can engage in physical activities and enjoy the outdoor space safely without wandering too far or getting lost. Other programmes have also been developed in the Khoo Teck Puat Hospital and the Salvation Army Peacehaven Nursing Home (Wee 2012). In these establishments, occupational therapists take patients to the outdoor gardens to soak in the sights, sounds and smells of greenery or to exercise their limbs and improve their motor skills by engaging in gardening. Moreover, as Singapore was once an agriculture-based society, many older people still remember growing their own food. Therefore, gardening as a form of activity for seniors can serve to stimulate the brain and memory.

Recognising the high potential of therapeutic horticulture being adopted by occupational therapists, NParks is collaborating with local educational institutions offering occupational therapy programmes to introduce a module on gardening into its curriculum. In the longer term, it would be beneficial for Singapore and other cities to develop a pool of certified horticulture therapists trained by experts and institutions. Both the design principles of therapeutic gardens and horticulture therapy programming can be scaled up and implemented at the community level to reach more beneficiaries.

The benefits of horticultural therapy are not limited to specific groups but are also broadly targeted towards the general public. Singapore, which envisions itself as a city nestled in an environment of trees, flowers, parks and rich biodiversity, has embarked on its plan to create an urban environment that supports ecologically healthy people and habitats. A growing initiative in this dense urbanised landscape is the community gardens movement. As part of the Community in Bloom (CIB) movement, over 1000 community gardens have been realised in neighbourhoods and organisations across Singapore. Programmes like these come under the definition of 'social horticulture', which the American Horticulture Therapy Association (2012) defines as 'a leisure or recreational activity related to plants and gardening, with neither treatment goals defined, nor therapist being present'. To integrate science to the programming, NParks has initiated research to quantify the physical and well-being benefits of exercising in parks and engaging in gardening activities. The studies are conducted in collaboration with medical researchers who designed the first randomised control trials in Southeast Asia to test the therapeutic effects of greenery with clinical evidence. Alongside the research, NParks is also working with the local Ministry of Health to progressively incorporate more senior-friendly amenities and therapeutic gardens in its network of parks (Ministerial Committee of Ageing Report 2015). The latter will be designed with special treatments such as landscape richness, and elements of surprise to support the mental well-being of park users, including those with dementia.

Complementing the Park Prescription movement by medical practitioners, a new group of researchers are also looking at ways to use neuroscience to inform the design of the environment to benefit public health. Investigating brainwave patterns in people exposed to different landscapes can bring about interesting conclusions about specific human responses to different designs while helping to discover the healing potential of these spaces. According to a study by Olszewska et al. (2014), the settings deemed most contemplative had panoramic vistas with long-distance views of more than 400 m. These settings tended to include large empty spaces, natural asymmetry, clearings and stimulation to look at the sky. In contrast, the least contemplative settings usually lacked these features; instead, they have characteristics such as paths and enclosed spaces (small pocket gardens). Evidence has supported the role which parks play in providing a restorative environment. Understanding the specific design features of a park through nascent neuro physio-psychological research and developing evidence-based design bring us one step closer to eventually developing the 'ideal park' for contemplation.

5.4 Health and Social Benefits Derived from Park Exposure

Exposure to nature offers a range of health and social benefits that contribute greatly to well-being. These benefits, documented since the 19th century, showed that mere exposure to nature is associated with stress reduction and psychological restoration. But even before then, human life depended on nature for food and shelter. As society advanced, this relationship with nature changed. Most societies today may neither have direct use nor harvest nature to supply their basic needs, instead spend time in nature largely for leisure or non-consumptive uses.

Through public pressure and organised labour movements, the industrialised society was afforded leisure time in nature with time off from work so that families could spend time together. Weekends and holidays were created from labour laws that recognised the need for workers to restore their physical and mental well-being, as well as build social relations. As workers earned higher wages, families could spend more time in urban parks or take vacations to the rural areas where water-based activities were popular and people of all ages could enjoy a more natural setting that the city often did not provide. Sessoms and Henderson (1994) highlight that even with new labour laws that afforded weekend and vacation time from work, the individual or household had to be motivated to be outdoors and visit public or private places designed as parks. Barriers to park use and constraints of intrapersonal, interpersonal and structural reasons for not visiting parks play an important role in understanding park underutilisation and social avoidance of outdoor public spaces (Jackson 1988).

Another social change that has allowed for greater time for leisure and spending time outdoors is the modernisation of society through technology. An agrarian society had many home and farming responsibilities to tend to on a daily basis. There was often no time for leisure, play or enjoyment. The invention of mechanical appliances which reduced the amount of time a family needed to spend on food production and house cleaning afforded extra time each day for recreation, fitness, and play. Technologies thus increased efficiencies in menial tasks to free up time for leisure and tranquillity (Godbey 1997). More recent information technology advances in smart-phones and televisions are also enabling people to structure the timing of their media consumption and entertainment, rather than on scheduled programming. For example, media firms now provide subscriptions which allows the viewer to watch television shows, movies, and other forms of media whenever and wherever a person chooses, including on a smart-phone in a park. On the other hand, broadcast television restricted viewers to scheduled timings on a single home appliance.

Many cities were designed with a single park or several large parks as a prominent feature of the urban landscape (Cranz 1982). New York City is well-known for Central Park designed by Frederick Olmsted, the founder of the landscape architecture profession (Wellman and Propst 2004). On the other hand, Paris was designed with straight, tree-lined boulevards, diagonals, squares, parks, and vistas across its 20 arrondissements. New York City and Paris, as cities surrounded by rural areas, also provide urban dwellers alternatives outside the city. The state of New York preserved land by creating the Catskills and Adirondack preserves where natural resources could be ‘forever wild’ and city residents could vacation in. Paris is known for its private chateaus and rural villages, which afforded urban dwellers an escape.

Gobster (2001) documents the use of scale to achieve varying levels of environmental and social conditions in park and green spaces from rural to urban neighbourhoods. Select cities, such as New York City and Paris over 200 years ago and more recently Singapore, had early visions of green cities with public space. Singapore’s model, a city-state with much less space than New York City or Paris, has a layering of natural areas to achieve biodiversity and social outcomes (Fig. 5.1), ranging from primary and secondary forests to neighbourhood parks. The protected nature parks are managed by the national resource agencies (NParks and Public Utility Board), while the management of the ‘convenient everyday parks’ are shared amongst housing agencies, both public and private, and community groups. Parks are often centrally located to maximise access to neighbourhoods and commercial or business districts. Planners attempt to place parks along transportation lines that may have included subway, bus, road or foot traffic. A large centralised park could also attract higher levels of awareness of the park by residents and tourists in comparison to neighbourhood parks tucked into dense

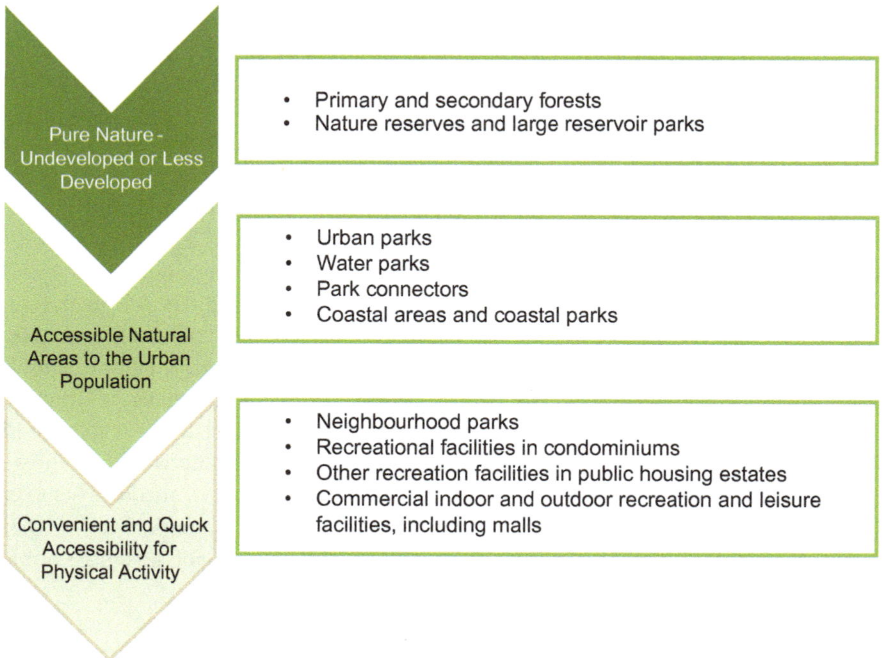

Fig. 5.1 Hierarchy of green spaces and other recreational facilities in Singapore

housing plots. Planners have found that the larger the park and the more prominent its location, the more likely people would use it (Giles-Corti et al. 2005). Events held at large parks serve as additional publicity. Music concerts, food festivals, art installations, and sport events like marathons are some of the contemporary forms of park uses that elevate the popularity of both the city and its park.

Urban planners are no longer the sole advocates for highly mobile and active cities as private corporations, such as Nike Inc., are participating in discussions that contribute to ways that a city can be mobile and active. A hotspot for testing these new sustainability approaches are global cities, which house a majority of the world's population. A city, therefore, becomes a place where benefits are maximised because of planning, policy, and promotions. The 2012 report, jointly produced by Sustran, Active Living Research and Nike (www.designedtomove.org), outlines that active cities have four successes—physical activity is a priority for all, existing resources are used, places are designed for people, and a legacy of lasting change. It promotes sustainability planning in cities around the globe to reinvigorate our thinking on where people desire to live, work and play. The report highlights research evidence of cities documenting less crime, reduced destructive effects of climate change, lower rates of depression and anxiety, better social cohesion, and more job growth from active living approaches. Patagonia is another corporation that has a long history of advocating for outdoor physical pursuits and supports conservation projects around the world.

Given the mounting scientific evidence that people need parks and urban greenery to achieve necessary health and social benefits, why is there then not a clear pathway for park investments in urban areas? The answer is often economics. Park development and management are often linked to the financial performance of cities. As one of many public services that compete for limited public funding, the budget of park departments are often in competition with safety departments like the police and fire which public administrations argue cities absolutely cannot do without. Despite funding challenges, park and urban planners appear to be moving forward with some new approaches and partners to deliver on quality outdoor spaces. Corporate partners, such as Nike and Patagonia, can help fill some of the gaps in funding and reach out to engage audiences. In the 1990s, a group of federal recreation staff and scholars created a new lens to promote conservation, parks and recreation. However, rather than highlighting the features of a place or the activities performed in a place, the lens shifted towards the achievement of goals or benefits. The benefit-based model places emphasis on social outcomes through the use of parks and recreation activities (Driver et al. 1991; Driver 2008). Driver, a United States Forest Service researcher, engaged local, state, and federal professionals to enumerate the extent of benefits and the park and greenery features that enabled these benefits to be realised. Driver and his co-authors used research to convince practitioners and policy makers that parks could be directly linked to improved human conditions. The results are an empirically based rationale that parks and their features are necessary for our well-being (Table 5.2).

Today's park professionals around the globe better understand how the management and marketing of park and greenery features produce specific and desirable outcomes. In densely populated cities like Singapore, park planners segment parks into activity areas to achieve a multitude of benefits. For instance, in Bishan-Ang Mo Kio Park, the Kallang River traverses through the park and activity areas such as playgrounds, seating areas and a lily garden for contemplation situated next to the river. Another example of parks delivering multiple uses and benefits is the Yishun Pond Park, which was designed to create a 'multi-generational and health-promoting garden' (Khoo Teck Puat Hospital et al. 2010). The pond's 2010 makeover surrounding the development of the then new Khoo Teck Puat Hospital saw the integration of newer health facilities (including the new Yishun Community Hospital which was recently completed in 2015), with an older park (Yishun Park), and the addition of a waterfront promenade with a 2.6 km trail. With the integrated trail, patients and their caregivers from the hospitals can make use of the place for recuperation and social interactions, whereas residents also use this trail for recreation and exercise. These examples demonstrate how well-planned and well-designed park features can create a multitude of social benefits. This suite of features that are directly linked to benefits is the new dialogue of park and planning professionals.

Table 5.2 Well-being benefits associated with park or greenery attributes

Benefits or outcomes	Park and greenery features
Physiological—human needs based on environment	Sun Fresh, clean air Trees Healthy ecosystem
Emotional-achieving restorative and mental health, as well as happiness	Beauty Quiet spaces within a city Presence of other urban dwellers Safety Presence of wildlife
Physical Activity—exercise, self actualisation	Trails Exercise areas
Social Gathering and Interaction—moving toward features with more engagement	Playgrounds Adventure activities (boating, ziplines and challenge courses) Benches Tables and pavilions for games and eating Barbeques Proximity to neighbourhoods Cafes Dog parks or bird cage areas
Human Development—learning, curiosity	Recreation activities with physical facilities Creative places like playgrounds for kids Programs of all types Exposure to living nature (animals, plants, weather)
Nourishment (Food)	Community gardens Fishing Gathering of people for community good

Adapted from Driver et al. (1991)

5.5 Leveraging the Park and Greenery Effects for Greater Human Well-Being

Thus far, this chapter has profiled many social and health outcomes that parks and greenery at various scales have produced across the globe. The evidence is compelling—those who use parks or are exposed to nature can benefit in a myriad of ways. So what is next for practitioners and scholars to employ to further leverage nature for greater human well-being?

Glover (2015), a leisure scholar, advocates that leisure research which is partly embedded within the parks and recreation field can illuminate opportunities for greater social innovation. He summarises many other scholar's work to show how parks can play a role in social justice in providing a safe place for all citizens, environmental justice as a place where physical activity can be offered for children and adults, and stewardship and opportunities for caring for nature and people through programmes held at parks. Recreational activities, park programmes, and the research that evaluates their impacts identify the stakeholders that are needed. These will help in enchancing the park experience and also measure the well-being impacts on the stakeholders. Knowledge about the contribution of programmes in parks needs to move us beyond status quo into action for even greater benefits for every one as societies struggle with many issues that diminish well-being including chronic health issues, overworked and stressed children and workers, and crime. A better informed planning model with social innovation as the core concept can inform government policies and the practices of private developers. It can also engage citizens on the multitude of possibilities that parks and green spaces can endow for the health and success of all ages, races, and ethnicities.

Recent efforts to align park, forestry, planning professionals, and scholars with the medical profession is another way to leverage applications and solutions for human betterment. In America, the foundation of the Johnson & Johnson health-care company, Robert Wood Johnson Foundation (www.rwjf.org), plays a leadership and funding role in igniting research and community practices that engage park and recreation, planning, health care, and public health fields to improved human outcomes. In Singapore, public agencies work across families and neighbourhoods, healthcare organisations, and employers to achieve high levels of health and physical activity across their society. As illustrated in this chapter, a more socially and physically networked approach to co-locate parks and medical facilities can provide greenery exposure to hospitalised patients; and trails and outdoor fitness equipment can provide for patients needing rehabilitation. Continued collaborations between parks, planning, and healthcare can effectively reach a broad cross-section of urban dwellers.

In conclusion, parks have a varied history of emphasising the aesthetic value of landscapes for human pleasure and contemplation. Some efforts are directed at conserving existing untouched wilderness and, at other times, directed at restoring 'used' landscapes. Today, parks continue to play a central role in urban planning for both conservation and restoration. Social innovation approaches and connectivity to health care require planners to deeply consider how humans can be exposed to maximum levels of nature, particularly in dense urban settings. While public agencies and public funding remain a key player in parks, the future is likely to have many more stakeholders from the private and non-profit sectors. A more engaged set of stakeholders will enable broader social responsibility for and greater use of and appreciation for parks and public natural resources.

References

Abdullah Z (2015) Singapore park connectors reach 300 km at 25-year mark. The Straits Times. http://www.straitstimes.com/singapore/singapore-park-connectors-reach-300km-at-25-year-mark. Accessed 26 Feb 2016

American Horticultural Therapy Association (2012) Definitions and positions. http://ahta.org/sites/default/files/DefinitionsandPositions.pdf. Accessed 7 Dec 2015

Antonovsky A (1996) The salutogenic model as a theory to guide health promotion. Health Promot Int 11(1):11–18

Bial R (2002) Tenement: immigrant life on the lower east side. Houghton Mifflin, Boston

Butterton H (1993) The derby arboretum. History Today 43(62). Academic OneFile. Accessed on 4 Jan 2016

Cheong YM (2015) Central CDC brings the community to IMH. IMHLink: Loving hearts, beautiful minds, Apr–Jun 2015. Institute of Mental Health. https://www.imh.com.sg/uploadedFiles/Publications/IMH_Link/IMH%20Link%20Apr%20-%20Jun%202015.pdf. Accessed 11 Jan 2015

Cranz G (1982) The politics of park design: a history of urban parks in America. MIT Press, Cambridge, Mass

Cranz G, Boland M (2004) Defining the sustainable park: a fifth model for urban parks. Landscape J 23(2):102–120

Driver BL (2008) Managing to optimize the beneficial outcomes of recreation. Venture Publishing, Inc, State College, PA

Driver BL, Brown PJ, Peterson GL (eds) (1991) Benefits of leisure. Venture Publishing, Inc, State College, PA

Giles-Corti B, Broomhall MH, Knuiman M, Collins C, Douglas K, Ng K, Lange A, Donovan RJ (2005) Increasing walking: how important is distance to, attractiveness, and size of public open space? Am J Prev Med 28:169–176

Glover TD (2015) Leisure research for social impact. J Leisure Res 47(1):1–14

Gobster PH (2001) Neighbourhood-open space relationships in metropolitan planning: a look across four scales of concern. Local Environ 6(2):199–212

Godbey G (1997) Leisure and leisure services in the 21st century. Venture Publishing, State College, PA

Hamlin J (2015) The nature cure. The Atlantic. http://www.theatlantic.com/magazine/archive/2015/10/the-nature-cure/403210/. Accessed 2 Mar 2016

Heller JC, Bhatia R (2007). The east bay greenway health impact assessment. http://www.urbanecology.org/greenway/GreenwayHealthImpactAssesment.pdf. Accessed 15 Dec 2015

Institute at the Golden Gate (2010) Park prescriptions: profiles and resources for good health from the great outdoors. http://www.parksconservancy.org/assets/programs/igg/pdfs/park-prescriptions-2010.pdf. Accessed 8 Jan 2016

Jackson EL (1988) Leisure constraints: a survey of past research. Leisure Sci 10(3):203–215

Jordan H (1994) Public parks. Gard Hist 22(1):85–113

Kaplan R, Kaplan S (1989) The experience of nature: a psychological perspective. Cambridge University Press, New York

Khoo Teck Puat Hospital, Housing Development Board, National Parks Board, Public Utilities Board (2010) Media release: Yishun pond makeover to create new green lung. https://www.ktph.com.sg/uploads/1284019944YishunPond_groundbreaking_mediarelease_WEB.pdf. Accessed 3 Mar 2016

Ministerial Committee on Aging (2015) $3 billion action plan to enable Singaporeans to age successfully. http://www.mfa.gov.sg/content/mfa/overseasmission/geneva/speeches_press_statements_and_other_highlights/2015/201508/press_20150826.html. Accessed 11 Jan 2015

Nike Inc (2012) Designed to move. http://e13c7a4144957cea5013-f2f5ab26d5e83af3ea377013dd602911.r77.cf5.rackcdn.com/resources/pdf/en/full-report.pdf. Accessed 20 Feb 2016

Olmsted FL (1881) A consideration of the justifying value of a public park. Tolman & White, Boston

Olszewska AA, Marques PF, Barbosa F (2014) Urban planning, neurosciences and contemplation for improving well-being in our cities. http://www.arcc-journal.org/index.php/repository/article/viewFile/253/195. Accessed 4 Mar 2016

Park SA, Son KC, Cho WK (2012) Practice of horticultural therapy in South Korea. Acta Hortic 954:179–185. doi:10.17660/ActaHortic.2012.954.24. http://dx.doi.org/10.17660/ActaHortic.2012.954.24. Accessed 9 Dec 2015

Parliament House of Commons (1833) Report from the Select Committee on public walks. Parliamentary Paper 15:1049–1059

Pasanen TP, Tyrväinen L, Korpela KM (2014) The relationship between perceived health and physical activity indoors, outdoors in built environments, and outdoors in nature. Appl Psychol Health Well-being 6(3):324–346

President's Commission of American Outdoors (1987) Report and recommendation. Reprinted as: American outdoors: the legacy, the challenge. US Government Printing Office, Washington, DC

Sessoms HD, Henderson KA (1994) Introduction to leisure services. 7th edn. Venture Publishing, Inc, State College, PA

Shoemaker CA (2004) Horticulture therapy: comparisons with other allied therapies and current status of the profession. Acta Hort (ISHS) 639:173–178

Sydney PB, Eva CW, Theodora MF, Jessica S (2014) Horticultural therapy. University of Florida. https://edis.ifas.ufl.edu/pdffiles/EP/EP14500.pdf. Accessed 8 Dec 2015

Tan K (2006) A greenway network for Singapore. Landscape Urban Plann 76:45–66

Taylor HA (1995) Urban public parks, 1840–1900: design and meaning. Gard Hist 23(2):201–221

Taylor DE (1999) Central park as a model for social control: urban parks, social class and leisure behaviour in nineteenth-century America. J Leisure Res 31(4):420–477

Thomas JD (2010) Patients' gardening bears fruits. IMHLink: loving hearts, Beautiful minds, Jan–Mar 2010. Institute of Mental Health. https://www.imh.com.sg/uploadedFiles/Publications/IMH_Link/IMH%20Link%20Apr%20-%20Jun%202015.pdf. Accessed 11 Jan 2015

Toyoda M (2012) Horticultural therapy in Japan-history, education, character, assessment. JAD 2. http://www.awaji.ac.jp/htcp/artis-cms/cms-files/20120703-164630-5665.pdf. Accessed 8 Dec

Ulrich R (1984) View through a window may influence recovery from surgery. Science 224 (4647):420–421 New Series

United Nations, Department of Economic and Social Affairs, Population Division (2013) World population aging 2013. ST/ESA/SER.A/348. http://www.un.org/en/development/desa/population/publications/pdf/ageing/WorldPopulationAgeing2013.pdf. Accessed 4 Jan 2016

Urban Redevelopment Agency (2008) Singapore's concept and master plans 2008. https://www.ura.gov.sg/uol/master-plan/View-Master-Plan/master-plan-2014/master-plan/Key-focuses/recreation/Recreation. Accessed 29 Dec 2015

Wagenfeld A, Atchison B (2014) Putting the occupation back in occupational therapy: a survey of occupational therapy practitioners' use of gardening as an intervention. Open J Occup Ther 2 (4), article 4. doi:10.15453/2168-6408.1128

Wee L (2012) Sowing the seeds of health. The Straits Times. Accessed 9 Dec 2015

Wellman JD, Propst DB (2004) Wildland recreation policy, 2nd edn. Krieger Publishing Company, Malabar, FL

Wilson EO (1986) Biophilia. Harvard University Press, Cambridge, MA

Young T (1995) Modern urban parks. Geogr Rev 85(4):535–551

Chapter 6
Urban Community Gardens as Multimodal Social Spaces

Jeffrey Hou

Abstract This chapter examines the social and collective dimensions of urban gardening. Through a review of recent literature and cases around the world, it examines urban community gardens in terms of their multiple modalities, specifically *as a convivial space, a cultural space, an inclusive space, a restorative space, a democratic space,* and *a resilient space*. As convivial spaces, urban gardens build and nurture agency of individuals as well as social ties in a community. As inclusive, cultural spaces, urban gardens can function as a place for cross-cultural learning and understanding and building of connections across social and cultural divides. As restorative space, urban gardens contribute to individual and community health and well-being. As democratic spaces, urban gardens serve as a vehicle to engage individuals and communities in efforts toward other social and environmental initiatives. As resilient space, urban gardens function as social safety nets and provide for the community in time of calamity and struggles. Through these different expressions and opportunities for active engagement by communities and citizens, the chapter argues that urban gardening can serve as a model for other urban greening strategies to incorporate considerations for multiple social, cultural, and economic goals.

Keywords Community gardens · Urban gardening · Social space · Convivial space · Restorative space

6.1 Introduction

Urban gardening has enjoyed a resurgence of interest around the world in recent years. In North America, once considered as a temporary use of vacant spaces, urban gardening has gained greater acceptance as a permanent part of the urban

J. Hou (✉)
University of Washington, Box 355734, Seattle, WA 98195-5734, USA
e-mail: jhou@u.washington.edu

P.Y. Tan and C.Y. Jim (eds.), *Greening Cities: Forms and Functions*, Advances in 21st Century Human Settlements,
DOI 10.1007/978-981-10-4113-6_6

landscapes. Such greater acceptance has been associated with a growing interest in environmental sustainability and urban food. The many benefits of urban gardening, including health and community development, have also been recognized through research. In Europe, where urban gardening has enjoyed a longer history and endurance, new sites have continued to emerge sometimes in unexpected places, such as the Tempelhof Airfield in Berlin and an abandoned railway in the city center of Stockholm. In East Asia, urban gardens have also emerged as results of both government initiatives and grassroots efforts. In South Korea, the Seoul Metropolitan Government has made an ambitious plan to become the world's capital of urban agriculture (Fig. 6.1). In Singapore, under the *Community in Bloom program*, hundreds of community gardens have emerged in the midst of tall apartment complexes. In Hong Kong, a temporary garden was recently installed in Kowloon Park, and *Rooftop Republic*, a social start-up, has been managing more than twenty sites of rooftop community gardens. In Taipei, the City has recently launched a *Garden City* programme to support community gardening and urban food production. Similar initiatives have also been developed in other cities in Taiwan, including the New Taipei City.

Compared with their counterparts in North America and Europe, urban gardening in dense mega-East Asian Cities represents a unique opportunity for greening the highly urbanized landscapes as well as integrating functions of food production, recreation, health and well-being, and environmental learning, along with other ecosystem- and social services. However, with scare land resource and

Fig. 6.1 Public farm on Seoul's Nodeul Island transformed a site previously slated for the construction of a new opera house into a urban farm for citizens (*Photo credit* J. Hou)

Fig. 6.2 Community gardens like this in Taipei came under criticism for taking over lands with high real estate value (*Photo credit* J. Hou)

high property values, urban gardening in East Asia is also faced with unique challenges and questions from the public and the skeptics who are not accustomed to food production in the city (Fig. 6.2). In Taipei, for instance, one of the most common criticisms for the city's recent Garden City initiative concerns why expensive urban lands are used for growing vegetables when they can be easily grown in agricultural areas and accessed in city markets, unlike the food deserts of North American cities. Another frequent question focuses on the safety of growing food in cities where there are many sources of pollution. Responses to these questions require policy makers and urban gardening advocates to better articulate the unique facets of urban gardening and to demonstrate their unique values and benefits to the public and the skeptics.

The purpose of this chapter is to unpack one particular category of the benefits of urban gardening—the social benefits, which include social interactions and building of social ties in dense urban settings through urban gardening. Specifically, through a review of predominantly recent literature, it examines the distinct social dimensions of community and urban gardening. As the current literature is primarily focused on the North American and European contexts, the chapter is meant to inform the current discussion on urban gardening as a strategy for city greening in dense urban contexts particularly in East Asia. Rather than prescribing specific approaches, it is intended to contribute to ongoing experimentation in planning, design and implementation of community gardens as well as the development of necessary discourse, organizational networks, and institutional support for community gardens.

6.2 Origins of Urban Gardening

Urban gardening is a broad term for describing a variety of gardening activities in the urban context. In North America, urban gardening is commonly traced to the allotment garden movement of the late nineteenth century developed 'in response to the dire needs of burgeoning lower-class population' (Loggins and Christy 2013: 14). Allotments at the time were given to the poor by factories, through schools, orphanage, and could be found or created in public parks (Loggins and Christy 2013). Later, in the early twentieth century, urban gardening efforts emerged again during wartime and economic depression to produce food and alleviate poverty (Lawson 2004, 2005). Beyond this predominant narrative, it was also argued that shared, food-producing urban open spaces have deep roots in the ethnic cultural heritage of Mexican-origin and other Latin American peoples (Mare and Peña 2010). In the 1970s, as inner-city decline spread throughout the U.S. due to white flights, suburbanization, and loss of tax revenues, community gardening became a way for many urban neighbourhoods to combat crimes and vandalism and improve neighborhood environment (Lawson 2004, 2005; Warner 1987). Specifically, in the context of the urban divestment in the 1970s, Warner (1987) argues that the community gardening movement reflects a politics of dignity and self-help.

In Europe, long traditions of allotment and community gardens can be found in countries such as Germany, Austria, the Netherland, and Norway (Francis et al. 1984). In Germany particularly, the Kleingarten represents a distinct form of community garden that has evolved over the past hundreds of years (Lewis 1979, cited in Francis et al. 1984). Starting as places for growing flowers and vegetables, these gardens have also become important centers for the social and cultural life for urban residents, especially Turkish immigrants (Francis et al. 1984). In Sweden, an allotment garden movement was developed in 1905, influenced by similar efforts in Denmark and Norway. The gardens have also become places where gardeners socialize with relatives and friends (Francis et al. 1984). Similar to experiences in the United States, urban gardening in Europe has also served as an important strategy for food production during the time of needs. In 1930, for example, 6% of Berlin's surface was covered with allotment gardens that produce food (Warneck et al. 2001, cited in Meyer-Renschhausen 2014).

In East Asia, urban gardening has emerged in recent years through both citizen movement and government initiatives. In Seoul, urban gardening was first promoted by Seoul Green Trust, a large grassroots nonprofit organization that was responsible for the development and co-management of the 297-acre (120 ha) Seoul Forest Park. In the face of declining municipal resources and diminishing land availability, urban gardening was viewed as a new strategy by the organization to pursue urban greening. The Seoul Municipal Government subsequently adopted an ambitious policy to promote urban agriculture and urban gardening. In Singapore, community gardening has been promoted under the government's Community in Bloom (CIB) initiatives. Launched in 2005, CIB currently has over 850 community gardens across Singapore, engaging over 20,000 residents.[1] In Taipei, the City Government recently adopted a Garden City initiative to promote urban gardening and food production effort, focusing on local schools and community garden plots. At the same time, a network of organizations including the nonprofit Community University system have also been promoting urban gardening through its curriculum to address interests in gardening and food learning. In Hong Kong, urban gardening has been spearheaded by social start-ups such as the Rooftop Republic that provides services to community and companies interested in rooftop gardening.

From Europe and North America to East Asia, urban gardening has served a variety of functions in the face of changing societal circumstances. Specifically, gardening has been interpreted and instrumentalized differently by different social groups and institutions for different purposes. These range from allotment gardens rented to individuals and sometimes commercial growing to community gardens that provide multiple benefits including food production and community building. Hancock (2001) makes an important distinction between allotment gardens and community gardens. He suggests that the community garden build social capital 'because, unlike an allotment garden, they are created and managed by the community itself and depend upon a cohesive social network to organize and manage the gardens' (Hancock 2001: 279). Indeed, the collective labour needed to maintain the garden as well as the informal social interactions occurring regularly in a garden all contribute significantly to the making of urban community gardens as an important and unique urban social space. In the following, the chapter examines aspects of the literature focusing specifically on the social and community building aspects of urban gardening.

6.3 Social Significance of Urban Gardening

Urban gardening, specifically community gardening, has been associated with a wide variety of personal and collective benefits as demonstrated in a growing number of empirical studies. These benefits range from personal well-being, health, and access to food, to community development and empowerment (Francis et al. 1984; Guitart et al. 2012; Hou et al. 2009; Lawson and Drake 2013; Ohmer et al. 2009; Okvat and Zautra 2011). Among these wide-ranging benefits, community building has ranked prominently in one study after another. In the 2012 survey conducted by Lawson and Drake (2013) for the American Community Gardening Association, 66% of the respondents identified social engagement, community

[1]See https://www.nparks.gov.sg/gardening/community-in-bloom-initiative.

building or neighbourhood revitalization as one of the most significant benefits resulting from community gardens, second to food production and access (at 74%). In a study that reviews the available research on community gardens, Guitart et al. (2012) found 'social development/cohesion' to be the top motivation, followed by 'to consume fresh foods' and 'improving health'. The most commonly demonstrated benefits were social benefits, 'such as community building/resilience and social interaction' (Guitart et al. 2012: 367). Similarly, Saldivar-Tanaka and Krasny (2004) found that much of the recent community gardening literature has focused on community development, and recognizes the role of gardens in creating a sense of community, economic opportunities, and an enhanced environment especially in poor, ethnically diverse neighbourhoods.

In the field of community psychology, in addition to empirical evidences, including those in individual well-being, cognitive benefits, affective benefits, community well-being, social network benefits, multicultural relations, community organizing and empowerment, crime reduction, nutrition and physical activity, economic benefits, environmental well-being, climate change mitigation, and other environmental benefits, Okvat and Zautra (2011: 378) cites a study by Kuo et al. (1998) that identifies a more distal benefit of green common spaces 'in that neighbourhood social ties significantly predicted a greater sense of safety and sense of adjustment to living in the neighbourhood'. 'Overall, green space appears to provide an opportunity for social contact and expansion of neighbourhood social networks, which serves as a foundation for building community' (Okvat and Zautra 2011: 378).

Aside from a benefit, community building in turn also serves as a key ingredient of successful community gardens. Through literature review and interviews with community garden leaders, Milburn and Vail (2010) identify community development as among four 'seeds' that contribute to successful community gardens, including secured land tenure, sustained interest, and appropriate design. Twiss et al. (2003), on the other hand, identify key elements for success to include volunteers and community partners and skill building opportunities, in addition to local leadership and staffing. In contrast, 'Gardens that are built by an external group and given to the community to maintain are often abandoned and vandalized as none of the residents feel responsible or have a sense of ownership' (Schmelzkopf 1995, cited in Milburn and Vail 2010, 79). Looking at the social benefits of urban gardening, Okvat and Zautra (2011) found that community gardens led to more neighbour-to-neighbour assistance. 'When one member was ill, injured, or busy, other members would tend their plots' (Okvat and Zautra 2011: 379).

In addition to contributing to the success of gardens itself, community involvement through gardening can have other positive spillover effects in a local community. In a study on community gardens in Toronto, Alaimo et al. (2010) examined associations between participation in community gardening/beautification projects and neighbourhood meetings with perceptions of social capital at both the individual and neighbourhood levels. They found that having a household member participate in community gardening/beautification and/or neighbourhood meetings was associated with more positive perceptions of bonding social capital, linking social capital, and the existence of positive neighbourhood norms and values (Alaimo et al. 2010). Glover (2004: 143) goes further to note, 'community gardens are less about gardening than they are about community'. The social interactions facilitated by gardens can foster norms of reciprocity and trust—conventional forms of social capital, and the development of social networks seemed to lead to further socializing outside of the garden project (Glover 2004: 150). Indeed, social capital has been a consistent theme in a growing number of studies on community gardens.

The community garden's unique capacity in fostering social interactions and building communities has been contrasted with other forms of urban open space. In a comparative study of community gardens and public parks in Sacramento, California, Francis (1987: 110) characterizes public parks as passive, publicly-controlled places which people often use alone and 'like'. On the other hand, gardens are active places that people make themselves, use for work and socializing, and can 'love'. Similar to the different characteristics of community gardens and public parks, garden organizers and institutions tend to have different views toward community gardens. Jamison (1985) noted that whereas bureaucracies tended to see gardening as personal activity, movement organizers stressed the increased self-worth and confidence mainly from participants' involvement in group action. 'This included both collective gardening and other activities related to the establishment and operation of gardening projects-activities which usually were unavailable to participants in bureaucratic gardening projects' (Jamison 1985: 476).

6.4 Urban Gardens as Multimodal Social Spaces

It is clear from the literature review above that socialization and community building constitute an important facet of urban community gardening. They are also critical to the success and uniqueness of urban gardens as community open spaces. In the following, the chapter seeks to define the multiple social dimensions of urban gardens, based on a review of literature and sampling of cases around the world.

6.4.1 Garden as Convivial Space

At the most fundamental level, through interactions at social events and everyday gardening activities, urban gardens serve as a convivial space for gardeners and non-gardeners alike. Everyday activities and occasional social events strengthen social ties and facilitate interactions among both gardeners and non-gardeners. They make the gardens function more as social spaces, rather than just garden plots. Through these interactions, urban gardening engages a wider audience beyond the gardeners, resulting in greater community benefits as well as greater support for urban gardening.

In a study on community gardens in New York City, Saldivar-Tanaka and Krasny (2004: 399) found that, aside from production of vegetables and herbs, the gardens also host numerous social, educational, and cultural events, including 'neighbourhood and church gatherings, holiday parties, children's activities, school tours, concerts, health fairs, and voter registration drives'. They found that gardeners and garden members view gardens 'more as social and cultural gathering places than as agricultural production' sites (Saldivar-Tanaka and Krasny 2004: 407). Gardens are seen as 'cultural and social neighbourhood centers, where people go to meet with friends, family, neighbours, newcomers, and visitors'; people of all ages get together to play cards, domino, etc., relax, exercise, cook and share food, chat, and 'find out what is going on in the community' (Saldivar-Tanaka and Krasny 2004: 404).

Similarly, community gardens in Seattle have been used as places for yoga exercises, concerts, BBQ, Wedding, and other more everyday forms of social events activities (Hou 2014; Hou et al. 2009) (see Table 6.1 and Fig. 6.3). Many of these events engage not only gardeners, but also neighbours, volunteers, and visitors alike. They represent a key ingredient to the success of the gardens (Hou et al. 2009). In Perth, Australia, Middle et al. (2014) also note that many community gardens actively facilitate community interactions through organized social events.

Table 6.1 List of social and community activities in selected community gardens in Seattle (*Source* Hou et al. 2009)

Sites	Activities
Interbay P-Patch	• Annual salmon Barbeque • Seasonal gathering: Christmas wreath-making, New Year's dinner; solstice party, Fourth of July party • Special events: Jazz in the Garden, Big Fat Greek Picnic • Fundraising: dahlia plant sale, honey sale • Tours, workshops, and teaching • Compost/Saturday soup specials • Friday night potlucks
Thistle P-Patch	• Volunteer workdays
Danny Woo International Community Garden	• Tours and workshops • Volunteer workdays (1–2 per month) • Annual pig roast
Bradner Gardens Park	• Tours, workshops, and demonstrations • Children's garden activities • Plant sales • Occasional concerts • Annual events: New Year's Eve burn, Halloween party, Fourth of July party
Marra Farm	• Volunteer workdays • Annual Fall Fest
Magnuson P-Patch	• Open-house events (first year) • Children's garden activities • Amphitheater events: music, drama, etc. • Volunteer workdays

Fig. 6.3 Annual pig roast at the Danny Woo International Community Gardens in Seattle brings together gardeners, volunteers, and neighborhood residents to enjoy the garden and the company (*Photo credit* J. Hou)

Similarly, these social events and activities make the gardens function more than a place to grow food and facilitate social interactions and building of social bonds between gardeners and non-gardeners.

Besides social events and activities, gardening itself can be a form of social activity and facilitates social interactions. In Amsterdam, Lin (2015) studies the case of Buurtmoestuin Dijkgraafplein where gardening is an important medium for social interaction in the neighbourhood. In Toronto, Baker (2004) finds that 'working bees', a regular monthly communal activity for the gardeners helps bind people together. In Seattle, Hou et al. (2009) noted cases in which gardeners prefer to join community gardens even though they could also garden in their backyards because of the social nature of community gardening and opportunities to meet with neighbours and other gardeners.

6.4.2 Garden as Cultural Space

For immigrants and refugees in particular, urban gardens provide important opportunities for continuing traditional cultural practices. The gardens are also used as sites for a variety of ethnic social events and cultural practices. As such, urban gardens function as a cultural space for specific communities and as a site for the expression of cultural identities and belonging.

Eizenberg (2011: 771) notes that the diversity of gardens in New York City enables gardeners to express and experience their culture collectively, rather than privately—'a rich experience that engages aesthetic and culinary preferences, rituals, customs, artistic expressions, and social interactions'. In their study of community gardens in New York City, Saldivar-Tanaka and Krasny (2004) found that the gardens provide a connection between immigrants and their cultural heritage. Particularly, the Latino gardens seemed to be particularly important as sites for maintaining Puerto Rican farming culture in an urban environment (Saldivar-Tanaka and Krasny 2004: 409). But rather than just agricultural practices, they found that in Latino cultures, 'agricultural practices are tightly intertwined with community celebrations, which often include dance, music, and food' (Saldivar-Tanaka and Krasny 2004: 409).

In Seattle, Hou et al. (2009) examined several cases of community gardens and found them to serve also as important venues for continued cultural practices. At Thistle P-Patch, Marra Farm, and the Danny Woo International Community Gardens, which all support primarily gardeners from ethnic communities, the gardens provide important opportunities for continuing the agrarian traditions and knowledge of immigrant and refugee gardens in particular. The gardens enable them to grow ethnic vegetables that could not be easily found in grocery markets. Growing and consuming these vegetables are part of important cultural practices for ethnic communities (Hou et al. 2009). In their study of Marra Farm in Seattle and South Central Farm in Los Angeles, Mare and Peña (2010) looked at the gardening efforts as a way of retaining cultural identity and create community. In both cases, they argue that the gardens serve as a site for immigrant gardeners to reconstruct their sense of place (Mare and Peña 2010).

6.4.3 Garden as Inclusive Space

For communities and gardens with ethnically diverse gardeners, the garden can serve particularly as a space for cross-cultural interactions and learning as well as building of social ties across ethnic and cultural boundaries. These occur through both the act of gardening as well as the social and community events occurring on the site.

For instance, in examining community gardens in Australia, Middle et al. (2014) suggest that community gardens have great potential for bringing socially and culturally diverse individuals together. The gardens 'can provide a similar community building function to other formal group activities such as sport, but for different and perhaps more varied demographic groups' (Middle et al. 2014: 4). In the case of Transvaal Buurttuinen, a multicultural community in Amsterdam, because every participant had to participate in the daily affairs, more social interactions happened in the neighbourhood. The garden is no longer only a place for growing vegetables and flowers, but also for growing social awareness, neighbourhood relationships and environmental justice (Lin 2015).

Though a case study of three gardens in Toronto, Baker (2004) found that through the gardening activities, the gardeners are actively shaping their community and connecting cross-culturally. Similarly, in another study in Toronto, Wakefield et al. (2007: 100) found that, ‘the gardens were seen by many as a place where communicating with people from other cultures could begin, using food and shared experience as a starting point for understanding’. This is echoed by a study of a community garden in Oakland, California, in which growing and sharing food also serves as a medium for cross-cultural learning and understanding between different cultural and ethnic groups (Prince 2013). Examining the community gardens in St. Louis, Shinew et al. (2004) further suggest that community gardens can serve as potential sites for racial integration.

The function and benefit of gardens to serve as an inclusive space are not only limited to ethnic groups. Rather, it can be extended to other demographic groups as well. For instance, Krasny and Tidball (2009: 3) studied the work of Garden Mosaics, an international educational programme taking place in urban community gardens in the United States and abroad to connect youths and elders ‘to investigate the mosaics of plants, people, and cultures in gardens’. In Minneapolis, Myers (1998: 181) finds that ‘gardening activities undertaken by persons with psychiatric disabilities can provide opportunities for empowerment and increased competence, while building bridges to naturally occurring supports and resources within the broader community’. As an activity that can engage a wide variety of urban residents, gardening can serve as a bridge for diverse individuals and communities.

6.4.4 Garden as Restorative Space

In recent years, the literature concerning health benefits of nature and access to nature has experienced a phenomenal growth (e.g., Cooper Marcus and Sachs 2014; Scouter-Brown 2015; Winterbottom and Wagenfeld 2015; Ward Thompson et al. 2010; Cooper Marcus and Barnes 1999; Talbot and Kaplan 1991). The recent literature has built on earlier work including Attention Restoration Theory (Kaplan 1995) and Stress Recovery Theory (Ulrich et al. 1991) that have contributed to a greater understanding of the health benefits of access to natural environment. Although much of the research has been specific to the context of hospitals and patients in health facilities, growing evidence has also been collected in the context of gardening, including community gardening.

At the individual level, working with plants and in the outdoors has been found to benefit the mental health, mental outlook, and personal wellness of individuals (Brown et al. 2004; Matsuo and Relf 1994, cited in Bellows et al. 2004). Cultivation activities are said to trigger both healing responses and illness prevention (Bellows et al. 2004), and enhance nutrition and physical activity (Twiss et al. 2003). Similarly, Middle et al. (2014: 638) find that community gardens provide a venue 'for an alternative and more accessible form of physical activity—gardening—and a restorative park environment…'. They also find that 'even as passive destinations, community gardens can be effective restorative environments. […] Community gardens, regardless of the activity undertaken therein, are effective forms of urban nature for providing restorative services' (Middle et al. 2014: 641).

Various case studies have supported these findings. For example, in a report on California Healthy Cities program, Twiss et al. (2003) found increase physical activity and increased consumption of fruits and vegetable being associated with school gardening programs. In a study on South-East Toronto, Wakefield et al. (2007: 92) found that community gardens were perceived by gardeners to provide numerous health benefits, 'including improved access to food, improved nutrition, increased physical activity and improved mental health'. In a study on community support programme project in ten garden sites in Chester County, Pennsylvania, Myer (1998) found gardening activities providing persons with psychiatric disabilities opportunities for empowerment and increased competence, while building bridges to supports and resources within the broader community.

Besides gardening activities and their health benefits, the restorative aspect of community gardening also stems from the social interactions taking place in the gardens. For example, Teig et al. (2009) argue that community gardens have the potential to promote public health not only through physical activity and improved nutrition, but also through social engagement. They further argue that the place-based social processes found in community gardens support collective efficacy, 'a powerful mechanism for enhancing the role of gardens in promoting health' (Teig et al. 2009: 1121). In the Toronto study, gardens were also perceived as promoting social health and community cohesion (Wakefield et al. 2007). In Seattle, the author found that gardeners valued the social interactions in the garden as much as the activity of gardening. In particular, immigrant gardeners have reported that the gardens have helped them with their adaptation to the new environment through interactions and friendships developed with fellow gardeners in the neighborhood garden (Hou 2013).

At both individual and community levels, community gardening also relieves stresses from traumatic experience. Winterbottom and Wagenfeld (2015: 261) suggest that community gardens offer safe, supportive places 'to confront fear and alleviate disturbing memories'. Similar to the experience in Seattle, gardeners and survivors of wars befriend each other through gardens and find it to be a place for relaxation and learning (Winterbottom and Wagenfeld 2015).

6.4.5 Garden as Democratic Space

Through collective decision-making and sharing of responsibilities, garden can serve as a space for democratic practices. Internally, community gardens can serve as a medium through which democratic values are practiced and reproduced

(Glover et al. 2005). Through organizing and community building through gardening activities, the engagement can also extend beyond the gardens into related social and political movements.

In the case study of a community garden in Melbourne, Australia, Kingsley and Townsend (2006: 525) found that the garden provides benefits including increased social cohesion in terms of the sharing of values that enables 'identification of common aims and the sharing of codes of behavior governing relationships'. In Perth, Western Australia, Middle et al. (2014) found that gardens could facilitate bridging interactions between different social groups and at the same time provide opportunities for local residents to participate actively in green space planning processes.

In terms of social and political engagement beyond the gardens, Saldivar-Tanaka and Krasny (2004) found that the activities of community gardeners in New York City served to catalyze community organizing. Specifically, 'in order to be more effective and to help each other, gardeners and garden support groups have formed coalitions to work on fund raising, publications, workshops, rallies, outreach, and support of other local campaigns' (Saldivar-Tanaka and Krasny 2004: 408). In some cases, organizing and leadership experiences gained through participation in community gardens can lead to engagement in the political process, such as voter drives and rallies (Saldivar-Tanaka and Krasny 2004). In Amsterdam, Lin (2015) suggests that community-based urban farming on micro-urban public space functions to framing a powerful grassroots movement to reclaim the right to the city.

In Sacramento, California, de la Peña (2015: 37) studied the case of Soil Born, 'an organization that has helped established networks that have mobilized citizens, urban farmers, gardeners, and gleaners to remake Sacramento as an edible city'. Similarly, in Toronto, Baker (2004) finds that through gardening activities, gardeners are being drawn into broader social movements such as the Community Food Security movement, through their association with local nongovernmental organizations. In New Orleans, Kato et al. (2014) examine the case of Lower Ninth Ward Food Access Coalition (LNWFAC) and how its self-directed struggle for food access has led neighbourhood residents to engage with legacies of racism and segregation, and to collaborate with a diverse array of actors to achieve their stated goals. They further looked at how political gardening in marginalized neighbourhoods represents an apparatus through which citizens can engaged in politicized conversations—'even if somewhat tacitly' (Kato et al. 2014: 3).

In Berlin, Rosol (2010) studied the case of an urban farm for children as an example of a grassroots project as part of a broader social movement developed in confrontation with urban planners, as well as other cases of community gardens that were developed in cooperation with urban planners. In all of these instances, regardless of conflicts or collaboration, community gardens serve as important sites for social mobilization and political engagement—a critical component of active democracy.

6.4.6 Garden as Resilient Space

Finally, urban gardens can serve as a community safety net in the face of calamity as well as economic despair. Through building of networks and social ties, and through the production of food, urban gardening can contribute to the resilience of a community in time of crisis.

In the context of disaster events, Tidball and Krasny (2007, cited in Okvat and Zautra 2011: 376) argue, 'community gardens can contribute to the creation of resilient urban neighbourhoods and facilitate a city's recovery when faced with a sudden crisis such as a natural disaster or human-made conflict, or a more gradual disturbance such as an economic downturn'. In the case of a community garden in Melbourne, Australia, Kingsley and Townsend (2006) find the garden provides social support in terms of having people to turn to in times of crisis. Another notable instance was the rescue and relief effort in Far Rockaway, Queens, following the Super Storm Sandy, in which a local community gardens served as a relief station for victims seeking. In this case, the relief effort also utilized both the space of the garden and the social networks that was built through the establishment and maintenance of the garden (Hou 2015).

As mentioned above, beyond instances of disasters, community gardens can also serve as a place for communities to respond to economic precariousness. In Chicago, Williams (2014: ii) studies how collective work in urban gardening enables the low-income African American community to actively transform 'the bleak circumstances into a places of opportunity for a better life'. In Seattle, again, growing food in the gardens also provide economic supplement for low-income families and especially for immigrants and refugees who can make use of their agrarian skills. For the immigrants and refugees, gardening provides a means to supplement limited household. The activity itself also contributes to the psychological health and well-being of individuals engaged in gardening (Hou et al. 2009).

6.5 Precautions and Conclusions

As discussed above, urban gardens can provide a variety of significant social benefits beyond food production and recreation. The benefits range from serving as convivial space for social and intercultural interactions, to a space for cultural and democratic practices, and a safety net against environmental and economic calamity. However, as much as the gardens themselves are often a work in progress, it is important to note these social benefits have to be actively nurtured and maintained, rather than being inherent or intrinsic.

In a case study in Melbourne, Australia, for instance, Kingsley and Townsend (2006) find that, at least in the early stages of development, the social benefits do not necessarily extend beyond the garden setting. In a case in Toronto, Baker (2004) finds that not all of the social capital generated in the garden was positive. In

Singapore, Tan and Neo (2009) find that community gardens developed under the Community in Bloom programme are viewed by some as exclusionary spaces due to their close association with government apparatus. While noting that those who are directly involved believe that the gardens have been effective in enhancing community bonding within the community they serve, 'presence of locks and fences around the garden has made the community garden a physical exclusionary space where access is determined by the managers and those directly involved with the garden' (Tan and Neo 2009: 535). Similarly, Chua (2000) found that when community projects in Singapore are linked with governmental bodies, a portion of the population would avoid participation. These evidences suggest the materialization and manifestation of social benefits are predicated on active engagement and conditions beyond the simple availability of gardens.

In growing cases around the world, it is clear that urban gardening that takes the form of community gardens provides unique opportunities for social interactions, cultural and democratic practices, and urban resilience. Unlike other forms of urban open spaces that preclude direct engagement of users and stakeholders in their implementation, programming, and management, urban gardens are direct products of both individual and collective efforts. As such, the gardens build and nurture agency of individuals as well as social ties in a community. As a collective space in the increasingly heterogeneous urban contexts, urban gardens can serve as a venue and vehicle for interactions between diverse users and stakeholders. As such, they can function as a place for cross-cultural learning and understanding and building of connections across social and cultural divides, where differences can be recognized and negotiated. Finally, as a place of social and democratic engagement, urban gardens can further serve as a vehicle to engage individuals and communities in efforts toward other social and environmental initiatives.

In democratic societies, urban greening initiatives require broad support of citizens and institutions. Through direct engagement and experiences of social and environmental benefits, urban gardens can support a broader movement for diverse urban greening strategies that are important in making contemporary cities more livable, just, and resilient. In the ever more complex urban systems today, urban greening in turn needs to embody multifaceted social, cultural, economic, and environmental goals and objectives. With its multimodal social outcomes and processes, urban gardening serves not only as an effective strategy but also as an exemplary model for other greening strategies to consider multiple social, cultural, and economic dimensions and active engagement of communities and citizens.

References

Alaimo K, Reischl TM, Ober Allen J (2010) Community gardening, neighborhood meeting, and social capital. J Community Psychol 38(4):497–514

Baker L (2004) Tending cultural landscapes and food citizenship in Toronto's community gardens. Geogr Rev 94(3):305–325

Bellows AC, Brown K, Smit J (2004) Health benefits of urban agriculture. Community Food Security Coalition's North American Initiative on Urban Agriculture, Portland, OR

Brown VM, Allen AC, Dwozan M, Mercer I, Warren K (2004) Indoor gardening and older adults: effects on socialization, activities of daily living, and loneliness. J Gerontological Nurs 30 (10):34–42

Chua BH (2000) The relative autonomies of state and civil society in Singapore: engaging the citizens. In: Ooi GL, Koh G (eds) State-society relations in Singapore. Institute of Policy Studies, Oxford University Press, Singapore, pp 92–96

Cooper Marcus C, Barnes M (eds) (1999) Healing gardens: therapeutic benefits and design recommendations. Wiley, New York

Cooper Marcus C, Sachs NA (2014) Therapeutic landscapes: an evidence-based approach to designing healing gardens and restorative outdoor spaces. Wiley, New York

de la Peña D (2015) Edible Sacramento: soil born farms as a community-based approach to expanding urban agriculture. Incite CHANGE/CHANGE insights: CELA 2015 conference proceedings, p 37–52

Eizenberg E (2011) Actually existing commons: three moments of space of community gardens in New York City. Antipode 44(3):764–782

Francis M (1987) Some different meanings attached to a city park and community gardens. Landscape J 6(2):101–112

Francis M, Cashdan L, Paxson L (1984) Community open spaces: greening neighborhoods through community action and land conservation. Island Press, Washington, DC

Glover T (2004) Social capital in the lived experiences of community gardens. Leisure Sci 26:143–162

Glover T, Shinew KJ, Parry DC (2005) Association, sociability, and civic culture: the democratic effect of community gardening. Leisure Sci 27:75–92

Guitart D, Pickering C, Byrne J (2012) Past results and future directions in urban community gardens research. Urban For Urban Greening 11(4):364–373

Hancock T (2001) People, partnerships and human progress: building community capital. Health Promot Int 16(3):275–280

Hou J (2013) Transcultural participation: designing with immigrant communities in Seattle's international district. In: Hou J (ed) Transcultural cities: border crossing and placemaking. Routledge, London and Seattle, pp 222–236

Hou J (2014) Making and supporting community gardens as informal urban landscapes. In: Mukhijia V, Loukaitou-Sideris A (eds) The informal American city: beyond taco trucks and day labor. The MIT Press, Cambridge MA, pp 79–96

Hou J (2015) Guerrilla resilience. Waterproofing New York, Urban research (UR), 2:90–99 (Terreform, New York)

Hou J, Johnson JM, Lawson LJ (2009) Greening cities, growing communities: learning from Seattle's urban community gardens. University of Washington Press, Seattle

Jamison MS (1985) The joys of gardening: collectivist and bureaucratic cultures in conflict. Sociol Q 26(4):473–490

Kaplan S (1995) The restorative benefits of nature: toward an integrative framework. J Environ Psychol 15:169–182

Kato Y, Passidomo C, Harvey D (2014) Political gardening in a post-disaster city: lessons from New Orleans. Urban Stud 51(9):1833–1849

Kingsley JY, Townsend M (2006) 'Dig In' to social capital: community gardens as mechanism for growing urban social connectedness. Urban Policy Res 24(3):525–537

Krasny ME, Tidball KG (2009) Community gardens as contexts for science, stewardship, and civic action learning. Cities Environ 2(1):18 p. Available via http://digitalcommons.lmu.edu/cate/vol2/iss1/8/. Accessed 27 Feb 2016

Kuo FE, Sullivan WC, Coley RL, Brunson L (1998) Fertile ground for community: Inner-city neighbourhood common spaces. Am J Com Psych 26:823–851

Lawson L (2004) The planner in the garden: a historical view into the relationship between planning and community gardens. J Plann Hist 3(2):151–176

Lawson L (2005) City bountiful: a century of community gardening in America. University of California Press, Berkeley
Lawson L, Drake L (2013) Community gardening organization survey 2011–2012. Community Greening Rev 18:20–47
Lewis GK (1979) The kleingarten: evolution of an urban retreat. Landscape 23:33–37
Lin Y-T (2015) Farming the rights to the city: case studies of community-based urban farming on micro-urban public space in Amsterdam. Paper presented at future of places III, Stockholm, Sweden, 29 June–1 July 2015
Loggins D, Christy L (2013) History of New York City open space. Community Greening Rev 18:14–19
Mare TM, Peña D (2010) Urban agriculture in the making of insurgent spaces in Los Angeles and Seattle. In: Hou J (ed) Insurgent public space: guerrilla urbanism and the remaking of contemporary cities. Routledge, London and New York, pp 241–254
Matsuo E, Relf PD (eds) (1994) Horticulture in human life, culture, and environment: international symposium 22 Aug 1994: 24th international horticultural congress, 21–27 Aug 1994, Kyoto, Japan, ISHS, Leuven, 19–29, ISHS Acta Horticulture, no. 391
Meyer-Renschhausen E (2014) Community gardening in Berlin and New York: a new eco-social movement. In: Viljoen A, Bohn K (eds) Second nature urban agriculture: designing productive cities. Routledge, London and New York, pp 146–153
Middle I, Dzidic P, Buckley A, Bennet D, Tye M, Jones R (2014) Integrating community gardens into public parks: an innovative approach for providing ecosystem services in urban areas. Urban Forestry Urban Greening 11(4):638–645
Milburn L-AS, Vail BA (2010) Sowing the seeds of success: cultivating a future for community gardens. Landscape J 29:71–89
Myer MS (1998) Empowerment and community building through a gardening project. Psychiatr Rehabil J 22(2):181–183
Ohmer ML, Meadowcroft P, Freed K, Lewis E (2009) Community gardening and community development: individual, social and community benefits of a community conservation program. J Community Pract 17(4):377–399
Okvat HA, Zautra AJ (2011) Community gardening: a parsimonious path to individual, community, and environmental resilience. Am J Community Psychol 47:374–387
Prince A (2013) Urban agriculture as 'agricultural' producer. In: Hou J (ed) Transcultural cities: border crossing and placemaking. Routledge, London and New York, pp 237–249
Rosol M (2010) Public participation in post-fordist urban green space governance: the case of community gardens in Berlin. Int J Urban Reg Res 34(3):548–563
Saldivar-Tanaka L, Krasny ME (2004) Culturing community development, neighborhood open space, and civic agriculture: the case of Latino community gardens in New York City. Agric Hum Values 21:399–412
Schmelzkopf K (1995) Urban community gardens as contested space. Geogr Rev 85(3):364–381
Scouter-Brown G (2015) Landscape and urban design for health and well-being. Routledge, London and New York
Shinew KJ, Glover TD, Parry DC (2004) Leisure spaces as potential sites of interracial interaction: community gardens in urban areas. J Leisure Res 36(3):336–355
Talbot JF, Kaplan R (1991) The benefits of nearby nature for elderly apartment residents. Int J Aging Human Dev 33(2):119–130
Tan LHH, Neo H (2009) 'Community in bloom': local participation of community gardens in urban Singapore. Local Environ 14(6):529–539
Teig E, Amulya J, Bardwell L, Buchenau M, Marshall JA, Litt JS (2009) Collective efficacy in Denver, Colorado: strengthening neighborhoods and health through community gardens. Health Place 15:1115–1122
Tidball KG, Krasny ME (2007) From risk to resilience: what role for community greening and civic ecology in cities? In: Wals AEJ (ed) Social learning towards a sustainable world: principles, perspectives and praxis. Wageningen Academic Publishers, Wageningen, pp 149–164

Twiss J, Dickinson J, Duma S, Kleinman T, Paulsen H, Rilveria L (2003) Community gardens: lessons learned from California healthy cities and communities. Am J Public Health 93 (9):1435–1438

Ulrich RS, Simons RF, Losito BD, Fiorito E, Miles MA, Zelson M (1991) Stress recovery during exposure to natural and urban environments. J Environ Psychol 11(3):201–230

Wakefield S, Yeudall F, Taron C, Reynolds J, Skinner A (2007) Growing urban health: community gardening in South-East Toronto. Health Promot Int 22(2):92–101

Ward Thompson C, Aspinall P, Bell S (eds) (2010) Innovative approaches to research landscape and health. Routledge, London and New York

Warneck P, Gröning G, Friedrich J (2001) Ein starkes Stück Berlin 1901–2001: 100 Jahre organisiertes Berliner Kleingartenwesen. Landesverband der Gartenfreunde e.V, Berlin

Warner SB (1987) To Dwell is to garden: histories and portraits of Boston's Community Gardens. Northeastern University Press, Boston

Williams DA (2014) Fertile ground: community gardens in a low-income inner-city Chicago neighborhood and the development of social capital among African Americans. Dissertation, University of Illinois at Urbana-Champaign

Winterbottom D, Wagenfeld A (2015) Therapeutic gardens: design for healing spaces. Timber Press, Portland and London

Chapter 7
Urban Green and Biodiversity

Peter Werner and John G. Kelcey

Abstract The chapter explores the relationship between urban green and biodiversity. Cities are home to a large number of native plant and animal species. Non-native species are an essential component for the species richness in the cities worldwide. The population of animal and plant species is not stable and the number of native species has been declining over the last decades and the portion of non-native species is increasing. Public and private gardening are main causes for the introduction of non-native species. The different contrasting attitudes towards non-native species between urban dwellers and nature conservationists are discussed. Three approaches are described representing different scales, namely the city in the region, the urban matrix and green patches. These three approaches offer a sophisticated view about the relationship between urban green and biodiversity and provide ways for a tailored management to improve biodiversity in urban areas. For each scale, recommendations for the management of urban green are presented. The chapter ends with basic principles for the development and management of green areas and green structures to enhance urban biodiversity and ecosystem services in cities.

Keywords Urban biodiversity · Native species · Non-native species · Species pool · Urban matrix · Patches

7.1 Introduction

The relationship between biodiversity and cities is ambiguous, or expressed in another way, cities are both the beast and the beauty with respect to biodiversity.

P. Werner (✉)
Institute for Housing and Environment, Darmstadt, Germany
e-mail: p.werner@iwu.de

J.G. Kelcey
Ceckovice, Bor U Tachova, 34802 Czech Republic
e-mail: johnkelcey@hotmail.com

P.Y. Tan and C.Y. Jim (eds.), *Greening Cities: Forms and Functions*,
Advances in 21st Century Human Settlements,
DOI 10.1007/978-981-10-4113-6_7

The beast: urban development destroys and fragments natural habitats, increases the abundance and distribution of alien and invasive species and strains the local, regional and global ecosphere. Urban development is responsible for one of the most serious global issues, that of biotic homogenization. The beauty: in relation to their surrounding landscapes, cities stand out because of their richness of species, a wide range of habitats and efficient use of land for human settlements. Cities have become the main habitat for several plant and animal species and a refuge for some rare species, as well as providing opportunities for most of the world's human population to integrate experiences of nature into everyday life (Müller et al. 2010; Elmqvist et al. 2013).

The Convention on Biological Diversity defines 'urban biodiversity' as: 'The variety and richness of living organisms (including genetic variation) and habitat diversity found in and on the edge of human settlements. This biodiversity ranges from the rural fringe to the urban core' (SCBD 2012).

On the one hand, it is reasonable to assert that the past, present and probably future total biodiversity of any city in the world is not known and never will be. On the other hand, the problem is that the complexity of determinants and the spatial and temporal dynamic of cities (Andersson 2006) preclude simple starting points and lines of argument to explain causal linkages between biological diversity and cities (Kinzig et al. 2005). Despite these difficulties the chapter examines some aspects of the relationship between urban green and biodiversity although it is appreciated that in doing so only partial explanations are possible at the present time.

However, in terms of a general principle, there is a simple message: more green = more biodiversity (Aronson et al. 2014; Beninde et al. 2015; Turrini and Knop 2015). Green areas such as parks other green elements and even a single tree provide the basic units that allow plants and animals to survive in urban areas.

To better understand the underlying processes, it is helpful to use three overlapping approaches, representing different eco-geographical scales. These approaches have been chosen to present the current knowledge about the relationship between the green components of cities and biodiversity as well as opportunities to increase both aspects of biodiversity by the management of green areas and other elements. The three approaches are:

1. City in the context of the region, because cities are embedded in biogeographic regions and are connected to the surrounding landscapes.
2. Urban fabric of the city as a whole, which focuses on the intimate mix of built-up and non-built-up areas and structures.
3. The local or site level, which looks more closely at green spaces or patches – green islands in the ocean of the concrete city.

After this introduction the chapter is divided into four sections starting with some general remarks about the origin and composition of the urban flora and fauna and followed by an examination of some aspects of non-native species. Then we present the above mentioned three approaches in the context of the development of biodiversity of cities. The final section contains the conclusion. Some key concepts and terms are defined and explained in Box 7.1.

Box 7.1

Definitions

Urban Biodiversity
The variety and richness of living organisms (including genetic variation) and habitat diversity found in and on the edge of human settlements. This biodiversity ranges from the rural fringe to the urban core (SCBD 2012).

Green Components
Green area is defined as land that consists predominantly of unsealed, permeable, 'soft' surfaces such as soil, grass, shrubs and trees. It includes all public and private areas of parks, gardens, play areas, wasteland, green roofs and so on as well as other green spaces with other origins like natural areas, agricultural land, orchards, nurseries. The term '(urban) green area' is used here as a short hand term and includes all these categories. Trees, hedgerows and other green single structures which support organisms in cities are described here as *green elements*.

Both green areas and green elements are summarized under the term green components.

Native Species
Native or indigenous species have originated in a given area with or without human involvement or have arrived without intentional or unintentional intervention of humans from an area in which they are native (Scholz 2007). In a strict sense, only such taxa can be defined as native if their genomes are part of the original regional species pool.

Non-native Species
Alien, non-native, non-indigenous, foreign, exotic means a species, subspecies, or lower taxon occurring outside of its natural range (past or present) and dispersal potential (i.e. outside the range it occupies naturally or could not occupy without direct or indirect introduction or care by humans), and includes any part, gametes or propagule of such species that might survive and subsequently reproduce (IUCN 2000).

7.2 Origin, Composition and Development of Plant and Animal Species in Urban Areas

It is often stated that the physical and chemical environment of cities, for example climate, water regime, soil structure, air, water and soil quality and vegetation cover are completely different from that of the surrounding landscape and that these altered conditions have a significant influence on the composition of the plant and animal populations (Sukopp and Wittig 1998; Douglas and James 2015) creating novel ecosystems (Kowarik 2011) of which green areas and structures are components.

Urban areas are very recent in the context of biogeographical history of the fauna, flora and other organisms on earth (Lenzin et al. 2004; Kelcey 2015). They are also highly dynamic because of continual changes that occur as the result of spatial expansion and changes in land use. Growing and shrinking, development and redevelopment of sites, technological innovations and changes in human and political priorities that influence values and therefore the planning, design and management of built-up and green areas. As a consequence the flora and fauna of cities lack stability (Sattler et al. 2010), including those that occur in public and private green areas. Only a subset of native species can cope with environmental changes, which occur with urbanization (Kark et al. 2007a; Williams et al. 2009).

Nevertheless many species representing various taxonomic groups such as vascular plants, birds and butterflies can be found in city areas (Table 7.1). However, this chapter concentrates mainly on terrestrial green issues in relation to vascular plant species and birds because most of the current knowledge is based on them. The total administrative area of a city includes parts of the surrounding landscape and various parks and green areas. As a result of their structural and habitat diversity, cities can accommodate a high level of species richness in terms of plants, animals and other organisms. It is essential to state the geographical context on which species-richness is measured, namely the total administrative area (the current baseline) or the area within the limits of development. With respect to both the total city area and vascular plants, cities like Berlin, Washington DC and Seoul are hotspots of biodiversity. At the more detailed level, closer to the city centre the environment becomes more inhospitable, green areas smaller and plants and animals more scarce (McKinney 2006a).

Although only relatively few species can survive in inner urban areas, some of them occur in high abundances. Many of these common species have migrated from their original natural habitats (especially rocks and cliffs) to urban centres. For example, in European cities the dominant breeding bird species include Rock Dove (*Columba livia domestica*), Collared Dove (*Streptopelia decaocto*), House Sparrow (*Passer domesticus*), Blackbird (*Turdus merula*), Starling (*Sturnus vulgaris*) and Black Redstart (*Phoenicurus ochruros*). Other species also breed in urban areas but feed (at least partially) outside it, for example Common Swift (*Apus apus*), Kestrel (*Falco tinnunculus*) and Eurasian Jackdaw (*Corvus monedula*) (Kelcey and Rheinwald 2005).

Table 7.1 Some basic data and species numbers of selected cities round the world

City	Country	Population	Population/km^2	City area in km^2	Developed area in km^2	Green space in km^2	Annual average temperature	Number of vascular plant species	Number of bird species	Number of Lepitoptera species
Adelaide	AU	1,203,873	659	1,826.90	n.a.	14.50	16.40	1,664	291	n.a.
Auckland	NZ	1,461,900	299	4,894.00	n.a.	n.a.	15.30	542	103	15[a]
Berlin	DE	3,510,032	3,927	891.70	70.29	22.50	8.90	1,393	148	n.a.
Curitiba	BR	1,851,215	4,056	434.97	n.a.	5.06	17.83	1,766	366	468
Edmonton	CA	817,498	1,168	699.80	29.73	n.a.	3.90	487	178	68
Hamburg	DE	1,802,041	2,349	755.30	44.40	23.30	9.00	1,086	160	n.a.
Montreal	CA	1,649,519	4,538	363.52	n.a.	5.44	6.00	1,063	124	274
Moskau	RU	11,503,501	4,583	2,510.00	n.a.	n.a.	5.00	1,647	152	n.a.
Nagoya	JP	2,267,048	6,945	326.43	56.53	10.28	16.10	1,000	231	75
New York	US	8,175,133	10,356	789.40	65.40	19.70	12.40	2,177	n.a.	87
Nitra	SK	78,875	785	100.48	n.a.	n.a.	8.40	608	144	n.a.
Seoul	KR	9,794,304	16,175	605.52	57.53	31.92	11.80	1,794	205	n.a.
Singapore	SG	5,183,700	7,257	714.30	n.a.	13.00	27.30	2,145	376	295
Toulouse	FR	440,204	3,721	118.30	n.a.	n.a.	12.60	767	92	n.a.
Zurich	CH	390,082	4,245	91.90	54.33	37.19	7.90	1,353	97	n.a.

[a]including migrants n.a. = not available

Sources Own data by P. Werner collected from local publications and websites, City Biodiversity Index fact sheets presented on COP 11 in Hyderabad 2012, Aronson et al. (2014)

The predominant urban birds are common species with a wide geographical range, being represented in temperate and Mediterranean cities by omnivores and granivores (Clergeau et al. 1998; Chace and Walsh 2006; Kark et al. 2007b) and by frugivoros and granivoros in tropical cities (Lim and Sodhi 2004).

Typical urban plants in Europe include those found in the following natural habitats (the English names follow Stace 2010):

- river banks, floodplains, woodlands and swamps: Ground Elder (*Aegopodium podagraria*), Hedge Bindweed (*Calystegia sepium*), Cleavers (*Galium aparine*);
- periodically flooded, nutrient-rich mud, sand and gravel surfaces of inland waters: Trifid Bur-marigold (*Bidens tripartita*), Creeping Cinquefoil (*Potentilla reptans*),
- strand lines, dunes and coastal rocks: Common Couch (*Elymus repens*), Perennial Sowthistle (*Sonchus arvensis*),
- areas of wind throw, clearings: Creeping Thistle (*Cirsium arvense*), Spear Thistle (*Cirsium vulgare*),
- open, disturbed ground, most of which has originated as the result of human activities, for example Annual Meadow Grass (*Poa annua*), Shepherd's Purse (*Capsella bursa-pastoris*) and Common Nettle (*Urtica dioica*).

The predominant characteristics of these urban plants are: the absence of hygromorphic leaves, increased leaf area index, wind-pollinated flowers, reproduction by seeds, short life-span or perennial hemicryptophytes.

When considering the habitats and species composition of the administrative area of a city, Kowarik (2005) identified four basic categories of nature (Table 7.2).

It is the variety of natural categories that is one of the reasons for the richness of species in urban areas compared to the lower number of species found in intensively managed arable land and forests.

A further reason for the species richness in relation to vascular plants is the presence of a large number of non-native species that have been introduced by human activities; they comprise 30–35% of the total number recorded in a European city. A similar situation applies throughout the world with exceptions of New Zealand and Australia. The same picture is true for birds but on a much lower level, only 10–15% of bird species in cities are non-native (Aronson et al. 2014).

In addition, many cities contain more native species than the surrounding landscape and, more surprisingly, a larger proportion of the regional species pool (La Sorte et al. 2014). An underlying cause of the species richness is that many

Table 7.2 Four basic forms of nature found in cities (from Kowarik 2005)

Nature 1	Old wilderness	Remnants of pristine nature
Nature 2	Traditional cultural landscape	Continuity of former agricultural or forested land
Nature 3	Functional greening	Urban parks, green areas and gardens
Nature 4	Urban wilderness	New elements by natural colonization processes particularly distinct on urban wastelands

cities, especially in Central Europe, have been established in productive and diverse landscapes combined with relatively stable environment conditions. This indicates that cities are often located in landscape settings that are naturally species-rich (Kühn et al. 2004). As a consequence, there is a special responsibility on central and local governments and others to protect this richness and to develop and manage green areas accordingly.

However, a comparison of the plant inventories of several cities on various continents has found a loss of native species ranging between 3% and 46% over a period of 100 to more than 150 years. The mean loss found in Europe is about 10–12% (Jackowiak 1998; Bertin 2002), for example a reduction of both native species and archaeophytes, but an increase of neophytes is reported in Frankfurt, Halle/Saale and Zurich. In Zurich the number of neophytes was higher than the number of native vascular plants that are thought to no longer occur in the city. The opposite has occurred in Frankfurt and Halle where the decrease of native species was larger than the increase of new species (Landolt 2000; Stolle and Klotz 2005; Gregor et al. 2012). It is instructive to compare those situations with the findings of Tait et al. (2005) and McKinney (2006b) who report the proportion and in many cases the absolute number of non-native species is still increasing

Successful ‘urban exploiters’ and ‘urban adapters’, which can be non-native or native species, are usually found in a large number of mapped grid fields or habitats of a city. However, during recent decades, the general trend appears to be that the proportion of species found in a larger number of grids, is declining. Conversely the proportion of species, including still common species, occurring in only a few sites is rising continually, which suggests a growing potential of threat (Chocholouskova and Pysek 2003).

7.3 Non-native Species

There is a voluminous literature on non-native species (many published the last few years), which it is impractical to review in this chapter, which focuses on some of the more important issues. Alien, non-native, non-indigenous, foreign, exotic means a species, or taxon of lower rank that occurs outside its natural range (past or present) and dispersal potential (i.e. outside the range it occupies naturally or could not occupy without direct or indirect introduction or care by humans), and includes any part, gametes or propagule of such species that might survive and subsequently reproduce (IUCN 2000).

As described earlier, cities are important centres for the importation, establishment and distribution of non-native species. So far as plants are concerned, deliberate introductions for horticulture, forestry and landscaping purposes play the major role while unintentional introductions in goods are of less importance (Müller et al. 2010).

It is apparent that early hominids started to select and cultivate plants from many localities for food, medication, dyes and the production of spices from different localities. Consequently, it is likely that the deliberate introduction of species into cities was contemporaneous with the building of the first cities in the Middle East. It is probable that even in these times there was at least some appreciation of the aesthetic qualities of plants, which was certainly the case by the time of the ancient Greeks. The Romans introduced fruit trees into the cities they conquered or built whilst monks created physic gardens for the production of medicines (many are now known as liquors). The building of castles and large houses with extensive grounds gave rise to plant hunters who scoured the world for 'new' species and varieties to satisfy the requirements of and competition between the aristocratic owners. These activities ultimately gave rise to the horticultural trade and plant breeding.

By the 19th century, interests in gardening were being pursued by the 'middle' classes, whilst wealthy industrialists were creating public parks. These interests resulted in the expansion of the horticultural trade to propagate plants in much greater numbers and produce more varieties by selection and/or breeding. Since the 1960s, gardening (including allotments and house plants) has become a very popular and fashionable recreational activity of ordinary people whilst the more affluent hire garden designers and contractors to create and maintain more exotic, complex gardens to compete with their peers. During the same period, local planning authorities have required and continue to require residential, retail, commercial and industrial developments to be accompanied by landscape schemes although the quality varies considerable. Other key drivers for the introduction of non-native plant and animal species include the management of public parks, land restoration, creation of sports fields, lawns and other areas of grassland using imported grass seeds of unknown origin.

As a consequence, the 'nursery and horticultural trade' has developed into a large commercial business operating from a relatively small number of countries producing (often by cloning) more and more plants of a limited pool of taxa for export. In addition, plant breeders are in 'hot pursuit' of new products (i.e. plant varieties) for their own commercial purposes and to satisfy the insatiable demands of the horticultural, landscape and agricultural industries. The result is thousands and an ever increasing number of varieties of trees, shrubs, 'flowers,' soft and hard fruits, vegetables and cereals, most of which are for planting in urban areas. The only change is the replacement of old varieties with new according to fashion. Foresters, horticulturalists, landscape architects, etc. have a theoretical view of what a good plant, of a specific taxon should look like. Producers of plants select for these preferred phenotypes and reject phenotypes that do not conform to the model. This selection is having a very serious detrimental effect on genetic diversity by reducing the amount of variations in genotypes.

It is estimated that since the Neolithic period, 12,000 species have been introduced into Central Europe for ornamental and cultural purposes, and approximately 10%, i.e. 1100 of those plants have become naturalized (Lohmeyer and Sukopp 2002). Stace and Crawley (2015) report that most non-native species have short 'life expectancies' in a particular locality. Consequently, there is likely to be a high turnover of non-native species in terms of losses and gains, replacement and changes in abundance and distribution.

The introduction of non-native species includes taxa that might become invasive. Cities are sources of non-native species and therefore of invasive species which colonize the surrounding landscape where they may out-compete native species, changing natural vegetation stands, causing loss of crop yield or damaging dams for flood control and much more. An invasive species is a species that has been introduced to an environment where it is non-native and whose introduction causes environmental or economic damage or harm to human health (IUCN 2015). However, only a small number of non-native species have become invasive and cause problems. For example, in Germany 115 non-native plants are rare, 559 occur occasionally and 609 have become naturalized, and of the latter about 30 are classified as invasive. In terms of non-native animals, it is estimated that nearly 5% of those that have established have the potential to become invasive (BfN 2016). In cities, invasive species are not a problem because the ecological situation prevents them from having an impact on the biological diversity.

There is a growing tension between the attitudes and values of some of the biological disciplines and nature conservationists and between them and people as a whole in respect of non-native species. Many urban dwellers like non-native plants species because they have attractive forms, foliage, flowers or fruits, useful for hedging and providing screens and are sometimes of wildlife value (e.g. Butterfly-bush *Buddleja davidii*). On the other hand most conservationists consider them detrimental to wildlife and wish to remove exotics or prevent them from being planted, even in cities. There are many conflicting, strongly held and often highly emotional opinions, the following are examples from Britain. Some nature conservationists are strongly opposed to the planting of non-native tree species, especially Sycamore (*Acer pseudoplatanus*), because they support fewer species of Arthropod compared with native species, especially Pedunculate Oak (*Quercus robur*). Others take the contrary view and argue that a large number of individuals of a few species is important (as is generally the case with the non-native species). In his book on urban ecology, Gilbert (1989) makes a special plea for the value of the alien species, Japanese Knotweed (*Fallopia japonica*), in providing cover for some woodland species in Britain. Especially for novel ecosystems, it is noted that non-native species can shelter the regrowth of native species and enhance ecosystem services from impacted or designed states (Morse et al. 2014). There is also a clash of cultures—science and aesthetics; for example, ecologists wish to remove *Rhododendron ponticum* because it suppresses the native flora whilst the public at large object because it is very attractive during the flowering season. Florists like some the Fleabanes (*Conyza* spp.) very much for flower arrangements whereas ecologist would like to see them exterminated.

In a strict although correct sense, a species is classified to be native when considered in the context of the whole country, but it may be non-native in a particular region(s) if it does not occur naturally in that region (Box 7.1). It is also essential to consider variation in the genotypes of the regional genetic pools—because of the structure and operations of the international horticultural and forestry trades native taxa used in landscape schemes, parks and gardens generally have non-native genomes. In Britain, at least, since the 1970s it has been fashionable (possibly an obsession) to create 'wildflower meadows' in some parks, road verges and land awaiting development or unused land. The proposals have three objectives, aesthetic and increasing floristic, and faunal diversity habitats for invertebrates. Although the objectives have been achieved to varying degrees, they have two negative aspects in terms of biodiversity. First, some of the species did not occur in the region or in the country. Second, at least in the early years, the seeds originated from outside Britain, often from agricultural strains from elsewhere in Europe, resulting in native British taxa with non-native genomes. Native species versus non-native genomes is often overlooked in landscape planting as a whole. Therefore, to save and enhance the gene pools of regional species, some institutions (for example, in Germany several federal states and local authorities) have decided to plant only regional genotypes of woody plants or use regional seed banks for meadows on public green areas. As mentioned above, sensu stricto, a taxon that is moved by human activity from a part of a region where it occurs naturally to a region in which it does not occur naturally is an alien taxon in that region just as much as if it was imported from another country. That approach raises two major matters to which scientists do not have the answers. First, detailed knowledge of the extent of variation in the genome of a taxon and its significance in evolutionary, pathological and other terms. Second, the taxonomy of an organism can vary across countries and continents.

7.4 Three Scales—Three Approaches

To analyze and understand the biodiversity that can be found in a city and to develop local guidelines for interventions to maintain and improve urban biodiversity in the green components, it is helpful to use three approaches representing different scales (Table 7.3).

7.4.1 City in the Region

Virtually all cities originated from small settlements mainly on trade routes, relatively few were developed de novo on 'green field' sites; those that were include Abuja, Brasilia, Canberra, Lelystad, Milton Keynes and St. Petersburg. The biodiversity of these cities before, during and 'after' the development has not been monitored and measured in qualitative or quantitative terms.

The plants and animals of urban areas are part of the regional species pools, which interact with the local pool. In relation to birds, the surrounding landscape is the dominant source and the urban habitats are sinks of many but not of all bird species (Schwarz and Flade 2000). Consequently, changes in the regional species

Table 7.3 Three scales of approaches to analyze urban biodiversity

Scale		
City in the region	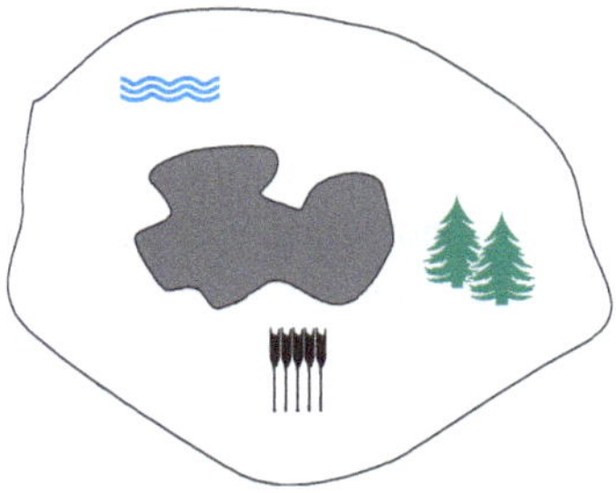	
Urban matrix	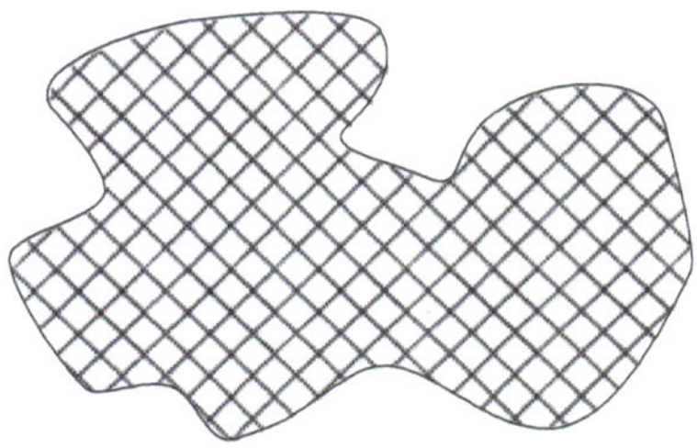	
Green patches	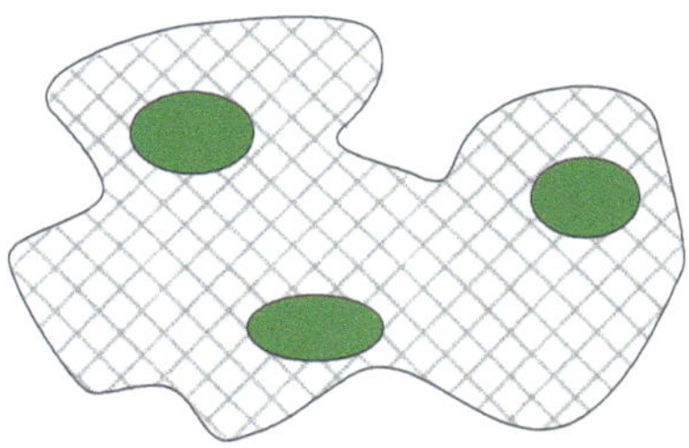	

pool will have an impact on the development of the bird populations in the city. Therefore, a reduction of the species diversity in the surrounding area of a city may cause significant changes in the plant and animal populations regardless of whether the habitat quality of green spaces in the city are suitable or not (Tait et al. 2005).

From a global perspective it is essential to appreciate that cities are embedded in different biogeographical regions and landscape settings, including cultural landscapes and wild lands. Consequently, there are many inter-relationships between cities and their surroundings. Urban areas encounter many different existing natural conditions and landscapes of the different biomes that have been more or less modified substantially by human activities (Palmer et al. 2008), for example biospheres of boreal forests and deserts.

Boreal forests are characterised by coniferous trees that cast a dense shade on the forest floor, relatively short of soil nutrients, moderate temperatures in the summer and a long harsh winter. As a result, only a few stress-tolerant species occur in them. It follows that the habitat characteristics of the boreal forests are in marked contrast to those in cities, especially during the winter months. Cities such as Oslo and Vancouver are remarkably warmer than surrounding environs and the growing season is noticeably longer. Urban green spaces are mostly open and productive landscapes influenced by human recreation activities and continuous management. Because of the distinct seasonality of summer and winter in the high latitudes, urban areas have a special function in winter, for example serving as a shelter for many species (Clergeau et al. 1998; Murgui 2009). Thus in cities, species whose natural distribution lie further south can selectively move the boundary further north (Luniak 2004; Kowarik and Säumel 2007).

The hot and dry deserts located in several parts of the world differ widely in their floristic composition, but their basic characteristics are similar: sparse vegetation cover, extremes of temperature, low soil fertility and scarcity of water. As a consequence, the regional flora and fauna is dominated by specialists adapted to these harsh conditions. The vegetation consists of dwarf trees and shrubs whilst many of the native animal species are nocturnal and confined to small habitats. The variation between dry and humid years exercises a major influence on the vegetation (Buyantuyev et al. 2010). Cities built in these extreme conditions such as Dubai and Phoenix are not as dry as the surrounding natural areas. They are more humid and have generally higher soil-water content because of numerous artificial water bodies and the irrigation of parks, gardens and even streets (Pickett and Cadenasso 2009). Irrigation partly compensates for differences in the water regime between drier and wetter seasons or years and with increasing green areas and 'woody' elements, generalist diurnal animal species occur in a sufficient amount and variety of habitats.

As some Australian cities demonstrate, additional factors such as the length of time following the cultural reshaping of the original (pristine) landscape and the time lapse since urban development processes also play an important role (Tait et al. 2005). The time period (duration of co-evolution) in which the regional pool of species is formed under the influence of silvi- and agricultural as well as urban development determines the relationships of the regional versus the urban pool of species. Over the course of the last 8000 years and more, man-made landscapes have been created throughout Europe, initially in the Mediterranean region. Both the rural and the urban landscapes are based on long adaptation processes to anthropogenic influence. Considerable similarity exists between the Mediterranean landscape and green structures in cities not only because both consist of open, sparsely forested land but also because the land was (and in part still is) compartmentalized. Rome is a good example of this; compared to cities of Central Europe it exhibits a relatively high percentage of native vascular plants of the region (Celesti-Grapow et al. 2006).

In the Central European environment, urban growth tends to take place primarily on agricultural land. While this also occurs in areas with rapid suburbanization (such as in the United States) or rapidly growing mega-cities, these scenarios are also likely to have a direct impact on areas of semi-natural wilderness (Grove et al.

2006). The development of urban areas directly in wilderness areas results in very different interactions between local and regional species pools, because there is a severe ecological break between urban and wilderness areas.

The urban fringe is the transition zone between the urban and surrounding areas. The designing of that zone has an impact on how species can enter or leave the urban areas. The zone, especially in relation to plants, form an important ecotone that supports a high level of species richness and diversity. These ‘edge of town’ areas are often used for the development of business parks and retail outlets, which can be of value for biodiversity strategies (Snep 2009).

Some studies have simulated the impact on local biodiversity by two varying scenarios of urban growth; first, a more compact growth resulting in large green areas and densely built houses with small gardens; second, urban sprawl comprising detached houses with large gardens. The simulations found that the compact growth has a smaller impact on the local species diversity than urban sprawl (Sushinsky et al. 2013; Varet et al. 2014).

The following recommendations for planning and management of green components of cities emerge from the analysis in this section:

- the regional species pool is the base for the local species pool and must be considered at all stages;
- the increase in species diversity of the surrounding areas has a positive impact on the local areas;
- improving and creating opportunities for the movement of species from the surrounding landscape to and from the urban areas. Such connections and transition zones are needed for both ‘generalists’ and ‘specialists’ plant and animal species;
- the spread of invasive species from urban areas into the surrounding semi-natural or natural areas should be prevented by management activities;
- compact growth including larger areas of green is preferable to urban sprawl.

7.4.2 Urban Matrix

A review of the literature shows that urban biodiversity is mainly assessed on floristic and faunistic studies conducted in parks and public green areas. The built-up areas (the urban fabric) were once described as unlivable habitats and therefore not recognized in biodiversity research (Franklin and Lindenmayer 2009). For the purpose of this chapter, built-up areas referred to as the ‘urban matrix’ are defined as a mix of densely and less densely built-up areas, comprising a relatively small-scale mosaic of buildings, streets, open and green spaces.

Only in the last few years has the biodiversity of the ‘urban matrix’ been studied and analyzed more intensively. These studies have demonstrated and confirmed studies made in previous decades that without the consideration of the urban matrix the knowledge of the biodiversity in a city would be incomplete (Lizee et al. 2011;

Conole and Kirkpatrick 2011). The influence of the vegetation of the whole city, including the urban matrix, is the key to increasing the species-richness and diversity in urban areas (Jones and Leather 2012; Pellisier et al. 2012).

The following are examples to illustrate the function of the urban matrix. Using presence-absence data for 16 bird species in the rural-urban interface, Taylor et al. (2015) demonstrated that the models fitted successfully when consideration was given to both patch and matrix characteristics. In this study the important matrix characteristic was the degree of tree cover. Several studies of the Barred Owl (*Strix varia*) found that suburban neighbourhoods with a well-developed stock of trees are similar to patches of forests (Dykstra et al. 2012). A study of the Red Squirrel (*Sciurus vulgaris*) showed that explanatory models of its occurrence in patches are significantly better if the permeability of the matrix is taken into account in the analysis (Verbeylen et al. 2003). Lizee et al. (2011) found that the configuration of the matrix had a significant influence on diversity of urban butterflies. It was concluded both matrix configuration and distance from the regional species pool overrides park size in contributing to variations in species richness. The amount of surrounding habitat areas and structures which correspond with the habitat functions of park and green spaces can minimize the substantial breaks between green patches and surrounding matrix. For that reason, the 'effective' size of parks may be larger than their actual size (Loeb et al. 2009).

There is an underestimation of the amount of small green areas inside the urban matrix. One reason is simply the fact that the matrix comprises mainly private property that cannot be accessed and studied as easily as public open space (Hodgson et al. 2007). In many cities these private green areas exceed the total amount of public green areas. For example, the developed city area of Hanau, Germany, contains 400 ha public parks and green spaces but the total amount of private gardens and private green areas, e.g. located between block of flats total nearly 800 ha (Werner 1999). If the green areas between blocks of flats were designed and managed to create a complex vegetation structure, they would provide valuable habitats for many more animal species.

It is estimated that in the United Kingdom, for example, between 19 and 27% of the area of cities is taken up by domestic gardens (Smith et al. 2006), providing high-quality habitats for plants and animals. As demonstrated by a study on Wood Mice (*Apodemus sylvaticus*), domestic gardens may further reduce the deleterious effects of fragmentation (Baker et al. 2003).

Meta-analysis and simulation models of fragmented terrestrial landscapes (such as urban landscapes) demonstrate the major influences of the matrix that surrounds green patches (Prugh et al. 2008) whilst lower matrix quality increased the requirements for the improvement of the habitat quality of the patches (Dunford and Freemark 2005). Taylor et al. (2015) came to the conclusion that the supporting function of the matrix increases with decreasing patch size and therefore the matrix can make a significant contribution to the importance of biodiversity in urban areas.

The analysis in this section gives rise to the following recommendations for planning and management of green components:

- improving the permeability of the urban matrix by increasing the vegetation cover and diversity by the creation of green areas, for example pocket parks, green roofs, green 'corridors' along streets, riversides and other green elements such as trees and green facades;
- increasing the vegetation cover around parks to enhance the interactions between the green and surrounding areas;
- combining the design and management of public and private areas to improve species mobility and interactions between public green areas and the surrounding private green components.

7.4.3 Green Urban Patches

The importance of urban parks and green spaces is clearly illustrated by a study in Flanders, Belgium. The investigated urban and suburban parks of Flanders comprise only 0.03% of the total area but contain about 29% of all wild plants and over 48% of all breeding birds in the region (Cornelis and Hermy 2004). Some 53% of bat species occurring in France that use green and forest areas can be found in urban parks (DeCornulier and Clergeau 2001). These few examples are sufficient to demonstrate that urban parks and green spaces are the backbones of species richness and diversity in urban areas.

Many scientists state that the habitat quality of green urban areas is determined by structural features (for example horizontal and vertical complexity and the diversity of microhabitats), size, age and the relationship with other habitats (Stenhouse 2004). The size and age of green spaces are related to the increase of structural diversity and diversity of microhabitats. To increase the species richness of birds and other animals, it is necessary to provide important features such as an open tree canopy with a cover of not less 30%, an age range of tress, including some with cavities, a shrub layer and a field layer with small structures offering shelter for small mammals, and good protection from feral cats and dogs (Garden et al. 2010; Taylor et al. 2015; Yang et al. 2015). Nielsen et al. (2014) reviewed the empirical findings of research of species richness in urban parks across all taxonomic groups that have been studied. They conclude that diversity of habitats and microhabitat heterogeneity contained in urban parks appears as the most decisive factor for the overall species richness in urban areas.

In nature conservation, there is a long lasting debate about the relationship between large and small areas, the so called SLOSS discussion—single large or several small (Tjorve 2010). Meanwhile, it is clear that both large tracts and small fragments have conservation value and that is also true for urban areas (Taylor et al. 2015). Large areas provide important reservoirs of species because they are able to support viable population while fragmented small areas have the function of stepping stones. However, cities have limited space and therefore there is a lack of large green areas and the creation of such areas is virtually impossible. The exception occurs if redundant large industrial or military areas can be converted to open spaces providing that there is no pressure for development or politicians have decided to leave them as open space. For these reasons the approach has to be changed to focus on small patches. Studies demonstrate that each small green space can provide different benefits for different species that can enhance species diversity if it is used as an integrated concept (Rega et al. 2015). Beyond that, small green patches distributed throughout the urban area offer more opportunities for people to have easier contact with nature.

Many rare and endangered species are found in urban semi-natural remnant areas. These remaining spaces do not only exist in the urban fringe but also in the middle of mega-cities. Notable examples of pristine habitats wholly or partly embedded within a city include the Mata Atlantica Forest in Rio de Janeiro, the remnant tropical evergreen forest in Singapore, the National Park El Avila with its rock faces in Caracas, various remnants of bushland in Perth, Sydney and Brisbane, natural forests in York (Canada) and Portland (USA), and rock faces and outcrops in Edinburgh (Heywood 1996; Miller and Hobbs 2002; Edinburgh Biodiversity Partnership 2008).

These examples demonstrate a further important point. The ‘urban standard’ of public and private managed green areas is short-mown grass and well-maintained (neat and tidy) sites. However, green areas that are hotspots of biodiversity in cities often do not comply with the ‘urban standard’; they include (Figs. 7.1, 7.2, 7.3 and 7.4):

- pristine remnants of native vegetation (Antos et al. 2006);
- parks and gardens with unchanged use and management over decades or even centuries (Kowarik 1998; Celesti-Grapow et al. 2006);
- unmanaged areas with natural succession and the emergence of differentiated vegetation structures (Hansen et al. 2005);
- large green areas of open, cultivated or forest land.

In summary it can be noted for green spaces that:

- structural diversity of the vegetation is one of the most important factors;
- size of habitat provides the opportunity to increase habitat structures and the variety of micro-habitats;
- habitat size is often linked to an increase in the number of species but the positive correlation between species richness and area is a gross simplification due to underlying processes that have to be considered;
- large and connected green areas are extremely rare in cities and should be protected;
- small patches have a function that is often underestimated;
- habitat age has a variety of aspects that comprise

Fig. 7.1 Remnants of a pristine tropical rain forest in the central catchment nature reserve of Singapore

  - pristine remnants, unchanged use and maintenance over decades or even centuries and
  - succession and the emergence of differentiated vegetation structure;
- quality of habitat networks can be described as spatial and functional connectivity.

More specifically, the more structurally complex, the larger, the older and the less isolated a habitat area is, the better are the chances for a high level of biological diversity (Angold et al. 2006; Chace and Walsh 2006).

7.5 Conclusion

This chapter has examined the provision and management of green spaces and structures that provide opportunities to enhance biodiversity and some ecosystem services in cities.

Cities have been created by human endeavours throughout the world for several millennia, mainly as economic and political centres. On the one hand they represent the pinnacle of human achievement whilst on the other they contain the depths of

Fig. 7.2 Garden of the temple 'Komyo zen-ji' in Dazaifu, Fukuoka

human depravation. Cities were often founded in landscapes that are naturally species rich. The expansion of cities has and is likely to continue to incorporate elements of the adjacent rural landscape with natural habitats and species to generate an urban-wilderness interface. The consequence is that pristine or semi-natural habitats survive in scattered patches of various sizes within the city.

The result is that the cities and their surroundings comprise spatially separated patches of high biodiversity in a matrix of simplified habitats with low biodiversity. This has given rise to a form of commensalism whereby the two elements rely on each other for the interchange of species. The most appropriate way of determining what positive action is required to improve biodiversity in cities is to examine the quality and species composition of the habitats and their connectivity in the context of a three-level eco-geographical hierarchy—the city in the region, the urban matrix and the green patches. In conceptual terms, the three domains represent overlapping sub-sets in multi-dimensional space.

The fundamental solution is the creation of a 'double' wheel whereby the primary centre (the city) is connected to the outer rim by spokes of green connectivity with HUBS (High Urban Biodiversity Sites) occurring within the primary centre and connected to its margins by secondary green 'spokes. The quantitative and qualitative structure of the connectivity is critical to the interchange of species between the three components. Aerial species are probably the least constrained followed by aquatic species (watercourses, including canals, are the only

Fig. 7.3 Südgelände, a ruderal area in Berlin

continuous habitat that flows through the landscape). The most disadvantaged are terrestrial species that do not have the advantage of continuity.

The planning, design and management of cities has resulted in some positive and negative impacts on biodiversity. From the earliest times, it is likely that people introduced plants into cities for food or aesthetic reasons. That trend has continued through the millennia resulting in the continual deliberate introduction of more and more species and varieties from other localities throughout the world. The consequence is that 30–35% of the plant species now recorded in cities are non-native; it also seems likely that the quantum of plants of non-native taxa is significantly higher than native taxa (of all ranks) Nevertheless, it is also suggested that cities may contain more native species than their surroundings. The combination of these two factors means that cities generally have a higher plant diversity than their surroundings. A similar principle applies to birds, although the figures are lower.

With respect to the above-mentioned three levels, the development and management of green areas and green structures to enhance urban biodiversity and ecosystem services in cities can be condensed in the following principles:

On the regional or landscape scale

- Improving the habitat quality of the surrounding landscape
- Connecting the regional species pool with the local species pool by green networks and permeable transition zones.

Fig. 7.4 The Tempelhofer Feld, a former airport area, in Berlin

On the city scale

- Improving the green cover and structures of the urban matrix

On the patch or habitat scale

- Improving the habitat quality of the patches
- Connecting the patches.

The implementation of these principles does not only increase urban biodiversity and ecosystem services but it is a big step to a sustainable, resilient and human-friendly city, too.

References

Andersson E (2006) Urban landscapes and sustainable cities. Ecol Soc 11(1):34. http://www.ecologyandsociety.org/vol11/iss1/art34/

Angold PG, Sadler JP, Hill MO, Pullin A, Rushton S, Austin K, Small E, Wood B, Wadsworth R, Sanderson R, Thompson K (2006) Biodiversity in urban habitat patches. Sci Total Environ 360:196–204

Antos MJ, Fitzsimons JA, Palmer GC, White JG (2006) Introduced birds in urban remnant vegetation: does remnant size really matter? Austral Ecol 31:254–261

Aronson MFJ, La Sorte FA, Nilon CH, Katti M, Goddard MA, Lepczyk CA, Warren PS, Williams NSG, Cilliers S, Clarkson B, Dobbs C, Dolan R, Hedblom M, Klotz S, Kooijmans JL, Kühn I, MacGregor-Fors I, McDonnell M, Mörtberg U, Pysek P, Siebert S, Sushinsky J, Werner P, Winter M (2014) A global analysis of the impacts of urbanization on bird and plant diversity reveals key anthropogenic drivers. Proc Royal Soc B 281:20133330

Baker PJ, Ansell RJ, Dodds PAA, Webber CE, Harris S (2003) Factors affecting the distribution of small mammals in an urban area. Mammal Rev 33:95–100

Beninde J, Veith M, Hochkirch A (2015) Biodiversity in cities needs space: a meta-analysis of factors determining intra-urban biodiversity variation. Ecol Lett 18:581–592

Bertin RI (2002) Losses of native plant species from Worcester, Massachusetts. Rhodora 104 (920):325–349

BfN—Bundesamt für Naturschutz (2016). https://www.bfn.de/fileadmin/MDB/documents/themen/artenschutz/IAS_DZN.pdf. Accessed 15 Mar 2016

Buyantuyev A, Wu J, Gries C (2010) Multiscale analysis of the urbanization pattern of the Phoenix metropolitan landscape of USA: time, space and thematic resolution. Landscape Urban Plan 94:206–217

Celesti-Grapow L, Pysek P, Jarosik V, Blasi C (2006) Determinants of native and alien species richness in the urban flora of Rome. Divers Distrib 12:490–501

Chace JF, Walsh JJ (2006) Urban effects on native avifauna: a review. Landscape Urban Plan 74:46–69

Chocholouskova Z, Pysek P (2003) Changes in composition and structure of urban flora over 120 years: a case study of the city of Plzen. Flora 198:366–376

Clergeau P, Savard J-PL, Mennechez G, Falardeau G (1998) Bird abundance and diversity along an urban-rural gradient: a comparative study between two cities on different continents. Condor 100:413–425

Conole LE, Kirkpatrick JB (2011) Functional and spatial differentiation of urban bird assemblages at the landscape scale. Landscape Urban Plan 100:11–23

Cornelis J, Hermy M (2004) Biodiversity relationships in urban and suburban parks in Flanders. Landscape Urban Plan 69:385–401

De Cornulier T, Clergeau P (2001) Bat diversity in French urban areas. Mammalia 65:540–543

Douglas I, James P (eds) (2015) Urban ecology. An Introduction. Routledge, Oxon and New York

Dunford W, Freemark K (2005) Matrix matters: effects of surrounding land uses on forest birds near Ottawa, Canada. Landscape Ecol 20:497–511

Dykstra CR, Simon MM, Daniel FB, Hays JL (2012) Habitats of Suburban Barred Owls (*Strix varia*) and Red-Shouldered Hawks (*Buteo lineatus*) in Southwestern Ohio. J Raptor Res 46:190–200

Edinburgh Biodiversity Partnership (2008). http://www.ukbap.org.uk/lbap.aspx?ID=381. Accessed 28 May 2008

Elmqvist T, Fragkias M, Goodness J, Güneralp B, Marcotullio PJ, McDonald RI, Parnell S, Schewenius M, Sendstad M, Seto KC, Wilkinson C (eds) (2013) Urbanization, biodiversity and ecosystem services: challenges and opportunities. A global Assessment. Springer, Dordrecht, New York, Heidelberg, London

Franklin JF, Lindenmayer DB (2009) Importance of matrix habitats in maintaining biological diversity. PNAS 106:349–350

Garden JG, McAlpine CA, Possingham HP (2010) Multi-scaled habitat considerations for conserving urban biodiversity: Native reptiles and small mammals in Brisbane, Australia. Landscape Ecol 25:1013–1028

Gilbert OL (1989) The ecology of urban habitats. Springer, Dordrecht

Gregor T, Bönsel D, Starke-Ottich I, Ziska G (2012) Drivers of floristic change in large cities—A case study of Frankfurt/Main (Germany). Landscape Urban Plan 104:230–237

Grove JM, Troy AR, O'Neil-Dunne JP, Cadenasso ML, Pickett STA (2006) Characterization of households and its implications for the vegetation of urban ecosystems. Ecosystems 9:578–597

Hansen AJ, Knight RL, Marzluff JM, Powell S, Brown K, Gude PH, Jones K (2005) Effects of exurban development on biodiversity: patterns, mechanisms and research needs. Ecol Appl 15:1893–1905

Heywood VH (1996) The importance of urban environments in maintaining biodiversity. In: di Castri F, Younes T (eds) Biodiversity, science and development: towards a new partnership. CAB International, Wallingford, Oxon, pp 543–550

Hodgson R, French K, Major RE (2007) Avian movement across abrupt ecological edges: differential responses to housing density in an urban matrix. Landscape Urban Plan 79:266–272

IUCN—International Union for Conservation of Nature (2000) Guidelines for the prevention of biodiversity loss caused by alien invasive species. IUCN Council, Gland, Switzerland

IUCN—International Union for Conservation of Nature (2015). http://www.iucn.org/about/work/programmes/species/our_work/invasive_species/. Accessed 15 Mar 2016

Jackowiak B (1998) The city as a centre for crystallization of the spatio-floristic system. Phytocoenosis 10 (N.S.), Supplementum Cartographiae Geobotanica 9 Warszawa-Bialowieza, pp 55–67

Jones EL, Leather SR (2012) Invertebrates in urban areas: a review. Eur J Entomol 109:463–478

Kark S, Allnutt TF, Levin N, Manne LL, Williams PH (2007a) The role of transitional areas as avian biodiversity centres. Glob Ecol Biogeogr 16(2):187–196

Kark S, Iwaniuk A, Schalimtzek A, Banker E (2007b) Living in the city: can anyone become an 'urban exploiter'? J Biogeogr 34:638–651

Kelcey JG (2015) Prologue. In: Kelcey JG (ed) Vertebrates and invertebrates of European cities: selected non-avian fauna. Springer, New York, Heidelberg, Dordrecht, London, pp 1–24

Kelcey JG, Rheinwald G (eds) (2005) Birds in European cities. Ginster Verlag, St. Katharinen

Kinzig AP, Warren P, Martin C, Hope D, Katti K (2005) The effects of human socioeconomic status and cultural characteristics on urban patterns of biodiversity. Ecol Soc 10(1):23. http://www.ecologyandsociety.org/vol10/iss1/art23/

Kowarik I (1998) Auswirkungen der Urbanisierung auf Arten und Lebensgemeinschaften - Risiken, Chancen und Handlungsansätze. Bundesamt für Naturschutz, Schriftenreihe für Vegetationskunde 29:173–190

Kowarik I (2005) Wild urban woodlands: towards a conceptual framework. In: Kowarik I, Körner S (eds) Wild urban woodlands. New perspectives for urban forestry. Springer, Heidelberg, pp 1–32

Kowarik I (2011) Novel urban ecosystems, biodiversity, and conservation. Environ Pollut 159:1974–1983

Kowarik I, Säumel I (2007) Biological Flora of Central Europe: *Ailanthus altissima* (Mill.) Swingle. Perspec Plant Ecol Evol Syst 8(4):207–237

Kühn I, Brandl R, Klotz S (2004) The flora of German cities is naturally species rich. Evol Ecol Res 6:749–764

Landolt E (2000) Some results of a floristic inventory within the city of Zürich (1984–1988). Preslia 72:441–445

La Sorte FA, Aronson MFJ, Williams NSG, Celesti-Grapow L, Cilliers S, Clarkson BD, Dolan RW, Hipp A, Klotz S, Kühn I, Pyšek P, Siebert S, Winter M (2014) Beta diversity of urban floras among European and non-European cities. Glob Ecol Biogeogr 23:769–779

Lenzin H, Erismann C, Kissling M, Gilgen AK, Nagel P (2004) Häufigkeit und Ökologie ausgewählter Neophyten in der Stadt Basel (Schweiz). Tuexenia 24:359–371

Lim HC, Sodhi NS (2004) Responses of avian guilds to urbanization in a tropical city. Landscape Urban Plan 66:199–215

Lizee M-H, Manel S, Mauffrey J-F, Tatoni T, Deschamps-Cottin M (2011) Matrix configuration and patch isolation influences override the species–area relationship for urban butterfly communities. Landscape Ecol. doi:10.1007/s10980-011-9651-x

Loeb SC, Post CJ, Hall ST (2009) Relationship between urbanization and bat community structure in national parks of the Southeastern US. Urban Ecosyst 12:197–214

Lohmeyer W, Sukopp H (2002) Agriophyten in der Vegetation Mitteleuropas. Braunschweiger Geobotanische Arbeiten 8:179–220

Luniak M (2004): Synurbization—adaptation of animal wildlife to urban development. In: Shaw et al. (eds) Proceedings 4th international urban wildlife symposium, pp 50–55

McKinney ML (2006a) Correlated non-native species richness of birds, mammals, herptiles and plants: scale effects of area, human population and native plants. Biol Invasions 8:415–425

McKinney ML (2006b) Urbanization as a major cause of biotic homogenization. Biol Conserv 127 (3):247–260

Miller JR, Hobbs RJ (2002) Conservation where people live and work. Conserv Biol 16:330–337

Morse NB, Pellissier PA, Cianciola EN, Brereton RL, Sullivan MM, Shonka NK, Wheeler TB, McDowell WH (2014) Novel ecosystems in the Anthropocene: a revision of the novel ecosystem concept for pragmatic applications. Ecol Soc 19(2):12. doi:10.5751/ES-06192-190212

Müller N, Werner P, Kelcey JG (eds) (2010) Urban biodiversity and design. Conservation science and practice no. 7, Wiley-Blackwell, Hoboken NJ

Murgui E (2009) Influence of urban landscape structure on bird fauna: a case study across seasons in the city of Valencia (Spain). Urban Ecosyst 12:249–263

Nielsen AB, van den Bosch M, Maruthaveeran S, van den Bosch CCK (2014) Species richness in urban parks and its drivers: a review of empirical evidence. Urban Ecosyst 17:305–327

Palmer GC, Fitzsimons JA, Antos MJ, White JG (2008) Determinants of native avian richness in suburban remnant vegetation: implications for conservation planning. Biol Conserv 141:2329–2341

Pellisier V, Cohen M, Boulay A, Clergeau P (2012) Birds are also sensitive to landscape composition and configuration within the city centre. Landscape Urban Plan 104:181–188

Pickett STA, Cadenasso ML (2009) Altered resources, disturbance and heterogeneity: a framework for comparing urban and non-urban soils. Urban Ecosyst 12:23–44

Prugh LR, Hodges KE, Sinclair ARE, Brashares JS (2008) Effect of habitat area and isolation on fragmented animal populations. PNAS 105:20770–20775

Rega CC, Nilon CH, Warren PS (2015) Avian abundance patterns in relation to the distribution of small urban greenspaces. J Urban Plan Dev 141:A4015002

Sattler T, Borcard D, Arlettaz R, Bontadina F, Legendre P, Obrist MK, Moretti M (2010) Spider, bee, and bird communities in cities are shaped by environmental control and high stochasticity. Ecology 91:3343–3353

SCBD—Secretariat of the Convention on Biological Diversity (2012) Cities and biodiversity outlook: action and policy. CBD, Montreal. http://www.cbd.int/doc/health/cbo-action-policy-en.pdf

Scholz H (2007) Questions about indigenous plants and anecophytes. Taxon 56:1255–1260

Schwarz J, Flade M (2000) Ergebnisse des DDA-Monitoringprogramms. Teil I: Bestandsänderungen von Vogelarten der Siedlungen seit 1989. Vogelwelt 121:87–106

Smith RM, Warren PH, Thompson K;. Gaston KJ (2006) Urban domestic gardens (VI): environmental correlates of invertebrate species richness. Biodiv Conser 15:2415–2438

Snep R (2009) Biodiveresity conservation at business sites. Options and opportunities. Alterra Scientific Contributions 28, Wageningen

Stace CA (2010) New flora of the British Isles, 3rd edn. Cambridge University Press, Cambridge

Stace CA, Crawley MJ (2015) Alien plants. William Collins, London

Stenhouse RN (2004) Fragmentation and internal disturbance of native vegetation reserves in the Perth metropolitan area, Western Australia. Landscape Urban Plan 68:389–401

Stolle J, Klotz S (2005) Flora der Stadt Halle(Saale). Calendula, 5. Sonderheft, Hallesche Umweltblätter, Halle(Saale)

Sushinsky JR, Rhodes JR, Possingham HP, Gill TK, Fuller RA (2013) How should we grow cities to minimize their biodiversity impacts? Glob Change Biol 19:401–410

Sukopp H, Wittig R (1998) Stadtökologie, 2nd edn. Gustav Fischer, Stuttgart

Tait CJ, Daniels CB, Hill RS (2005) Changes in species assemblages within the Adelaide metropolitan area, Australia, 1836–2002. Ecol Appl 15:346–359

Taylor JJ, Lepczyk CA, Brown DG (2015) Patch and matrix level influences on forest birds at the rural-urban interface. Landscape Ecol. doi:10.1007/s10980-015-0310-5

Tjrove E (2010) How to resolve the SLOSS debate: lessons from species-diversity models. J Theor Biol 264:604–612

Turrini T, Knop E (2015) A landscape ecology approach identifies important drivers of urban biodiversity. Glob Change Biol 21:1652–1667

Varet M, Burel F, Petillon J (2014) Can urban consolidation limit local biodiversity erosion? Responses from carabid beetle and spider assemblages in Western France. Urban Ecosyst 17:123–137

Verbeylen G, deBruyn L, Adriaensen F, Matthysen E (2003) Does matrix resistance influence Red squirrel (Sciurus vulgaris L. 1758) distribution in an urban landscape? Landscape Ecol 18:791–805

Werner P (1999) Why biotope mapping in populated areas? Deinsea 5:9–26

Williams NS, Schwartz MW, Vesk PA, McCarthy MA, Hahs AK, Clemants SE, Corlett RT, Duncan RP, Norton BA, Thompson K, McDonnell MJ (2009) A conceptual framework for predicting the effects of urban environments on floras. J Ecol 97:4–9

Yang G, Xua J, Wanga Y, Wang X, Peic E, Yuanc X, Lia H, Dinga Y, Wang Z (2015) Evaluation of microhabitats for wild birds in a Shanghai urban area park. Urban For Urban Greening 14:246–254

Chapter 8
Urban Agriculture as a Productive Green Infrastructure for Environmental and Social Well-Being

Brenda B. Lin, Stacy M. Philpott, Shalene Jha and Heidi Liere

Abstract Urban agricultural (UA) systems appear in many forms, from community farms and rooftop gardens to edible landscaping and urban orchards. They can be productive features of cities, providing important environmental and social services that benefit and support urban communities. These benefits include the provision of high levels of biodiversity and ecosystem services that contribute to urban nature and environmental processes as well as a range of social benefits, such as food and nutrition, cultural resources and recreational benefits. However, there are a number of challenges that prevent UA from expanding despite various acknowledged benefits. Increasing competition for space and environmental constraints often limits the ability to establish UA systems in many city areas, and negative spillover from UA to urban areas can create hazards to the natural environment and the local community. Further expansion and development of UA as a productive green infrastructure will require win-win strategies that maximize environmental and social benefits while taking advantage of vacant or under-utilized pockets of urban land.

B.B. Lin (✉)
CSIRO Land & Water, Aspendale, VIC, Australia
e-mail: Brenda.Lin@csiro.au

S.M. Philpott
Environmental Studies Department, University of California, Santa Cruz, CA, USA
e-mail: sphilpot@ucsc.edu

S. Jha
University of Texas at Austin, Austin, TX, USA
e-mail: sjha@austin.utexas.edu

H. Liere
Reed College, Portland, OR, USA
e-mail: heliere@reed.edu

P.Y. Tan and C.Y. Jim (eds.), *Greening Cities: Forms and Functions*, Advances in 21st Century Human Settlements,
DOI 10.1007/978-981-10-4113-6_8

Keywords Green space · Urban food · Biodiversity · Land use · Public good · Food security

8.1 Introduction

Of the range of green infrastructure types that are often studied and acknowledged in cities, one type gaining worldwide interest is urban farming or urban agriculture. Urban agricultural (UA) is common across continents with urban gardens covering hundreds of hectares in Amsterdam, Montreal, Beijing and Barcelona, amongst many other cities (reviewed in Lovell 2010), and such green spaces serve many environmental and social uses for urban citizens. UA is regarded as an important feature of the overall urban support systems at long-term and global scales (Barthel and Isendahl 2013), and thus important to the sustainability and resilience of cities. Additionally, because of the benefits to cities, urban policy and development have been increasingly adopted to introduce and maintain such systems (McClintock et al. 2012). However, some challenges are associated with agricultural systems in cities such as competition for land use. In this chapter, both the benefits and the challenges are reviewed for the on-going establishment and persistence of UA as an integral part of urban green infrastructure.

8.1.1 What Is Urban Agriculture?

UA is defined as the production of crop and livestock goods within cities and towns (Zezza and Tasciotti 2010), generally integrated into the local urban economic and ecological systems (Mougeot 2010). Conceptualizing what ‘urban’ precisely means remains a challenge in the urban green infrastructure literature (Montgomery 2008). Broadly speaking, urban areas consist of predominantly human-made surfaces, with high concentrations of people and are the hub of economic activities (Martezello et al. 2014). UA also often includes peri-urban agricultural areas around cities and towns, which may provide products and services to the local urban population (Mougeot 2010).

UA activities in and of themselves are diverse and can include the cultivation of vegetables, medicinal plants, spices, mushrooms, fruit trees and other productive plants, as well as keeping livestock for eggs, milk, meat, wool and other products (Lovell 2010). The different types of UA allow for a diverse set of ecosystem structures to contribute to the edible landscape in a range of community types and provide a broad array of services based on community desires (McLain et al. 2012). UA systems are highly heterogeneous in size, form and function and can be found in different types of urban green spaces.

8.1.2 Typology of Urban Agriculture

The list of UA examples below highlight how diverse urban farming can be. This diversity is based on some important factors including land tenure, management, types of food and service provision, and scale of production.

- **Community or allotment gardens** often represent small-scale, highly-patchy and qualitatively rich (vegetatively complex and species rich) semi-natural ecosystems that are usually located in urban or semi-urban areas for food production (Colding et al. 2006).
- **Private gardens** are primarily located in suburban areas and may be the most prevalent form of urban agriculture in cities (Loram et al. 2007). Privately owned gardens cover an estimated 22–27% of the total urban area in the UK (Loram et al. 2007), 36% in New Zealand (Mathieu et al. 2007), and 19.5% in Dayton, Ohio, USA (Sanders and Stevens 1984).
- **Easement gardens** are located within private or community properties, but are often regulated by the local government (Hunter and Brown 2012). Urban easements are established with the purpose of improving water quality and erosion control (Forman and Alexander 1998), but they can include a wide array of biodiversity, including food plants, depending on management type (Hunter and Hunter 2008). Gardening on verges may also be done as a form of 'guerrilla gardening' where local communities garden on small patches of soil when few unpaved spaces are available.
- **Rooftop gardens or green roofs** are any vegetation established on the roof of a building and can be used to improve insulation, create local habitat, provide decorative amenity, and cultivate food plants (Whittinghill and Rowe 2012).
- **Urban orchards** are tree-based food production systems that can be owned and run privately or by the community. Increasingly, schools and hospitals are establishing fruit trees that provide crops, erosion control, shade and wildlife habitat, and producing food for the local community (Drescher et al. 2006).
- **Peri-urban agriculture** usually exists at the outskirts of cities that largely serves the needs of the nearby urban population (Zasada 2011). Typically, these are multifunctional agricultural systems that include a large variety of activities and diversification approaches and contribute to environmental, social and economic functions.

Many UA systems may fit into more than one category. For example, both private gardens and community gardens may exist as rooftop gardens, and orchards may exist within community gardens. See Fig. 8.1 for photographs of examples of UA types.

Fig. 8.1 Photographs of different types of UA. **a** Community garden in Toledo, Ohio, **b** Allotment garden in Salinas, California, **c** Private garden in Toledo, Ohio, **d** Easement garden in Melbourne, Australia, **e** Rooftop garden in New York City, **f** Urban orchard in San Jose, California. Photos courtesy of P. Bichier (**a**, **b**, **f**), P. Ross (**c**), G. Lokic (**d**), and K. McGuire (**e**) (From Lin et al. 2015)

8.2 Urban Agriculture: Contribution to Liveability, Sustainability and Resilience

UA offers multiple contributions to the liveability, sustainability and resilience of cities. Besides local food production, urban agricultural systems provide a place for recreation and social interaction, community engagement, biodiversity, and a range of ecosystem services to the community (Owen 1991; Baker 2004; Saldivar-Tanaka and Krasny 2004; Miotk 1996; Smith et al. 2006a; Matteson et al. 2008; Blaine et al. 2010; Aubry et al. 2012). Additionally, urban farming systems may be considered important as a means of maintaining or developing local employment and incomes and even landscape-scale environmental quality (Aubry et al. 2012). The range of benefits is reviewed below to show the contributions to: (1) local food production, health and nutrition, (2) biodiversity and environmental services, and (3) social and cultural services.

8.3 Local Food Production, Health and Nutrition

Urban planners are increasingly interested in maintaining agriculture within and around cities due to food security concerns. Several US cities contain 'food deserts', where access to fresh produce is limited due to reduced proximity to markets, financial constraints, or inadequate transportation (Thomas 2010; ver Ploeg et al. 2009). For example, in Oakland, CA, positioned at the heart of the Bay Area's 'foodie' culture, 87% of school children receive free or reduced lunch due to financial need, and one third of Alameda County residents are food insecure (Beyers et al. 2008; OFPC 2010). Various assessments of the Oakland food system have underscored that affordability is the most important factor that influences where low-income residents shop for food (Wooten and Ackerman 2011), and limited access to transportation is another fundamental constraint to accessing healthy food (Treuhaft et al. 2009). In New Haven, CT, limited access to urban supermarkets co-varies with socio-economic indicators, thus highlighting the social justice implications of food deserts specifically for minority communities and the urban poor (Russell and Heidkamp 2011).

In response to food insecurity, UA in the US has expanded by >30% in the past 30 years, especially in under-served communities (Alig et al. 2004). UA has rapidly increased in developing countries all over the world, especially after the 2008 increase of global food prices (FAO 2014). In many African nations, for example, the percentage of low-income urban population participating in UA has grown from 20% in the 1980s to about 70% (Bryld 2003). A recent FAO report indicates an increasing number of Latin America cities are promoting and incentivizing UA through national governments, city administrations, civil society and non-governmental organizations (FAO 2014). This is because UA can be very productive, providing an estimated 15–20% of the global food supply (Hodgson et al. 2011; Smit et al. 1996). For example, UA provides 60% of the vegetables and 90% of the eggs consumed by residents in Shanghai, 47% of the produce in urban Bulgaria, 60% of vegetables in Cuba, and 90–100% of the leafy vegetables in poor households of Harare, Zimbabwe (Lovell 2010). Furthermore, with structural connectivity and governance, cities can provide good infrastructure, access to labour, and low transport costs for cost-effective local food distribution (Hodgson et al. 2011).

Additionally, as urban crop cultivation can also provide significant dietary contributions, communities around the world are using it to improve the health of urban residents (Beniston and Lal 2012) (Box 8.1). For example, there is an increasing desire to transform vacant land in post-industrial cities to address nutrition and childhood obesity issues in disadvantaged urban neighbourhoods (Yadav et al. 2012). Further, UA enhances food availability and quality across nations and economies; community members participating in UA in both developed and developing nations have been documented to exhibit greater dietary nutrition compared to non-participating community members (Zezza and Tasciotti 2010).

Many successful UA programmes have increased the food security of local residents. Existing UA programmes in Philadelphia produce over 900,000 kg of vegetables per year, worth more than US$4 million (Vitiello and Nairn 2009), and farms in Milwaukee gross more than US$200,000 per acre (4047 m^2) (Lovell 2010). New York City's (NYC) Green Thumb has become the largest community gardening programme in the US, with more than 600 gardens that support 20,000 urban residents (Lovell 2010). They are located in ethnically and culturally diverse neighbourhoods where a wide range of community members cultivate and manage the gardens. Ongoing expansion in Detroit's urban gardening scene is expected to produce 31% of the vegetables and 17% of the fruits currently consumed by city residents on just 100–350 ha of land (Colasanti and Hamm 2010). Cuba now has 383,000 urban farms, producing enough to supply 40–60% of fruits and vegetables to Havana and nearby cities (Funes et al. 2009), and the city of Quito currently has 140 community gardens, 800 family gardens, and 128 school gardens (FAO 2014).

Private gardens can also contribute significantly to local food production and food security. In Chicago, of the large number of community gardening projects reported by non-government organizations and government agencies, only 13% could be identified as food production sites via satellite image analysis, suggesting that many public spaces are supporting urban gardening projects without making notable physical changes to the environment. However, the food production area of home gardens identified by the study was almost threefold that of community gardens. This suggests that home food gardens can contribute significantly to enhancing community food sovereignty (Taylor and Lovell 2012) although it may be more difficult to regulate or incentivize.

Box 8.1

Unequal access to the available dietary diversity and calories leads to nutritional inequalities and diet-related health inequities in rich and poor cities alike. Three case studies presented by Dixon et al. (2007) illustrate how food insecurity can exist in cities regardless of the economic context of the city.

Case Study 1: In Nairobi, Kenya's capital, the poor constitute 55% of the population. Poverty and a reduction in agricultural production means that about 47% of the population is food-insecure. As in many parts of Africa, low- and medium-income households spend about three-quarters of their income on food. In urban areas, food is usually available but a nutritionally adequate diet is too costly for at least one third of households. In a context of low national GDP, under-nutrition is the major result of food insecurity, with 20% of Kenyan children underweight and 31% stunted. Anaemia and vitamin A deficiencies are also prevalent among children and women.

Case Study 2: Approximately 20% of Thailand's 65 million population lives in Bangkok, and per capita income differentials between the national capital and the rest of Thailand remain wide: 229,000 Bhat (US$6830) per annum compared to 74,600 Bhat (US$2225). Bangkok contains 70% of the country's supermarkets and superstores, whereas the rest of Thailand accesses food largely through Thai–Chinese shop houses, street stalls and wet markets. Urban wet markets cannot compete with supermarkets on price or perceived food safety, but they cater to the Thai population that is considered poor, of low education (55% of population in 2000), and who value a traditional diet. The major dietary issues in Thailand include undernutrition in rural areas, and growing over-nutrition or obesity in children amongst both rich and poor populations in urban areas.

Case Study 3: Australia is the world's most urban nation and has a population of 20 million people with a per capita GDP of US$25,353 in 2003. More than 75% of Australian women with families have paid employment, and nearly 27% of household food expenditure is on takeaway, fast foods and restaurant foods. On average 13% of total energy intake in the Australian diet comes from foods prepared outside the home. In some households this can be as high as 60%. Australians are among the most overweight and obese populations in the developed world. Obesity is more prevalent among poorer women and among richer men.

8.3.1 Biodiversity and Ecosystem Services

Urban green spaces such as UA can bring diverse green infrastructure back into the urban system by providing vegetative structure and biodiversity for ecosystem functions and services across fragmented habitats and spatial scales. UA provides many opportunities for re-vegetating the landscape at the local scale within a vegetatively depauperate urbanized landscape. Further, UA has the potential to support not only in situ biodiversity, but also nearby areas due to a landscape-mediated 'spillover' of energy, resources and organisms across habitats. Such spillover may be an important process for the persistence of wildlife populations in human-dominated landscapes because it allows for resource acquisition and re-colonization events (Blitzer et al. 2012). Movement of species between landscape elements can allow organisms to carry out functions at different points in space and time and maintain services that would otherwise be isolated (Lundberg and Moberg 2003). Thus, UA that provide landscape elements supportive of multiple species across time periods may be critical for the persistence of biodiversity in cities. Readers can refer to Chap. 7 in which Peter Werner and John Kelcey evaluated the relationship between urban greenery and biodiversity.

Vegetative diversity: The wide variety of UA types in practice allow for considerable variation in vegetative complexity and diversity. Domestic gardens vary widely in features that may promote plant biodiversity, such as ponds, moss, groundcover and varied vascular vegetative structures (Smith et al. 2005). For example, tropical home gardens have stratified vegetation similar to those seen in the multi-layered vertical structure of agroforestry systems (sensu Moguel and Toledo 1999) and can thus provide a large amount of planned and associated biodiversity (WinklerPrins 2002). The diversity of vegetation types within 21 home gardens has been documented in Santarem, Brazil, where 98 plant species including a large diversity of fruit trees and shrubs (comprising 34% of garden cover), ornamental plants (10%), vegetable or herb plants (13%) and medicinal plants (45%) were identified (WinklerPrins 2002). In Leon, Nicaragua, 293 plant species belonging to 88 families were recorded across 96 surveyed home gardens (González-García and Gómez-Sal 2008). In Hobart, Australia, 12 distinctly different garden types with different species, habits, and canopy heights were documented in front and backyard gardens (Daniels and Kirkpatrick 2006a), and a similar survey conducted in Toronto found an average of 25 woody plant species and 17 different herbaceous plant species per backyard garden (Sperling and Lortie 2010). In an example from five UK cities, more than 1000 species were recorded in 267 gardens, exceeding that recorded in all other local urban and semi-natural habitats (Loram et al. 2008).

Allotment and community gardens also provide substantial levels of vegetative biodiversity. In Stockholm, allotment gardens are older than many backyard gardens, often representing lush, well-managed flower-filled spaces covering large areas (3450–70,000 m^2). Such areas are often extremely rich in plant diversity, with more than 440 different plant species recorded in a single 400 m^2 allotment garden

(Colding et al. 2006). In Toronto, besides the typical local vegetables (cabbage, tomatoes, peppers and eggplant), farmers grew an additional 16 vegetable crops to supply the local community with foods unavailable in local grocery stores. These crops included Asian vegetables, such as bok choy, long bean, hairy gourd and edible chrysanthemums, and these plants substantially increased the vegetative diversity of the urban garden system (Baker 2004).

Arthropod diversity: In general, plant diversity is a principal predictor of arthropod diversity at small spatial scales (Southwood et al. 1979). Plant diversity and small-scale structural complexity have been shown to be important for tree-dwelling arthropods (Halaj et al. 2000), ground-dwelling arthropods (Byrne et al. 2008), web spiders (Greenstone 1984), grasshoppers (Davidowitz and Rosenzweig 1998), bees (Jha and Vandermeer 2010), and ground-dwelling beetles (Romero-Alcaraz and Ávila 2000) in natural and managed ecosystems.

Many studies have also shown that in urban systems plant diversity is highly correlated with insect diversity. For example, in urban backyard gardens in Toronto, invertebrate abundance and diversity was enhanced as the number of woody plant structures and plant species diversity increased, and backyard gardens had higher abundances of winged flying invertebrates when compared to urban grasslands and forests (Sperling and Lortie 2010). Likewise, within domestic gardens in the UK, invertebrate species richness was positively affected by vegetation complexity, especially by the abundance of trees (Smith et al. 2006b). In Pennsylvania, butterfly diversity increased with native plantings within suburban gardens (Burghardt et al. 2009), and parasitoid diversity increased with floral diversity within urban sites (Bennett and Gratton 2012).

Because of a rich abundance of flowering plants that prolongs the season for nectar supply, allotment gardens can support urban pollinators for long periods of time (Colding et al. 2006). In a survey of 16 allotment gardens in Stockholm, the number of bee species observed per allotment garden ranged between 5 and 11, including a large number of bumble bees, which were observed on a total of 168 plant species, especially those in the Lamiaceae, Asteraceae, Fabaceae, Boraginaceae and Malvaceae (Ahrne et al. 2009). However, bumble bee diversity decreased with increasing urbanization, from around eight species on sites in more rural areas to between five and six species in urban allotment gardens (Ahrne et al. 2009). In a survey of different garden types in Vancouver, a mean richness of 23 bee species was found (Tommasi et al. 2004). Similarly, community gardens in NYC provide a range of ornamental plants and food crops that supported 54 bee species, including species that nest in cavities, hives, pith and wood (Matteson et al. 2008). In another study in NYC community gardens, butterflies and bees responded to sunlight and floral area, but bee species richness also responded positively to garden canopy cover and the presence of wild or unmanaged areas in the garden (Matteson and Langellotto 2010). In Ohio, bee abundance in private, backyard gardens increased with native plantings, increases in floral abundance and taller herbaceous vegetation (Pardee and Philpott 2014). Additionally, a study of wild bee pollination of tomato plants in urban agricultural systems within San Francisco showed that wild bee pollination significantly increased overall production from the plants (Potter and LeBuhn 2015). This finding reinforces the idea that vegetation complexity within UA that can bring in more biodiversity can be beneficial to food production.

Overall, these studies support the idea that UA management with high vegetation diversity can have positive effects on invertebrate biodiversity in urban systems.

Vertebrate diversity: Wildlife friendly features implemented in UA can increase vertebrate diversity (Goddard et al. 2012). Practices such as planting fruit or seed-bearing plants, limiting the use of pesticides and herbicides, and constructing compost heaps and bird tables increase bird and vertebrate abundance and diversity (Good 2000). For avian diversity, garden heterogeneity that includes native plant species may be particularly important. Numerous avian studies show that gardens with sufficient native vegetation can support large populations of both native and exotic bird species at the local level (Daniels and Kirkpatrick 2006b) and at the landscape level, and garden heterogeneity can increase the overall diversity of insectivorous birds (Andersson et al. 2007). Heterogeneity that includes native plant species may be particularly important, as studies of suburban gardens in Australia show that nectarivorous birds prefer native genera over exotic ones as foraging sites (French et al. 2005).

For non-avian vertebrates, garden size, management style, and vegetation structure are critical for population persistence in urban areas. Baker and Harris (2007) reported 22 mammalian species or species groups in garden visitation surveys within the UK; however, mammal garden use declined as housing became more urbanized (e.g. more impervious habitat) and garden size and structural complexity decreased. Key findings from a range of garden studies show that in addition to high cultivated floral diversity, the three dimensional structure of garden vegetation is an important predictor of vertebrate abundance and diversity (Goddard et al. 2010). Increases in the vegetation structure and genetic diversity of domestic garden habitats have been shown to improve the connectivity of native populations currently limited to remnants (Doody et al. 2010) and aid conservation of threatened species (Roberts et al. 2007). For example, one study in Latin America documented that garden area and tree height were positively related to the presence and abundance of iguanas within urban areas, and increased patio extent allowed for greater iguana movement across the urban landscape (González-García et al. 2009). These studies show that garden management practices that provide food and nesting resources or movement corridors can be important strategies for maintaining vertebrate diversity in cities.

Ecosystem services: Ecosystem services are often a function of biodiversity levels (Loreau et al. 2001), thus the composition, diversity and structure of plant and animal communities within and around UA are important to consider for the delivery of urban ecosystem services. Specifically, biodiversity may enhance vital ecosystem services that city planners value—including energy efficiency, stormwater runoff, air pollution removal, carbon storage and sequestration, and water quality provision (McLain et al. 2012). Additionally, comparable to agricultural systems, where ecosystem services like water storage, pollination, and pest control increase US crop production resilience and protect production values by over US$57 billion per year (Daily 1997; Losey and Vaughan 2006), UA may strongly depend on biodiversity-mediated ecosystem services. However, there remains a large knowledge gap around the provisioning of services in UA systems. The key issues include increasing global food demands, climate-related crop failure, and consistent limitations in fresh food access within urban centres (Aubry et al. 2012; Thomas 2010; ver Ploeg et al. 2009).

Successful management and maintenance of ecosystem services within a city may need to extend beyond the city limits. For example, due to its large spatial extent, peri-urban agriculture can also play a key role in the management of the social, aesthetic and environmental functions of urban agglomerations nearby (Davoudi and Stead 2007). Depending on the type and intensity of the farming practise, peri-urban agriculture provides abiotic resources and ecosystem functions for the nearby urban areas. For instance, with its high water infiltration rates, pasture and arable land possess capacities for groundwater replenishment (Haase and Nuissl 2007) and flood control (Kenyon et al. 2008; Wheater and Evans 2009). Along with forest and wetlands, farmland including peri-urban agriculture, also contributes to urban-climate moderation (Lamptey et al. 2005) and carbon sequestration (Freibauer et al. 2004; Hutchinson et al. 2007) and thus should be incorporated into the large-scale management plans for sustainable cities.

8.3.2 Social and Cultural Resilience

In a number of ways, UA can enhance social and cultural resilience within cities. Urban gardening and urban social movements can build local ecological and social response capacity against major collapses in urban food supplies, helping to ensure food security in times of crisis (Barthel et al. 2013). Such systems allow for redundant food production solutions as a response to uncertain environmental, economic, or political futures. Hence, they should be incorporated as central elements of sustainable urban development. Additionally, communal garden spaces like allotment gardens can serve as conduits for transmitting collective social-ecological memories of food production. Specifically, they provide a venue for discussing roles and strategies for protecting urban green space, thus allowing communities to maintain local knowledge in the face of global change.

Urban agriculture can provide social safety nets to combat food insecurity, allowing healthy foods to be produced and shared by individuals and communities. Urban agriculture mapping initiatives, such as Fallen Fruit, created in Los Angeles, CA, where artists have mapped fruit trees in their neighbourhoods, create resources for the public to easily find and benefit from the local and free produce (Fallen Fruit 2014). This initiative, which has created over 60 neighbourhood and city maps from all over the US, is a great example of how UA can be combined with art, community strengthening activities, and neighbourhood beautification projects to help communities question and process themes such as public versus private land and the representation of ownership (Fallen Fruit 2014) in order to improve community well-being.

Social and cultural resilience of potentially underserved communities can particularly benefit from UA by making culturally relevant food and medicinal plants more accessible than traditional pathways. For example, as discussed above, in Toronto, surveys showed that besides the typical local vegetables (cabbage, tomatoes, peppers and eggplant), urban farmers grew an additional 16 vegetable crops (e.g. bok choy and hairy gourd) to supply the local community with foods unavailable in local grocery stores (Baker 2004).

In many cities, wet market stallholders and street vendors, principally women, have lost income as more commercial markets have expanded (FAO 2014), and the subsequent rise of income inequity acts with food insecurity to exacerbate diet-related health inequities. The phasing out of fresh produce markets, largely because of urban development pressures and the entry of supermarket and convenience store chains diminishes food access for poorer communities (Dixon et al. 2007). In contrast, the development of a vigorous UA system can provide enhanced opportunities for selling produce and accessing more nutritious foods. Specifically, aside from the act of cultivation and harvesting, the many employment opportunities associated with urban farming systems can boost the local economy by providing thousands of employed positions, from local food processing initiatives, to food distribution centres, to the establishment and management of healthy food market services (Dixon et al. 2007). For example, the proportion of the income coming from UA for the poorest communities can be as high as 30% in Africa, 20% in SE Asia and 10% in Latin America (Zezza and Tasciotti 2010). The provision of reliable flows of household income, via UA development, also improves access to nutritious foods in the cities via trade with other informal and small food producers.

Urban farming may also provide recreation and leisure opportunities that contribute to the quality of life (Antrop 2004). As inner cores of urban regions reach their limitations in complying with the increasing demand in green urban areas, the open spaces within and around cities, including urban and peri-urban farmland, provide valuable potential to deliver these services and functions and become increasingly important as the level of urbanisation increases (de Vries et al. 2003). Even if agricultural production represents the dominating land use in the peri-urban area, it still provides a ‘breathing space’ for the city nearby (Bryant and Johnston 1992) and access to the peri-urban landscape to enjoy open-space activities (Boulanger et al. 2004; Sharpley and Vass 2006). Matsuoka and Kaplan (2008) also found in their review of people’s needs in the urban landscape, that individuals greatly prefer urban landscapes that are dominated by naturalistic features and elements. Particularly organic farming is highly appreciated by urban residents, as argued by Brink (2003). Similarly, in the Brussels metropolitan region, more than half of the population support the protection of agricultural land use in the peri-urban fringe as a mechanism to preserve green space in the face of development (Boulanger et al. 2004).

Furthermore, urban gardening has been suggested as an effective tool for enhancing social cohesion and bridging racial divides by bringing people from different ages, races and income levels together (Shinew et al. 2004; Blaine et al. 2010). As such, UA can be linked to crime reduction, maintenance of cultural diversity, community empowerment, and promotion of civic participation (Warner and Hansi 1987; Murphy 1999). Community gardens can also serve as vehicles for education outreach programs for children and adults where they can learn about ecological processes, biodiversity and food production (Blaine et al. 2010). Learning gardens, in particular, have gained increased attention as efficient outdoor classrooms to foster healthy eating habits, increase physical activity, and demonstrate the importance of land stewardship and biological diversity (Williams and Brown 2013).

8.4 Challenges and Strategies in Promoting Urban Farming

During World War II, the US Department of Agriculture promoted Victory Gardens, which supplied in 1944 40% of the country's vegetables and 8 million tons of food (Nordahl 2009). Victory Gardens also produced self-reliance, self-respect, economic independence, community, and financial, physical and spiritual well-being (Nordahl 2009). However, these goals and benefits were not carried on after the war, and many of these benefits eroded in the face of industrialized agriculture and have not returned. Currently, urban farming challenges exist in urban areas, even in places that used to have significant support for urban food production.

Though public and scientific interest in UA has re-emerged and grown dramatically in the past two decades, there are significant challenges for integrating UA in an increasingly competitive urban landscape (Rural 2006). Much of the debate is centred around land-use trade-offs of UA versus other types of urban development, environmental constraints of the urban environment, and ecosystem dis-services that may come with UA.

8.4.1 Space Availability

Increased urbanization has led to greater competition for space in cities, making it difficult to make an argument to set aside land for urban farming. The question is how to best take advantage of the limited space available for urban gardens and maximize the benefits within these areas. We present a number of possibilities below.

Private yards. Private yards make up a significant proportion of green space in a city and do not require the acquisition of new space for urban farming. Even at a small-scale, private gardens can provide significant area for gardens and support complex vegetation structures in the urban matrix (Sperling and Lortie 2010). Strategies to incentivize wildlife-friendly gardening activities such as the National Wildlife Federation's certification for 'wildlife-friendly' gardens (National Wildlife Federation 2013) help encourage urban gardening in private households. More research is needed to understand the effectiveness of these incentives to support larger scale delivery of social and environmental benefits, such as food production, as many of the techniques are focused on the augmentation of ornamental or floral plants rather than food crops. Furthermore, due to lack of land tenure in poor communities, this type of gardens may not help in combating food insecurity for those who suffer from it.

Public spaces. Because greater housing density has been linked to smaller garden sizes, there is an acute need to better understand how UA can be supported within public green spaces, such as community gardens and easements (Smith et al. 2009). Although, zoning regulations often serve as obstacles to UA's expansion, a number of cities are working to understand how to move beyond these obstacles. For example, the Oakland Food Policy Council (OFPC) (in Oakland, CA) is fostering UA's expansion in public spaces by developing specific recommendations for urban agriculture zoning (McClintock et al. 2012). The city of Chicago, IL has recently passed a municipal code allowing community gardens, indoor, outdoor and rooftop operations in public, civic and commercial areas (Mayor Emanuel 2011). Additionally, the Chicago Urban Agriculture Mapping Project has inventoried and mapped all the metro area urban farms, community gardens, residential vegetable gardens, school gardens, etc. in order to assess current distributions. San Francisco, CA, created the first Urban Agricultural Zone allowing and promoting by means of tax reductions, plots between 0.1 and 3 acres (405 m^2 and 1.21 ha) to be converted to agricultural purposes for at least 5 years. This new ordinance requires public benefit to help knit the community together and give residents access to local produce. However, many compact cities are still facing land use debates about keeping urban gardens versus converting the land to much needed low-income housing in city centres, thus magnifying the environmental justice issues surrounding lower socio-economic communities and their access to green spaces and the benefits gained from them.

Vacant lots. Vacant lots provide opportunities to create functional green spaces where industrial redevelopment is not likely to happen (Beniston and Lal 2012). UA in these areas can be utilized to provide physical and psychological health for people in cities (Tzoulas et al. 2007). However, a better understanding of how to successfully rehabilitate vacant lots is needed in order to promote these spaces as options for urban farming. For example, creating gardens in abandoned lots has implications for urban land tenure for garden management, and it would be helpful to investigate whether temporary gardens can make positive contributions to the social and environmental health of cities in the same ways that more permanent gardens do. In many cases, the use of vacant lots for urban farming will add substantially more to the city than leaving the lot as an unused piece of land. In Buffalo, Metcalf and Widener (2011) showed that urban gardens on vacant lots were able to motivate community based farming initiatives and contribute to the social movement of 'local food' and 'healthy food' in the urban centre.

Peri-urban areas. In peri-urban areas, farming has to compete on the land market with other non-agricultural land uses, such as housing with its higher rents (Robinson 2004). As the price for a piece of farmland with an associated building permit rises dramatically, there is a strong financial incentive for farmers to sell land for purposes of urban development. Thus there is a decreasing amount of land reserved for peri-urban farming under many urban growth scenarios (Munton 2009); however, peri-urban areas remain as necessary areas to feed the local food economy and contribute to the urban metabolism of the city.

Rooftops. Although previously regarded as unusable space, the landscape of rooftops is being reclaimed for productive and sustainable purposes across many highly compact cities (Dunnett and Kingsbury 2008; Luckett 2009; Weiler and Scholz-Barth 2009). In Chap. 11, C.Y. Jim discusses the development of green roofs in cities and associated technological advances. Rooftops can be replenished to provide open space for social interaction in an increasingly depleted public realm and densifying city (Pomeroy 2012). Their use as alternative social and amenity spaces should be included alongside the conventional urban spaces of the street and square (public) or alternative social spaces of the mall, arcade, court or hotel lobby (semi-public) in the broader open space infrastructure of city development. However, rooftops are treated differently in different neighbourhoods—often as forgotten spaces for the underprivileged while providing leisure and recreation spaces for the affluent (Pomeroy 2012). For example, in Hong Kong, growing concerns about environmental issues and the need to promote sustainable urban environment, have led to growing development of green roofs in recent years (Hui 2009; Urbis Limited 2007). It is believed that green roofs can help mitigate the adverse effects of urban heat island in the city by lowering urban temperatures, but they also bring nature back to urban areas and improve urban aesthetics while reducing pollutant concentrations and noise (Hui 2006). Additionally, in Singapore, a proposal to develop rooftop farming in public housing estates has been developed to address the issue of food security and reduce the carbon footprint associated with food imports (Lim and Kishnani 2010). If such a scheme was to be implemented extensively in Singapore, it could result in a 700% increase in domestic vegetable production, satisfying domestic demand by 35.5% (Lim and Kishnani 2010). An example of a rooftop garden that is serving multiple purposes in Singapore is the rooftop garden of Khoo Teck Phuat Hospital which opens its rooftop garden for community garden and school use, but it also serves as a green space for hospital patients to enjoy (http://www.greenroofs.com/projects/pview.php?id=1622).

8.4.2 Environmental Constraints

Environmental changes brought on urbanization affect the agroecological conditions for food production, such as water availability, nutrient supply, soil degradation and pest pressure (Eriksen-Hamel and Danso 2010; Kaye et al. 2006). Thus, we need to understand the particular environmental conditions needed to support the safe and sustainable production of food in urban areas.

Climatic extremes. Cities tend to have higher air and surface temperatures than their rural surroundings because urban form and materials store and trap heat. This is a phenomenon known as urban heat island (Oke 1997). Evyatar Erell reviewed in Chap. 4 the role of greenery in modifying urban climate. The presence and management of garden trees, shrubs, and other plants in urban farming systems influence air and surface temperatures and has the potential to lower energy use and costs in urban environments. Additionally, UA plantings could enhance carbon sequestration while allowing enough light for cultivating ground crops and could assist in reducing the carbon footprint of cities. However, very little is known about how different UA respond to climate change or climate extremes, and how the urban environment in which UA is embedded may exacerbate or buffer climate effects. Thus, more research is needed to understand how plants in UA will respond to increasing temperature and drought, and changing rainfall amount, nutrient deposition and weather extremes.

Water use. Research on environmental constraints related to water use is also needed in UA, as irrigation is often required to provide water necessary for urban farming, and local supplies of water may be highly dependent on regional water systems (Mawois et al. 2011). Rainwater or grey water can be used for garden irrigation, and it is cheaper and at times more available than potable water-based irrigation, but UA gardeners must be aware of the potential pathogens and heavy metal contaminants that can cause human and environmental health problems (Qadir et al. 2010). For example, concentrations of potentially toxic elements were measured in soils in five tropical leafy vegetables grown in contaminated urban agriculture sites in Kampala City, Uganda, with soil contamination from poor waste disposal practices leading to considerable metal uptake in the crops (Nabulo et al. 2012).

Soil ecology. Urban soils are usually compacted, have low levels of organic matter, altered soil moisture characteristics, and sometimes have lead or other heavy metal contamination due to urban environmental processes (Beniston and Lal 2012). A number of methods, such as cover cropping, mulching, producing in raised beds, and changing subsurface drainage through piping, can improve soil conditions to support food production (Beniston and Lal 2012). However, more research must be done to understand how to sustainably rehabilitate urban soils. Alternative methods, such as 'organoponics', where organic compost is used as a growing medium instead of existing soils, need to be further explored to develop farming methods that are successful in the urban environment (Drescher et al. 2006).

Pest control and pollination. Food production requires important ecosystem services provided by vertebrate and invertebrate animal species to be ecologically and economically sustainable. Pollinator density and diversity are essential for optimal fruit and seed setting of many crop species (Klein et al. 2007), while insectivorous birds and arthropod predators and parasites can keep crop pests below damaging levels (Letourneau et al. 2009). These ecosystem services are particularly important in UA systems, where most of the crops depend upon bee pollination (Matteson and Langellotto 2009; Oberholtzer et al. 2014), and urban gardeners and growers greatly rely on natural pest control since they often face severe restrictions in using chemical pesticides. Nevertheless, habitat fragmentation, lack of vegetation cover, and constant disturbances make cities inhospitable habitats for many animal species (McKinney 2006; Goddard et al. 2010) that mediate important ecosystem services to UA such as pest control and pollination. Furthermore, even though food-web dynamics of crops commonly found in UA have been widely studied in rural settings, human forces may alter environmental stressors and create unique interactions in urban ecosystems (Shochat et al. 2006). Consequently, there is a great need for research on animal population persistence and food-web dynamics for the successful management of animal-mediated ecosystem services in UA.

8.4.3 Potential Ecosystem Disservices and Tradeoffs of Urban Agriculture

Besides many potential benefits provided by urban farming, the potential negative impacts on ecosystem functioning and human health should be evaluated.

Spillover into natural systems. In some cases, there is the possibility of negative spillover from managed farms to natural systems or vice versa of weed, pathogen or pest populations, potentially harming native ecosystems and damaging ecosystem-service delivery from natural systems (Blitzer et al. 2012; Zhang et al. 2007). The juxtaposition of natural systems to urban farm systems also potentially leads to an increased opportunity for biological invasions and detrimental competition with native species (Niinemets and Penuelas 2008). Genetic introgression within natural ecosystems by urban garden plants can negatively alter the genetic composition of native vegetative patches and affect the long term viability of these systems (Whelan et al. 2006). At the same time, chemical, water, and animal movement is bi-directional, and intensified management implemented in backyards, such as pesticide application, extensive pruning, frequent mowing and other disturbances, can limit the capacity of gardens to maintain rare or sensitive insect species (Matteson and Langellotto 2011). The problem of chemical spillover may be especially prevalent in developing nations where there may be a lack of governmental support or where UA has been considered an illegal activity until recently (Smith 1996; Færge et al. 2001; Deelstra and Girardet 2000). Urban farmers are thus often forced to hide their gardens by planting only less conspicuous and short-cycled crops and using more chemical inputs to reduce the length of the growth cycle (Bryld 2003). These situations, combined with the lack of tenure and constant danger of evictions, make urban farmers in many developing countries less motivated to practise urban agriculture in a sustainable manner (Bryld 2003).

Negative impacts on human health. If managed carelessly, urban farm areas may also lead to increased human health issues and greater disease transmission to urban populations. For example, UA systems provide increased mosquito breeding sites due to the presence of standing water from irrigation or rainwater, and this may potentially increase the rate of mosquito borne diseases in certain areas of the city (Matthys et al. 2010). Additionally, in non-organic UA systems, there is the potential for spillover of chemicals into natural and human habitats, leading to environmental pollution and air or water borne health risks, as discussed in the previous sub-section (Robbins et al. 2001). Additionally, UA in many countries remains largely unregulated, with very little official support or technical assistance provided by local governments. This creates environmental and health hazards due to frequent use and misuse of chemical fertilizers and pesticides (Smith 1996), and irrigation with contaminated water (FAO 2014).

Gentrification of low-income neighbourhoods. Recognizing many local benefits to the community, cities around the world have implemented strategies to increase urban green space, especially in lower socioeconomic neighbourhoods where the supply is usually inadequate. However, these actions can actually exacerbate the existing problem. While the creation of new green space to address environmental injustice can make neighbourhoods healthier and more aesthetically attractive, it also can increase housing costs and property values, forcing residents to find cheaper housing elsewhere. Ultimately, this can lead to gentrification and a displacement of the very residents the green-space strategies were designed to benefit (Wolch et al. 2014). Given that urban community gardening has been sometimes been linked to gentrification of urban areas (Martinez 2010), their development has been received with skepticism in many poor and minority communities (Shinew et al. 2004). Thus, the development of urban gardens across communities must be considered carefully to avoid displacing the extant communities.

8.5 Conclusion

Despite multiple environmental and social benefits to promoting urban agriculture within cities, maintaining and increasing this specific type of green space remain challenging in the face of other urban processes. Identifying win-win areas for urban farming, where environmental and social benefits can be maximized in otherwise unused land, will be necessary to build support and acceptance for these urban farming systems both socially and politically. On the research side, projects that map and collate data on urban farming systems (e.g. Taylor and Lovell 2012) will help develop a database of farm attributes and the benefits provided by each system. On the policy side, major changes to zoning regulations that allow communities to take advantage of otherwise unused land to develop urban farms will be necessary to transform vacant lots or overlooked public spaces into active UA systems. Understanding how the transformation of land into UA systems affects other social and economic processes in surrounding communities will also be necessary to prevent unintended consequences of chemical spillover or displacement. We posit that such knowledge will be required to develop local urban agriculture systems that allow people the opportunity to interact with the natural landscape around them while improving the environment and social health of the communities around them.

References

Ahrne K, Bengtsson J, Elmqvist T (2009) Bumble bees (*Bombus* spp.) along a gradient of increasing urbanization. PLoS ONE 4(5):e5574

Alig RJ, Kline JD, Lichtenstein M (2004) Urbanization on the US landscape: looking ahead in the 21st century. Landscape Urban Plan 69(2):219–234

Andersson E, Barthel S, Ahrné K (2007) Measuring social-ecological dynamics behind the generation of ecosystem services. Ecol Appl 17(5):1267–1278

Antrop M (2004) Landscape change and the urbanization process in Europe. Landscape Urban Plan 67:9–26

Aubry C, Ramamonjisoa J, Dabat MH et al (2012) Urban agriculture and land use in cities: an approach with the multi-functionality and sustainability concepts in the case of Antananarivo (Madagascar). Land Use Policy 29(2):429–439

Baker LE (2004) Tending cultural landscapes and food citizenship in Toronto's community gardens. Geogr Rev 94(3):305–325

Baker PJ, Harris S (2007) Urban mammals: what does the future hold? An analysis of the factors affecting patterns of use of residential gardens in Great Britain. Mammal Rev 37(4):297–315

Barthel S, Isendahl C (2013) Urban gardens, agriculture, and water management: Sources of resilience for long-term food security in cities. Ecol Econ 86:224–234

Barthel S, Parker J, Ernstson H (2013) Food and green space in cities: a resilience lens on gardens and urban environmental movements. Urban Stud 57:1321–1338

Beniston J, Lal R (2012) Improving soil quality for urban agriculture in the North Central U.S. In: Lal R, Augustin B (eds) Carbon sequestration in urban ecosystems. Springer, Netherlands, pp 279–313

Bennett AB, Gratton C (2012) Local and landscape scale variables impact parasitoid assemblages across an urbanization gradient. Landscape Urban Plan 104(1):26–33

Beyers M, Brown J, Cho S et al (2008) Life and death from unnatural causes: health and social inequity in Alameda County. Alameda County Public Health Department, Oakland, CA

Blaine TW, Grewal PS, Dawes A, Snider D (2010) Profiling community gardeners. J Extension 48 (6):1–12

Blitzer EJ, Dormann CF, Holzschuh A et al (2012) Spillover of functionally important organisms between managed and natural habitats. Agric Ecosyst Environ 146(1):34–43

Boulanger A, Meert H, Van Hecke E (2004) The societal demand for public goods in peri-urban areas: a case from the Brussels urban region. In: Conference proceedings 90th EAAE seminar multifunctional agriculture, policies markets: understanding the critical linkage, Rennes, France, 28–29 Oct 2004

Brink A (2003) Welche Rolle spielt der ökologische Landbau für eine Großstadt. In: Rahmann G, Nieberg H (eds) Ressortforschung für den ökologischen Landbau. Bundesforschungsanstalt für Landwirtschaft, Braunschweig, pp 80–83

Bryant CR, Johnston TRR (1992) Agriculture in the city's countryside. Belhaven Press, London

Bryld E (2003) Potentials, problems, and policy implications for urban agriculture in developing countries. Agric Hum Values 20(1):79–86

Burghardt KT, Tallamy DW, Shriver GW (2009) Impact of native plants on bird and butterfly biodiversity in suburban landscapes. Conserv Biol 23(1):219–224

Byrne L, Bruns M, Kim K (2008) Ecosystem properties of urban land covers at the aboveground–belowground interface. Ecosystems 11(7):1065–1077

Colasanti KA, Hamm MW (2010) Assessing the local food supply capacity of Detroit, Michigan. J Agri Food Syst Commun Dev 1(2):41

Colding J, Lundberg J, Folke C (2006) Incorporating green-area user groups in urban ecosystem management. Ambio 35(5):237–244

Daily G (1997) Nature's services: societal dependence on natural ecosystems. Island Press, Washington, DC

Daniels GD, Kirkpatrick JB (2006a) Comparing the characteristics of front and back domestic gardens in Hobart, Tasmania, Australia. Landscape Urban Plan 78(4):344–352

Daniels GD, Kirkpatrick JB (2006b) Does variation in garden characteristics influence the conservation of birds in suburbia? Biol Conserv 133(3):326–335

Davidowitz G, Rosenzweig ML (1998) The latitudinal gradient of species diversity among North American grasshoppers (Acrididae) within a single habitat: a test of the spatial heterogeneity hypothesis. J Biogeogr 25(3):553–560

Davoudi S, Stead D (2007) Urban–rural-relationships: an introduction and brief history. Build Environ 28:269–277

Deelstra T, Girardet H (2000) Urban agriculture and sustainable cities. In: Bakker N, Dubbeling M, Gündel S, Sabel-Koshella U, de Zeeuw H (eds) Growing cities, growing food. Urban agriculture on the policy agenda. Zentralstelle für Ernährung und Landwirtschaft (ZEL), Feldafing, Germany, pp 43–66

de Vries S, Verheij RA, Groenewegen PP, Spreeuwenberg P (2003) Natural environments—healthy environments? An exploratory analysis of the relationship between greenspace and health. Environment and Planning A 35:1717–1731

Dixon J, Omwega AM, Friel S, Burns C, Donati K, Carlisle R (2007) The health equity dimensions of urban food systems. J Urban Health 84(1):118–129

Doody B, Sullivan J, Meurk C, Stewart G, Perkins H (2010) Urban realities: the contribution of residential gardens to the conservation of urban forest remnants. Biodivers Conserv 19 (5):1385–1400

Drescher A, Holmer R, Iaquinta D (2006) Urban homegardens and allotment gardens for sustainable livelihoods: management strategies and institutional environments. Tropical Homegardens, Springer, The Netherlands, pp 317–338

Dunnett N, Kingsbury N (2008) Planting green roofs and living walls, revised and updated edition. Timber Press, Oregon

Eriksen-Hamel N, Danso G (2010) Agronomic considerations for urban agriculture in southern cities. Int J Agricu Sustain 8(1–2):86–93

Færge J, Magid J, de Vries FW (2001) Urban nutrient balance for Bangkok. Ecol Model 139 (1):63–74

Fallen Fruit Biography (2014). Retrieved 25 Sept 2014, from http://fallenfruit.org/about/

FAO (2014) Growing green cities in Latin America and the Caribbean. Food and Agriculture Organization of the United Nations, Rome

Forman RT, Alexander LE (1998) Roads and their major ecological effects. Annu Rev Ecol Syst 29:207–231

French K, Major R, Hely K (2005) Use of native and exotic garden plants by suburban nectivorous birds. Biol Conserv 121:545–559

Freibauer A, Rounsevell MDA, Smith P, Verhagen J (2004) Carbon sequestration in the agricultural soils of Europe. Geoderma 122:1–23

Funes F, Altieri MA, Rosset P (2009) The Avery diet: the Hudson Institute's misinformation campaign against Cuban agriculture. http://www.landaction.org/IMG/pdf/The-Avery-Diet.pdf

Goddard MA, Dougill AJ, Benton TG (2010) Scaling up from gardens: biodiversity conservation in urban environments. Trends Ecol Evol 25(2):90–98

Goddard MA, Dougill AJ, Benton TG (2012) Why garden for wildlife? Social and ecological drivers, motivations and barriers for biodiversity management in residential landscapes. Ecol Econ 86:258–263

González-García A, Belliure J, Gómez-Sal A, Dávila P (2009) The role of urban greenspaces in fauna conservation: the case of the iguana *Ctenosaura similis* in the 'patios' of León city, Nicaragua. Biodivers Conserv 18(7):1909–1920

González-García A, Gómez-Sal A (2008) Private urban greenspaces or 'patios' as a key element in the urban ecology of tropical Central America. Human Ecol 36(2):291–300

Good R (2000) The value of gardening for wildlife-what contribution does it make to conservation? Br Wildlife 12(2):77–84

Greenstone M (1984) Determinants of web spider species diversity: vegetation structural diversity vs. prey availability. Oecologia 62(3):299–304

Haase D, Nuissl H (2007) Does urban sprawl drive changes in the water balance and policy? The case of Leipzig (Germany) 1870–2003. Landscape Urban Plan 80:1–13

Halaj J, Ross DW, Moldenke AR (2000) Importance of habitat structure to the arthropod food-web in Douglas-fir canopies. Oikos 90(1):139–152

Hodgson K, Campbell MC, Bailkey M (2011) Urban agriculture: growing healthy, sustainable places. Am Plann Assoc, Washington, DC

Hui SCM (2006) Benefits and potential applications of green roof systems in Hong Kong. In: Proceedings of the 2nd megacities international conference 2006, 1–2 Dec 2006, Guangzhou, China, pp 351–360

Hui SCM (2009) Study of thermal and energy performance of green roof systems: final report. Department of Mechanical Engineering, The University of Hong Kong, Hong Kong

Hunter MCR, Brown DG (2012) Spatial contagion: gardening along the street in residential neighborhoods. Landscape Urban Plan 105(4):407–416

Hunter MCR, Hunter MD (2008) Designing for conservation of insects in the built environment. Insect Conserv Div 1(4):189–196

Hutchinson JJ, Campbell CA, Desjardins RL (2007) Some perspectives on carbon sequestration in agriculture. Agric For Meteorol 142:288–302

Jha S, Vandermeer JH (2010) Impacts of coffee agroforestry management on tropical bee communities. Biol Conserv 143(6):1423–1431

Kaye JP, Groffman PM, Grimm NB, Baker LA, Pouyat RV (2006) A distinct urban biogeochemistry? Trends Ecol Evol 21(4):192–199

Kenyon W, Hill G, Shannon P (2008) Scoping the role of agriculture in sustainable flood management. Land Use Policy 25:351–360

Klein AM, Vaissiere BE, Cane JH et al (2007) Importance of pollinators in changing landscapes for world crops. Proc Royal Soc B: Biol Sci 274(1608):303–313

Lamptey BL, Barron EJ, Pollard D (2005) Impacts of agriculture and urbanization on the climate of the Northeastern United States. Global Planet Change 49:203–221

Letourneau DK, Jedlicka JA, Bothwell SG, Moreno CR (2009) Effects of natural enemy biodiversity on the suppression of arthropod herbivores in terrestrial ecosystems. Annu Rev Ecol Evol Syst 40:573–592

Lim YA, Kishnani NT (2010) Building integrated agriculture: utilising rooftops for sustainable food crop cultivation in Singapore. J Green Build 5(2):105–113

Lin BB, Philpott SM, Jha S (2015) The future of urban agriculture and biodiversity ecosystem-services: challenges and next steps. Basic Appl Ecol 16(3):189–201

Loram A, Tratalos J, Warren P, Gaston KJ (2007) Urban domestic gardens (X): the extent & structure of the resource in five major cities. Landscape Ecol 22(4):601–615

Loram A, Warren PH, Gaston KJ (2008) Urban domestic gardens (XIV): the characteristics of gardens in five cities. Environ Manage 42(3):361–376

Loreau M, Naeem S, Inchausti P et al (2001) Biodiversity and ecosystem functioning: current knowledge and future challenges. Science 294(5543):804–808

Losey JE, Vaughan M (2006) The economic value of ecological services provided by insects. Bioscience 56(4):311–323

Lovell ST (2010) Multifunctional urban agriculture for sustainable land use planning in the United States. Sustainability 2(8):2499–2522

Luckett K (2009) Green roof construction and maintenance. McGraw-Hill, New York

Lundberg J, Moberg F (2003) Mobile link organisms and ecosystem functioning: implications for ecosystem resilience and management. Ecosystems 6(1):0087–0098

Martellozzo F, Landry JS, Plouffe D et al (2014) Urban agriculture: a global analysis of the space constraint to meet urban vegetable demand. Environ Res Lett 9(6):064025

Martinez MJ (2010) Power at the roots: gentrification, community gardens, and the Puerto Ricans of the Lower East Side. Lexington Books, Plymouth

Mathieu R, Freeman C, Aryal J (2007) Mapping private gardens in urban areas using object-oriented techniques and very high-resolution satellite imagery. Landscape Urban Plan 81(3):179–192

Matsuoka RH, Kaplan R (2008) People needs in the urban landscape: analysis of landscape and urban planning contributions. Landscape Urban Plan 84:7–19

Matteson KC, Ascher JS, Langellotto GA (2008) Bee richness and abundance in New York City urban gardens. Ann Entomol Soc Am 101(1):140–150

Matteson KC, Langellotto GA (2009) Bumble bee abundance in New York City community gardens: implications for urban agriculture. Cities Environ 2(1):5

Matteson KC, Langellotto GA (2010) Determinates of inner city butterfly and bee species richness. Urban Ecosyst 13(3):333–347

Matteson KC, Langellotto GA (2011) Small scale additions of native plants fail to increase beneficial insect richness in urban gardens. Insect Conserv Divers 4(2):89–98

Matthys B, Koudou BG, N'Goran EK et al (2010) Spatial dispersion and characterisation of mosquito breeding habitats in urban vegetable-production areas of Abidjan, Côte d'Ivoire. Ann Trop Med Parasitol 104(8):649–666

Mawois M, Aubry C, Le Bail M (2011) Can farmers extend their cultivation areas in urban agriculture? A contribution from agronomic analysis of market gardening systems around Mahajanga (Madagascar). Land Use Policy 28(2):434–445

Mayor Emanuel R (2011) Amendment of Chapter 17-2 of municipal code regarding urban agriculture uses. City Council Document Tracking Sheet, City of Chicago, Office of the Chicago City Clerk

McClintock N, Wooten H, Brown A (2012) Toward a food policy 'First Step' in Oakland, California: a food policy council's efforts to promote urban agriculture zoning. J Agri Food Syst Commu Dev 2(4):15–42

McKinney ML (2006) Urbanization as a major cause of biotic homogenization. Biol Conserv 127:247–260

McLain R, Poe M, Hurley PT et al (2012) Producing edible landscapes in Seattle's urban forest. Urban For Urban Greening 11(2):187–194

Metcalf SS, Widener MJ (2011) Growing Buffalo's capacity for local food: a systems framework for sustainable agriculture. Appl Geogr 31(4):1242–1251

Miotk P (1996) The naturalized garden—a refuge for animals? first results. Zoologischer Anzeiger 235:101–116

Montgomery MR (2008) The urban transformation of the developing world. Science 319:761–764

Moguel P, Toledo VM (1999) Biodiversity conservation in traditional coffee systems of Mexico. Conserv Biol 13:11–21

Mougeot LJ (ed) (2010) Agropolis: the social, political and environmental dimensions of urban agriculture. Routledge, London

Munton R (2009) Rural land ownership in the United Kingdom: changing patterns and future possibilities for land use. Land Use Policy 26:S54–S61

Murphy C (1999) Cultivating Havana: urban agriculture and food security in the years of crisis. Food First, Oakland, CA

National Wildlife Federation (2013) Garden for wildlife: making wildlife habitat at home. http://www.nwf.org/How-to-Help/Garden-for-Wildlife/Create-a-Habitat.aspx?campaignid=WH10BGHF&adid=54466
Niinemets U, Penuelas J (2008) Gardening and urban landscaping: significant players in global change. Trends Plant Sci 13(2):60–65
Nordahl D (2009) Public produce: the new urban agriculture. Island Press, Washington, DC
Oakland Food Policy Council [OFPC] (2010) Transforming Oakland food system: a plan for action. Oakland Food Policy Council and Food First, Oakland, CA. Available at http://www.oaklandfood.org
Oberholtzer L, Dimitri C, Pressman A (2014) Urban agriculture in the United States: characteristics, challenges and technical assistance needs. J Extension 52(6):61
Oke TR (1997) Urban climates and global environmental change. In: Thompson RD, Perry AH (eds) Applied climatology: principles and practice. Routledge, New York, pp 273–287
Owen J (1991) The ecology of a garden: the first fifteen years. Cambridge University Press, Cambridge
Pardee G, Philpott S (2014) Native plants are the bee's knees: local and landscape predictors of bee richness and abundance in backyard gardens. Urban Ecosyst 17(3):641–659
Pomeroy J (2012) Room at the top: the roof as an alternative habitable/social space in the Singapore context. J Urban Des 17(3):413–424
Potter A, LeBuhn G (2015) Pollination service to urban agriculture in San Francisco, CA. Urban Ecosyst 18:885–893
Qadir M, Wichelns D, Raschid-Sally L et al (2010) The challenges of wastewater irrigation in developing countries. Agric Water Manag 97(4):561–568
Robbins P, Polderman A, Birkenholtz T (2001) Lawns and toxins: an ecology of the city. Cities 18(6):369–380
Roberts DG, Ayre DJ, Whelan RJ (2007) Urban plants as genetic reservoirs or threats to the integrity of bushland plant populations. Conserv Biol 21(3):842–852
Robinson GM (2004) Geographies of agriculture: globalisation, restructuring and sustainability. Pearson Education, Harlow, UK
Romero-Alcaraz E, Ávila JM (2000) Landscape heterogeneity in relation to variations in epigaeic beetle diversity of a Mediterranean ecosystem: implications for conservation. Biodivers Conserv 9(7):985–1005
RURAL (2006) Rural areas under pressure. Case studies of rural–urban relationships across Europe. In: Overbeek MM, Terluin IJ (eds), European report. Agricultural Economics Research Institute, Den Haag, The Netherlands
Russell SE, Heidkamp CP (2011) 'Food desertification': the loss of a major supermarket in New Haven, Connecticut. Appl Geogr 31(4):1197–1209
Saldivar-Tanaka L, Krasny ME (2004) Culturing community development, neighborhood open space, and civic agriculture: the case of Latino community gardens in New York City. Agric Hum Values 21:399–412
Sanders RA, Stevens JC (1984) Urban forest of Dayton, Ohio: a preliminary assessment. Urban Ecol 8(1–2):91–98
Sharpley R, Vass A (2006) Tourism, farming and diversification: an attitudinal study. Tour Manag 27:1040–1052
Shinew KJ, Glover TD, Parry DC (2004) Leisure spaces as potential sites for interracial interaction: community gardens in urban areas. J Leisure Res 36(3):336
Shochat E, Warren PS, Faeth SH et al (2006) From patterns to emerging processes in mechanistic urban ecology. Trends Ecol Evol 21(4):186–191
Smit J, Nasr J, Ratta A (1996) Urban agriculture: food jobs and sustainable cities. United Nations Development Programme (UNDP), New York
Smith C, Clayden A, Dunnett N (2009) An exploration of the effect of housing unit density on aspects of residential landscape sustainability in England. J Urban Des 14(2):163–187

Smith D (1996) Urban agriculture in Harare: socio-economic dimensions of a survival strategy. In: Grossman D, van den Berg LM, Ajaegbu HI (eds) Urban and peri-urban agriculture in Africa: proceedings of a Workshop: Netanya, Israel, 23–27 June 1996, Ashgate, Aldershot, UK
Smith RM, Gaston KJ, Warren PH, Thompson K (2005) Urban domestic gardens (V): relationships between landcover composition, housing and landscape. Landscape Ecol 20 (2):235–253
Smith RM, Thompson K, Hodgson JG et al (2006a) Urban domestic gardens (IX): composition and richness of the vascular plant flora, and implications for native biodiversity. Biol Conserv 129:312–322
Smith RM, Warren P, Thompson K, Gaston KJ (2006b) Urban domestic gardens (VI): environmental correlates of invertebrate species richness. Biodivers Conserv 15(8):2415–2438
Southwood TRE, Brown VK, Reader PM (1979) The relationships of plant and insect diversities in succession. Biol J Linn Soc 12(4):327–348
Sperling C, Lortie C (2010) The importance of urban backgardens on plant and invertebrate recruitment: a field microcosm experiment. Urban Ecosyst 13(2):223–235
Taylor JR, Lovell ST (2012) Mapping public and private spaces of urban agriculture in Chicago through the analysis of high-resolution aerial images in Google Earth. Landscape Urban Plan 108(1):57–70
Thomas BJ (2010) Food deserts and the sociology of space: distance to food retailers and food insecurity in an urban American neighborhood. Int J Human Social Sci 5(6):400–409
Tommasi D, Miro A, Higo HA, Winston ML (2004) Bee diversity and abundance in an urban setting. Can Entomol 136(06):851–869
Treuhaft S, Hamm MJ, Litjens C (2009) Healthy food for all: building equitable and sustainable food systems in Detroit and Oakland. PolicyLink, Oakland, CA
Tzoulas K, Korpela K, Venn S et al (2007) Promoting ecosystem and human health in urban areas using green infrastructure: a literature review. Landscape Urban Plan 81(3):167–178
Limited Urbis (2007) Study on green roof application in Hong Kong, final report. Architectural Services Department, Hong Kong
ver Ploeg M, Breneman V, Farrigan T et al (2009) Access to affordable and nutritious food: measuring and understanding food deserts and their consequences. Report to Congress on access to affordable and nutritious food: measuring and understanding food deserts and their consequences. USDA Economic Research Service, Washington, DC
Vitiello D, Nairn M (2009) Community gardening in Philadelphia: 2008 harvest report. Penn planning and urban studies. University of Pennsylvania, PA
Warner SB, Hansi D (1987) To dwell is to garden: a history of Boston's community gardens. Northeastern University Press, Boston
Weiler S, Scholz-Barth K (2009) Green roof systems: a guide to the planning, design and construction of building over structure. Wiley, Hoboken, NJ
Wheater H, Evans E (2009) Land use, water management and future flood risk. Land Use Policy 26:S251–S264
Whelan RJ, Roberts DG, England PR, Ayre DJ (2006) The potential for genetic contamination vs. augmentation by native plants in urban gardens. Biol Conserv 128(4):493–500
Whittinghill LJ, Rowe DB (2012) The role of green roof technology in urban agriculture. Renew Agric Food Syst 27(04):314–322
Williams D, Brown J (2013) Learning gardens and sustainability education: bringing life to schools and schools to life. Routledge, London
WinklerPrins AGA (2002) House-lot gardens in Santarém, Pará, Brazil: linking rural with urban. Urban Ecosyst 6(1–2):43–65
Wolch JR, Byrne J, Newell JP (2014) Urban green space, public health, and environmental justice: the challenge of making cities 'just green enough'. Landscape Urban Plan 125:234–244
Wooten H, Ackerman A (2011) Seeding the city: land use policies to promote urban agriculture. Public Health Law & Policy/NPLAN, Oakland, CA. Available at http://changelabsolutions.org/publications/seeding-city

Yadav P, Duckworth K, Grewal PS (2012) Habitat structure influences below ground biocontrol services: a comparison between urban gardens and vacant lots. Landscape Urban Plan 104 (2):238–244

Zasada I (2011) Multifunctional peri-urban agriculture: a review of societal demands and the provision of goods and services by farming. Land Use Policy 28(4):639–648

Zezza A, Tasciotti L (2010) Urban agriculture, poverty, and food security: Empirical evidence from a sample of developing countries. Food Policy 35(4):265–273

Zhang W, Ricketts TH, Kremen C et al (2007) Ecosystem services and dis-services to agriculture. Ecol Econ 64(2):253–260

Chapter 9
Urban Nature and Urban Ecosystem Services

Wendy Y. Chen

Abstract Worldwide more and more people live and work in cities, where urban nature and their ecosystem services are the basis for economic development and social wellbeing. Therefore, how to ensure that cities in the present and future can provide a whole range of ecosystem services to meet urban dwellers' needs become a front and center issue in urban resilience and sustainability on the global agenda. This article presents a literature review that explores our understanding about various natural elements in cities (including urban green and blue spaces) and their diverse ecosystem services and some disservices. While the importance of urban nature and urban ecosystem services has been increasingly recognized, the integration of ecological, social and economic understanding of urban ecosystem services into relevant policy making processes is still at an embryonic stage. Some pertinent challenges are highlighted for the theorization and governance of urban ecosystem services.

Keywords Ecosystem services · Urban nature · Urban sustainability · Urban green spaces · Urban disservices · Review

9.1 Diverse Nature in Cities

As a human invention of more recent origin, modern cities are usually stacked with paved streets, concrete buildings, and other man-made structures for residential, commercial, industrial and infrastructural purposes. Over the course of their development throughout the world, much pristine natural features have been removed or replaced (Lundholm and Richardson 2010). For example, original soils are moved and reconfigured and usually the vertical stratification of A and B horizons (where most flora and fauna live) becomes nonexistent (Schaefer 2009). Plants and animals are directly or indirectly extirpated (Pincetl 2012). Thus, cities

W.Y. Chen (✉)
University of Hong Kong, Pokfulam, Hong Kong
e-mail: wychen@hku.hk

P.Y. Tan and C.Y. Jim (eds.), *Greening Cities: Forms and Functions*,
Advances in 21st Century Human Settlements,
DOI 10.1007/978-981-10-4113-6_9

are often thought to be separate from nature, a word calling to mind open spaces with lush vegetation, wild animals, or bodies of water. However, even though they frequently escape our notice cities present a mosaic of fragmented habitats and harbour diverse ecological niches ranging from remnant natural, modified natural, semi-natural, emulated natural or entirely human-created habitats with different conditions and resources (Kowarik 2011; Vilisics and Hornung 2009). They retain a surprisingly large amount of species of different taxa (Meffert and Dziock 2013), and sometimes serve as refuges for many endangered species, and thus play significant role in biodiversity conservation (Vilisics and Hornung 2009).

Although urbanization has notable effect on biotic homogenization (McKinney 2006), flora in cities has long been recognized as considerably rich in species (Breuste et al. 2008; Kowarik 2011). For example, it has been found that there are 671 angiosperm species in Brussels (the capital of Belgium), accounting for half of the total number of species present in the country (Godefroid 2001). The occurrence of rich flora in cities is mainly attributed to a high heterogeneity of urban environments (McKinney 2008) and associated heterogeneous geological substrates. While the majority of urban vegetation (including trees, shrubs and grass) are situated in domestic gardens, public parks and residual woodlands, it is not surprising that a wide variety of floral species have been found to inhabit roadside verges (Lundholm and Marlin 2006), playgrounds and sport fields (Schulman and Peters 2008), cemeteries (Uslu et al. 2009), building walls (Jim and Chen 2011), roof tops (Dvorak and Volder 2010), golf courses (Colding and Folke 2009), intensively cultivated allotments and orchards (Andersson et al. 2007), river and stream corridors (Sung et al. 2011), wetlands (Baldwin 2004), as well as derelict lands (Venn and Niemelä 2004). Additionally, urban flora could also be significantly enriched by alien, exotic and introduced plant species imported by landscape managers, horticulturists, gardeners and foresters. For example, a total of 617 plant species has been introduced into Adelaide city (Australia), leading a significant increase (about 46%) in plant species richness since its foundation in 1836 (Tait et al. 2005). Please also see Chap. 7 on biodiversity in other cities.

Associated directly with diverse floral communities as well as natural and artificial habitat niches within urban matrix, rich terrestrial faunas have also been recorded in many cities. The most-studied group is birds (McKinney 2008). Urban environments have been likened to a cliff/organic detritus zone (Lundholm and Richardson 2010) and a number of birds have forsaken their rock faces and adapted to breeding on tall buildings, such as feral pigeon, kestrel, black redstart, starling, house sparrow, etc. (Savard et al. 2000). Ortega-Álvarez and MacGregor-Fors (2009) investigated bird diversity, structure, and composition pattern in Mexico City and recorded 58 landbird species including three endangered ones. A rich diversity of mammals has also been found in urban areas and received much research attention. Possum, glider, and bandicoot species are common Australian urban mammals (Garden et al. 2006). Comparatively, invertebrates, aquatic fauna, reptiles and amphibians are under studied (McIntyre 2000) and our knowledge on these urban faunal groups is far from complete. Yet sporadically, the occurrence of various faunal groups and influencing factors in various cities have been examined, such as arthropods in Sydney (Gibb and Hochuli 2002), soil isopods in Budapest (Vilisics and Hornung 2009), and reptiles and amphibians in Adelaide (Tait et al. 2005). Urban woodlands are important for faunal species composition and diversity in cities. Large woodlots, high spatial heterogeneity, complex vertical structure, diverse species composition of vegetation, and presence of exotics could be associated with high faunal species richness (Meffert and Dziock 2013). In contrast, the replacement of vegetation by impervious surfaces will reduce animal diversity by the loss of habitable area and the deterioration of remaining vegetation (McKinney 2008).

Diverse urban nature has been an integrated and indispensable component of modern cities. However, urban nature has long been ignored by scientists and systematic studies only started several decades ago (Vilisics and Hornung 2009). Yet, a wealth of knowledge on diverse nature in cities has been produced. Much of the knowledge is from ecological perspectives (Breuste et al. 2008), covering key issues on the ecological characteristics of the biota of cities, such as ecological structure and function of habitats or organisms (e.g. Sandström et al. 2006), biodiversity and conservation (e.g. Kowarik 2011; Savard et al. 2000), invasion and role of exotic species (e.g. Dos Santos et al. 2010), dynamics of urban biota (e.g. Clarke et al. 2013), etc. While current knowledge has provided a basis and useful guidelines for urban planning and management of diverse natural resources within cities (Breuste et al. 2008), an essential step to managing urban nature more effectively and cities more resiliently and sustainably is a fuller understanding of the complex urban ecosystems by integrating social and economic dimensions into mainstream ecological studies (Gómez-Baggethun and Barton 2013). As an emerging, interdisciplinary and unique theme that is increasingly capturing the attention of scholars, policy makers and practitioners, the quantification and economic valuation of urban nature's ecosystem services can serve as a new lens with which to observe changes in our cities and safeguard the well-being of city inhabitants.

9.2 Ecosystem Services of Nature in Cities

Ecosystem services, defined as all benefits that human population obtains, directly or indirectly, from natural ecosystems (Costanza et al. 1997; De Groot et al. 2010), provide a conceptual framework and practical tool for further understanding the human-nature interface in urban areas (Luederitz et al. 2015; Tobias 2013). In 1997, Costanza and others quantified and evaluated a total of 17 ecosystem services from 16 biomes. Unfortunately, urban ecosystem services (those ecosystem services produced by urban ecosystems and their diverse natural components, according to Breuste et al. 2013; Luederitz et al. 2015) were missing in this milestone paper mainly due to the unavailability of relevant data (Costanza et al. 2014). Hitherto, although what constitutes an urban ecosystem and urban ecosystem services remains a question to be answered (Luederitz et al. 2015), scholars have begun to develop a body of literature expounding on important ecosystem services arising from or being appropriated within urban regions. Such studies have advanced our understanding of urban ecosystem services from biophysical, economic, and socio-cultural dimensions (Gómez-Baggethun and Barton 2013; Kremer et al. 2015).

A wide range of urban ecosystem services has been identified (e.g. Gómez-Baggethun and Barton 2013). Even though urban ecosystems are profoundly affected by extensive anthropogenic disturbances, and resultant changes in ecological conditions and functions associated with urbanization (such as soil contamination and sealing, species homogenization, etc.) might lead to the type, quality and quantity of urban ecosystem services to differ strongly from more natural ecosystems (Larondelle and Haase 2013), the classification framework developed by the Millennium Ecosystem Assessment (MEA 2005) and more recently by TEEB (2011) is still useful for studies on urban ecosystem services. This classification system has been widely adopted in diverse empirical studies (e.g. Jim and Chen 2009; Tobias 2013), and is recruited in this article to group urban ecosystem services into four categories: 'provisioning services' (the production of products from ecosystems), 'regulating services' (benefits generated through the regulation of ecosystem functions), 'cultural services' (the nonmaterial benefits humans gain from a contact with ecosystems through spiritual enrichment, cognitive development, reflection, recreation, and aesthetic experiences), and 'supporting services' (the processes that are essential for the production of all provisioning, regulating and cultural services).

9.2.1 Provisioning Services

Even though urban ecosystems mainly rely on energy and materials imported from other ecosystems (Folke et al. 2011), it is not impossible to nurture urban food forests, community vegetable and fruit gardens, and allotment gardens to improve urban sustainability via increased food security (Andersson et al. 2007; Barthel and Isendahl 2013) and landscape multi-functionality (Barthel et al. 2015; Lovell 2010). It is estimated that 15–20% of food is produced within cities or at peri-urban areas globally (Armar-Klemesu 2000). A survey in Sheffield, United Kingdom, indicated that the provision of fruits and vegetables has been considered as one of the major benefits of having a private garden, and 19% and 23% of the respondents grew vegetables and fruits in their private gardens for the taste, aroma and freshness of home-grown products (Dunnett and Qasim 2000). In urban Stockholm, garden holders are urged to grow traditional flowers, fruits, berries and vegetables (Barthel et al. 2010). In New York City, the overwhelming majority of community gardens use their space to produce food (McPhearson et al. 2013). In Seattle, US, fruit-tree species, as a source of food, are applied to private yards, developed parks, vacant lots, street planting strips, and public lands with remnant orchards (McLain et al. 2012). An experiment in Melbourne, Australia, demonstrated that urban rain gardens (which are designed and installed to reduce the volume and frequency of urban runoff) can yield a range of vegetables including beetroot, onion, spinach, tomato and broad bean (Richards et al. 2015). In addition to fruit and vegetable production, other products are increasingly provided, such as medicinal plants, spices, mushrooms, as well as livestock keeping for eggs and milk (Lovell 2010). Similar trends have been observed in developing countries. For example, an urban wetland in Uganda, African was extensively used for crop cultivation, papyrus harvesting, brick making and fish farming (Schuyt 2005). In Qingdao city (Shandong province, China), urban residents used rooftops, residential gardens, street greenbelts for cultivating vegetables, herbs, and fruits (Wang 2014).

Water scarcity is a severe problem facing many cities throughout the world, particular those with dry climates (Richards et al. 2015). While the generation of stormwater within cities is beginning to be seen as a valuable resource (Berndtsson 2010), urban waterbodies (Lundy and Wade 2011), urban vegetation and associated soil's capability of stormwater retention, storage and groundwater recharge (Speak et al. 2013) would contribute to meeting local water needs. However, how urban nature might affect the quantity of available water and secure water supply for drinking and other human uses hitherto received little scholarly attention.

The supply of genetic information is another provisioning service (MEA 2005). For example, veteran trees serve as living specimens and seed banks and a gene pool for enhancing biological diversity in urban landscapes and aid in the conservation of threatened species (Lin et al. 2015). Different species able to tolerate highly polluted urban habitats also offer interesting opportunities for bioremediation of these habitats (Lundy and Wade 2011). Unfortunately, there has been little empirical investigation on this particular provisioning service in urban areas.

9.2.2 *Regulating Services*

Regulating services are the most commonly examined category of ecosystem service in urban areas, which might be attributed to the availability of primary and secondary data to quantify and evaluate these services (Luederitz et al. 2015). This can also be attributed to the intimate dependence of urban residents' quality of life on regulating services, such as via the moderation of micro/meso-environmental conditions and quality that have usually been severely degraded in urban settings.

9.2.2.1 Air Quality Regulation

The abatement of atmospheric pollution is one of the key regulating services provided by urban vegetation (Nowak et al. 2006), as air pollution is ubiquitous and a major environmental concern in most major cities worldwide (Irga et al. 2015). The capacity of urban vegetation to regulate urban air quality is through two mechanisms. Firstly, plants can intercept and accumulate atmospheric particles through leaf pubescence and by providing large rough surfaces (such as branches, twigs and foliage) on which dry deposition can occur (Irga et al. 2015). Various gaseous pollutants (such as CO, NO, NO_2, O_3, HNO_3 and SO_2) can be absorbed by urban plants through the stomata and then diffuse into intercellular spaces. Urban plants thus serve as significant sinks for gaseous pollutants (Nowak et al. 2014). It was estimated that via dry deposition, urban trees in 55 US cities can remove 711,000 tonnes of air pollutants (including CO, NO_2, SO_2 and PM_{10}) annually (Nowak et al. 2006). Secondly, urban vegetation can significantly mitigate the urban heat island and cool down the ambient temperature. Reduction in energy use leads to decreased emission of various air pollutants from power plants (Akbari 2002) and the lower temperature slows the smog formation process (Nowak et al. 2014), hence improving urban air quality indirectly. For instance, it was found that the cooling effect of urban greening in Stuttgart, Germany, can decrease ozone concentrations by 5–8% on average (Fallmann et al. 2016).

The extent of air pollutant removal in a city is dependent on plant species (such as canopy structure, leaf area, morphology and biomass), planting density, amount of vegetation, pollutant concentrations, atmospheric precipitation and other meteorological factors affecting tree transpiration and the deposition velocity of air pollutants (Setälä et al. 2013).

9.2.2.2 Climate Regulation

It has been widely recognized that urban nature, mainly urban vegetation together with soils, contributes to regulating climate from microscale, mesoscale to macroscale (Chen 2015). At the microscale, trees close to building affect local microclimate through reducing the penetration of solar radiation and wind shielding (Akbari 2002), thereby altering the heat exchange between buildings and their surroundings (Wang et al. 2016). Trees and other vegetation also absorb latent heat from ambient atmosphere via evapotranspiration (Hedquist and Brazel 2014; Jim and Chen 2009). Due to the combined effect of shade and evapotranspiration, air temperature reductions of 1–3 °C can be achieved under the canopy in urban green areas, depending on the season, climate, and soil conditions (Duarte et al. 2015). A recent empirical study demonstrated that an urban park (Kensington Gardens) in London can lower ambient air temperature by 1.1 °C on average, with a maximum of 4 °C cooling, over a 440 m distance from the park (Doick et al. 2014).

At the mesoscale (city/regional level), vegetation cover would influence incoming solar radiation, relative humidity, surface roughness and albedo, and heights of boundary-layer, hence leading to changes in various aspects of regional meteorology (Cavan et al. 2014). When vegetation is applied throughout the city, modifying the albedo in process, the energy balance is also modified, which then produces citywide changes in climate (Duarte et al. 2015). An increase of green area within a city may reduce temperatures significantly at the city scale (Norton et al. 2015). In Berlin (Germany), it was demonstrated that urban greening can effectively mitigate heat waves in summer (Schubert and Grossman-Clarke 2013).

At the global level, urban vegetation and soils have the potential to mitigate carbon dioxide (CO_2) emission, one of the major greenhouse gases that cause global warming. Urban plants can reduce atmospheric CO_2 directly through photosynthesis during daytime and storing carbon as biomass in the form of stems, branches or roots (Nowak et al. 2013; Weissert et al. 2014). Empirical evidence also suggests the process of carbon sequestration and storage by urban plants can be more effective than plants in natural environments. This can be attributed to many factors, such as regular irrigation and fertilization, lack of natural parasites and enemies, wide spacing, higher temperatures, etc., which facilitate the growth of urban vegetation and higher rates of photosynthesis compared to plants in natural environments (Niinemets and Peñuelas 2008). It is estimated that urban trees in the 50 US states store a total of 643.2 million tons of carbon with an annual gross sequestration rate at 25.6 million tons (Nowak et al. 2013). Urban vegetation of 35 major Chinese cities stored a total of 18.7 million tons of carbon and carbon sequestration totaled 1.90 million tonnes per year (Chen 2015). Similarly, urban soils also act as passive carbon sinks (Weissert et al. 2014), an area which is now increasingly attracting scholarly attention. Urban vegetation can also reduce power demand for heating and cooling buildings, thus indirectly reducing CO_2 emissions related to the production of electric power (Akbari 2002; Donovan and Butry 2009).

In addition to urban vegetation and associated soils, other natural landscape elements in cities can also generate climate-regulating service. For example, water bodies in the city can act as heat sinks and help even out temperature deviations (Lundy and Wade 2011).

9.2.2.3 Water Regulation

Increase of impervious surface in cities during the process of urbanization often results in enhanced hydraulic efficiency, which substantially reduces the capacity for rainwater infiltration and lead to a concomitant increase in runoff generation and incidence and severity of flooding (Armson et al. 2013; Liu et al. 2015). Vegetation and associated soils have strong influence on urban hydrology (Mullaney et al. 2015). Plant canopies and stems intercept rainfall (Livesley et al. 2014), allow for evapotranspiration and groundwater recharge, and enhance storage and infiltration into the root and soil zones (Inkiläinen et al. 2013). All these processes help to reduce and attenuate surface water volumes and velocities, thereby reducing the occurrence of urban floods.

On an individual tree basis, it was estimated that annual reductions in stormwater runoff volume range from 3.2 to 11.3 m^3 per tree (Mullaney et al. 2015). Trees with higher plant area index (canopy density) could intercept a greater amount of gross rain fall (Livesley et al. 2014). Mature evergreen trees can intercept more than 4000 gallons per year (Cappiella et al. 2005). Collectively, it was found that in Manchester, trees and their associated tree pits in 9 m^2 plots reduced runoff from asphalt by as much as 62%, whereas grass almost totally eliminated surface runoff (Armson et al. 2013). In Raleigh, North Carolina, USA, urban residential forests with varying vegetation structure could potentially reduce stormwater runoff by 9.1–21.4% (Inkiläinen et al. 2013). At the city scale, it was estimated that a total of 97.9 million m^3 of excess surface runoff was retained by urban green space (22,998 ha, consisting of 64.6% tree canopy, 32.8% lawns, and 2.6% farmland) in Beijing, China (Yao et al. 2015). Although stormwater runoff reduction is influenced by many factors, such as amount of rainfall, soil type and condition, and urban morphology (Liu et al. 2015), urban vegetation and underlying soils regulate significantly and measurably stormwater runoff rates and volumes in urban catchments (Ossola et al. 2015).

9.2.2.4 Erosion Regulation

Although urban soils are mostly sealed, suggesting that soil erosion does not easily occur, exposed soils on steeply slopes are highly susceptible to erosion. In addition, bank erosion can be commonly observed in urban streams due to the clearance of riparian vegetation and replacement of deep-rooted vegetation. Vegetation strongly controls the frequency and magnitude of erosion. Vegetation communities affect soil properties directly through several mechanisms. The net rainfall and raindrop energy can be reduced by canopy interception and thus prevents some of the ground and soil displacement that cause erosion. The leaf litter underneath plants increases surface roughness and also serves as a sponge for the water. Trees can also absorb water in the soil by root uptake. Vegetation root can also enhance soil cohesion (Vanacker et al. 2014). Together, vegetation, roots and leaf litter stabilize soil and reduce erosion. For example, some weeds, such as Creeping Charlie (*Glechoma hederacea*), are kept by residents in their yards as low maintenance ground cover to prevent soil erosion in the Saint Paul-Minneapolis metropolitan area of Minnesota, USA (Dahmus and Nelson 2014). In Chongqing, China, communities of tall grass (*Saccharum spontaneum*) and trees (such as *Pterocarya stenoptera*) are planted at urban stream banks to control erosion (Xian et al. 2015).

9.2.2.5 Water Purification

All natural elements in cities can contribute to reducing the levels of pollutants in rainfall and surface runoff. Initially the removal of pollutants from rainfall depends upon the interception, adsorption and absorption of pollutants, by plant roots, soils, and associated microbial communities, which reduce the amount of harmful substances reaching ground or surface waters (Yang et al. 2015). Most pollutants absorbed by plants can be transformed into non-harmful forms. After rainfall, the movement of water through soils improves water quality via sedimentation, transformation/decomposition of persistent organic pollutants, sequestration and conversion of inorganic ions, and removal of disease-causing microbes (Jansson 2013). The observations in Yixing city, China, indicated that on average 17.29 kg total nitrogen, 3.10 kg ammoniacal nitrogen (NH_4–N), and 0.40 kg total phosphorus can be removed by a hectare of urban woodlands annually (Yang et al. 2015).

Moreover, improved urban watercourses and floodplain habitats would contribute to the physico-chemical purification of water and waste substances (via processes such as dilution, assimilation and chemical decomposition), as demonstrated by the restoration of Mayes Brook in east London, UK (Everard and Moggridge 2012). However, the interrelationship between vegetation, soils, urban rivers, and water quality in urban environs is still poorly understood.

9.2.2.6 Pest and Disease Regulation

Pests and pathogens have been major threats to not only urban ecosystem functioning, but also natural amenity and human health (Boyd et al. 2013; Tomlinson et al. 2015). Urban plants, especially where there is low diversity of species, are particularly vulnerable (Laćan and McBride 2008). Plants and insects in urban environments interact in a complex matrix. While it is argued that urban environments create opportunities for pests (and thus susceptibility to disease) to increase due to changes in host quality, natural enemy abundance and diversity, as well as microhabitats that may disrupt movement and colonization of pests and natural enemies (Raupp et al. 2010), empirical evidence suggests that diversifying plant species and structural complexity can provide favourable microhabitats and refuge from predators, as well as food sources for natural enemies to regulate pest population and function against pest and disease outbreaks (Lin et al. 2015). For example, less-well-manicured ponds in Manchester (UK) allotment gardens provide a habitat for frogs, which help control garden pest populations (Speak et al. 2015). In Stockholm, Sweden, the protection and improvement of habitats for insectivorous birds can be commonly found in allotment gardens, which can increase abundance of bird species and support pest regulation (Barthel et al. 2010).

9.2.2.7 Pollination

Urban green spaces, functioning as nodes of various sizes within larger ecological networks with abundant floral species, can promote pollinators' nesting and dispersal, and support a diverse assemblage of bees, butterflies, insects, birds and other pollinators (Lowenstein et al. 2015). Private and public gardens, which are usually designed to ensure abundant flowering (Verboven et al. 2014) and wide variety of colours as a result of ornamental planting (Radford and James 2013), could sustain the richness and diversity of pollinators and thus provide successful pollination service for the whole urban ecosystems (Biesmeijer et al. 2006). This condition can sometimes even spill over to surrounding areas (Barthel et al. 2010; Andersson et al. 2014). It is found that wild bees in urban habitats provide adequate pollination service to urban agriculture (in allotment and community gardens) in San Francisco, CA, (Potter and LeBuhn 2015). In reverse, in addition to pollen and nectar sources from vegetables and ornamental flowers grown on allotment gardens, spontaneous plant species found in allotment gardens attract various pollinators (Speak et al. 2015).

9.2.2.8 Natural Hazard Regulation

Cities, due to their high population concentration and intense economic activities, are extremely vulnerable to threats from natural hazards such as heat waves, floods, earthquakes and hurricanes, especially for those located in or near floodplains, earthquake fault zones, and hurricane-prone shorelines (Godschalk 2003). Fortunately, urban natural systems can provide valuable hazard mitigation functions. For example, empirical evidence has demonstrated that urban green spaces could cool down urban temperatures (see Sect. 9.2.2.2) and thus mitigate the impacts of heat waves. A comparison between open-site and below-canopy climatic conditions in Switzerland during a heat wave in 2003 suggested that the maximum temperatures were cooler under the canopy and deciduous and mixed forests can function better than coniferous forests (Renaud and Rebetez 2009). Urban waterbodies can also attenuate extreme temperatures (Everard and Moggtidge 2012). Additionally, urban rivers and wetlands, as well as urban forests (plants and associated soils) are capable of reducing surface runoff and thus prevent and reduce flooding (see Sect. 9.2.2.3). Mangroves and coral reefs can act as natural barriers that protect coastal cities from hurricanes and tsunamis (Gómez-Baggethun and Barton 2013). However, our understanding of urban nature's mitigation of various natural hazards is far from complete and the interpretation of this evidence in terms of ecosystem services is only beginning to emerge.

9.2.3 Cultural Services

Cultural services in urban contexts remain poorly explored (La Rosa et al. 2016), although some cultural services such as recreation, aesthetics and ecotourism are growing in importance to human well-being and are most frequently emphasized in urban planning and policy-making (Maraja et al. 2016). Increasingly, cities are places where most people have the most contact with nature. Various natural elements play an important role as providers of aesthetic, cultural, psychological and other non-material benefits, contributing in particular to human health and well-being.

Recreation and aesthetic benefits are amongst the highest valued type of cultural services (La Rosa et al. 2016). Urban green spaces (such as parks, gardens, greenbelts, rooftop gardens, vegetated streetscapes) and waterbodies (such as lakes, ponds and rivers) serve as spaces where a wide range of recreational and leisure activities, such as walking, jogging, cycling, picnicking, and sporting, can be pursued in order to relax and reduce stress, escape from city, and enjoy peacefulness and tranquility (Völker and Kistemann 2015). In contrast with grey and monotonous urban buildings, natural components may also improve the scenic quality of city neighborhoods, provide privacy, shelter residents from the negative effects of undesirable land uses, and thus confer aesthetic benefits to urban inhabitants (Panagopoulos et al. 2016).

Urban nature's cultural heritage, inspirational, spiritual/religious, and educational benefits have been mentioned in several studies (e.g. Gómez-Baggethun and Barton 2013). The peaceful and tranquil atmosphere and beautiful scenery of urban green spaces inspire reflection, meditation, and a general feeling of harmony between one self and the surrounding. Strong emotional bonds to gardens have been detected in Stockholm, leading to the development of sense of place and environmental stewardship (Andersson et al. 2007). Waterbodies act as an important element of emotional attachment to the place, as it evokes strong emotions and awakens 'creative' or 'spiritual' thoughts (Völker and Kistemann 2015). By providing meeting places where users develop and maintain emotional ties to local communities, urban natural spaces can enhance social cohesion (Kaźmierczak 2013). For example, it is found that allotment gardens can bring together people of different background but with a shared interest in gardening and thus promote social cohesion (Speak et al. 2015). Urban community gardens can also serve as 'pockets' of social–ecological memory in urban landscapes, and thus act as a cultural heritage for local residents (Barthel et al. 2015). Personal exposure to nature in everyday life plays major role in educating urban population about nature and environmental issues (Savard et al. 2000). However, these cultural services have received much less attention and only piecemeal evidence is available in the literature.

9.2.4 Supporting Services

According to MEA (2005), supporting services, including primary production, soil formation, water cycling and habitat provision, differ from the other categories of services in that their impacts on humans are indirect and/or occur over very long periods of time. In comparison with other categories of ecosystem services provided by urban nature, supporting services were investigated the least often, despite their critical importance and necessity to the generation of other types of services. Nevertheless, several empirical studies have estimated the primary productivity of urban green spaces, mainly with regards to the influence on carbon cycle (Jim and Chen 2009; Nowak et al. 2013). Moreover, cities as critical habitats for wildlife have been documented and they can serve as hotspots for nature conservation (O'Farrell et al. 2012), which in turn is associated with the provision of ecosystem services (Kowarik 2011). Other supporting services, such as the magnitude of terrestrial and aquatic vegetation's contribution to oxygen levels within the atmosphere and water bodies, as well as soil formation and maintenance, have yet to be quantified on a city scale.

9.2.5 Urban Ecosystem Disservices

Alongside various ecosystem services, urban ecosystems also generate some effects that are perceived as harmful, unpleasant or unwanted by citizenry, which are defined as ecosystem disservices (Lyytimäki 2015). For example, plants might contribute to air pollution in urban areas. Some species emit a considerable amount of biogenic volatile organic compounds (VOCs), which can react with nitrogen oxides (NOx) to form ozone, secondary organic aerosol and particulate matters (Curtis et al. 2014). In addition, in the flowering season, pollens might increase particulate concentration and induce allergenic symptoms (Mücke et al. 2014). It was found that roadside urban vegetation might reduce the ventilation of street canyons, and thus lead to increased pollutant concentrations (Salmond et al. 2013).

Other examples of urban ecosystem disservices include the increase of water/energy use related to urban plantings (Pataki et al. 2011), introduction of invasive plants (Escobedo et al. 2011), inhibition of human mobility and safety (Lyytimäki 2015), host pathogens or pests (Tomlinson et al. 2015), and risk of diseases transmitted by animals (Gómez-Baggethun and Barton 2013). A latest review of urban ecosystem disservices was given by von Döhren and Haase (2015).

9.3 Challenges for the Theorization and Governance of Urban Ecosystem Services

In comparison with other landscapes, urban ecosystem services/disservices are still rarely discussed and the theoretical foundation is less well developed. Whilst there is mounting well-recognized evidence indicating that urban nature, in particular urban vegetation, can offer a wide range of ecosystem services for urban residents (Dobbs et al. 2014; Irga et al. 2015), there are still major knowledge gaps and challenges which lay ahead in the theorization and governance of urban ecosystem services to achieve urban sustainability. These are discussed below.

9.3.1 Overlooked Dimensions

While ecosystem services provided by urban vegetation (in particular urban trees) have been extensively investigated, information concerning ecosystem services and values of urban fauna, soils and waterbodies is currently scarce. For example, there is missing linkage between urban soils and the provision of ecosystem services. Urban soils potentially provide the same ecosystem services as in wild environments, even though they usually experience serious depletion of basic functions (Ajmone-Marsan et al. 2016). Additionally, urban water spaces are amongst citizen's most preferred natural element, contributing significantly to the delivery of a full spectrum of ecosystem services (Lundy and Wade 2011; Völker and Kistemann 2015). However, there are important knowledge deficits with respect to ecosystem services rendered by urban waterbodies, and integrating urban plants, soils, waterbodies and grey infrastructure is necessary for facilitating the provision of multiple ecosystem services and developing nature-based solutions (Haase 2015).

9.3.2 Linkages Between Urban Biodiversity and Provision of Ecosystem Services/Disservices

Linkages between urban diversity and the provision of specific ecosystem services, and interactions of social, economic, institutional, ecological, and environmental subsystems within cities at various structural and functional levels need to be established in order to quantify, monitor and understand the contribution of individual species and overall biodiversity to the provision of ecosystem services and social welfare. However, quantitative relationships between terrestrial and aquatic biodiversity in urban areas, ecosystem functioning and the provision of ecosystem services/disservices are still poorly investigated. Although anecdotal evidence suggests that high species diversity might be helpful to optimize multiple ecosystem services (Morgenroth et al. 2016; Speak et al. 2015), the variation in ecological structure and the complexity of ecosystem functioning pose uncertainties about the role of biodiversity in the provision of ecosystem services/disservices, specifically within coupled social ecological systems (Gonzalea-Redin et al. 2016).

9.3.3 Assessment of Urban Ecosystem Services/Disservices

The assessment of ecosystem services has been a challenging issue (Lauf et al. 2014). Diverse and somewhat fragmented approaches for ecosystem service assessment are available in the literature. However, there is a lack of the standardized procedures to quantifying and evaluating urban ecosystem services/disservices at the various scales. Inconsistent research results derived via varying approaches has limited the inter-city, regional or national transferability and comparability of our knowledge accumulated and lesson learnt (Luederitz et al. 2015), which has hindered progress in utilizing ecosystem services in urban planning.

9.3.4 Synergies and Trade-Offs

Although not yet widely employed in practices of urban planning and environmental governance, the understanding of urban ecosystem services/disservices can provide an opportunity to develop land use plan for sustainable urban ecosystems and urban society (Lauf et al. 2014; Panagopoulos et al. 2016; von Döhren and Haase 2015). Identifying desirable ecosystem services and undesirable disservices (Pataki et al. 2011), as well as taking into account the full repertoire of both ecosystem services and disservices in urban regions are vital for preventing and solving controversies related to environmental management and planning (von Döhren and Haase 2015). Detecting trade-offs and synergies of competitive ecosystem services/disservices associated with particular urban designs will be necessary for informing decisions (Holt et al. 2015).

9.4 Concluding Remarks

While numerous studies show that urban dwellers depend on the productive and assimilative capacities of ecosystems well beyond city boundaries (e.g. Folke et al. 2011), increasing attention has been drawn to the potential for cities, where most people live and work, to remediate some of their own environmental impacts (Kowarik 2011) and to reduce reliance on far-flung natural resource imports (Pincetl 2012). How to understand diverse urban nature and their ecosystem services, and integrate this understanding into urban planning to ensure the provision of ecosystem services to their urban dweller in the present and future (Dobbs et al. 2014), and concurrently achieve sustainable development of cities where human society, economic activities and nature are tightly linked together, are front and center issues. They present a fertile ground for further research, and the challenges identified above will help to catalyze directed research in this burgeoning and important field.

References

Ajmone-Marsan F, Certini G, Scalenghe R (2016) Describing urban soils through a faceted system ensures more informed decision-making. Land Use Policy 51:109–119

Akbari H (2002) Shade trees reduce building energy use and CO_2 emissions from power plants. Environ Pollut 116:S119–S126

Andersson E, Barthel S, Ahrné K (2007) Measuring social-ecological dynamics behind the generation of ecosystem services. Ecol Appl 17(5):1267–1278

Andersson E, Barthel S, Borgström Sm Colding J, Elmqvist T, Folke C, Gren Å (2014) Reconnecting cities to the biosphere: stewardship of green infrastructure and urban ecosystem services. Ambio 43:445–453

Armar-Klemesu M (2000) Urban agriculture and food security, nutrition and health. In: Bakker N, Dubbeling M, Gundel S, Sabel-Koschella U, de Zeeuw H (eds) Growing cities, growing food: urban agriculture on the policy agenda. GTZ/DSE, Germany

Armson D, Stringer P, Ennos AR (2013) The effect of street trees and amenity grass on urban surface water runoff in Manchester, UK. Urban For Urban Greening 12:282–286

Baldwin AH (2004) Restoring complex vegetation in urban settings: the case of tidal freshwater marshes. Urban Ecosyst 7(2):125–137

Barthel S, Isendahl C (2013) Urban gardens, agricultures and waters management: sources of resilience for long-term food security in cities. Ecol Econ 86:215–225

Barthel S, Folke C, Colding J (2010) Social–ecological memory in urban gardens—retaining the capacity for management of ecosystem services. Glob Environ Change 20(2):255–265

Barthel S, Parker J, Ernstson H (2015) Food and green space in cities: a resilience lens on gardens and urban environmental movements. Urban Stud 52:1321–1338

Berndtsson JC (2010) Green roof performance towards management of runoff water quantity and quality: a review. Ecol Eng 36:351–360

Biesmeijer JC, Roberts SPM, Reemer M, Ohlemüller R, Edwards M, Peeters T, Schaffers AP, Potts SG, Kleukers R, Thomas CD, Settele J, Kunin WE (2006) Parallel declines in pollinators and insect-pollinated plants in Britain and the Netherlands. Science 313:351–354

Boyd IL, Freer-Smith PH, Gilligan CA, Godfray HCJ (2013) The consequence of tree pests and diseases for ecosystem services. Science 342:1235773

Breuste J, Niemelä J, Snep RPH (2008) Applying landscape ecological principles in urban environments. Landscape Ecol 23:1139–1142

Breuste J, Qureshi S, Li J (2013) Applied urban ecology for sustainable urban environment. Urban Ecosyst 16:675–680

Cappiella K, Wright T, Schueler T (2005) Urban forestry watershed manual—Part1: methods for increasing forest cover in a watershed. US Department of Agriculture

Cavan G, Lindley S, Jalayer F, Yeshitela K, Pauleit S, Renner F, Gill S, Capuano P, Nebebe A, Woldegerima T, Kibassa D (2014) Urban morphological determinants of temperature regulating ecosystem services in two African cities. Ecol Ind 42:43–57

Chen WY (2015) The role of urban green infrastructure in offsetting carbon emissions in 35 major Chines cities: a nationwide estimate. Cities 44:112–120

Clarke LW, Jenerette GD, Davila A (2013) The luxury of vegetation and the legacy of tree biodiversity in Los Angeles, CA. Landscape Urban Plann 116:48–59

Colding J, Folke C (2009) The role of golf courses in biodiversity conservation and ecosystem management. Ecosystems 12:191–206

Costanza R, d'Arge R, de Groot R, Farber S, Grasso M, Hannon B, Limburg K, Naeem S, O'Neill RV, Paruelo J, Raskin RG, Sutton P, van den Belt M (1997) The value of the world's ecosystem services and natural capital. Nature 387:253–260

Costanza R, de Groot R, Sutton P, van der Ploeg S, Anderson SJ, Kubiszewski I, Farber S, Turner RK (2014) Changes in the global value of ecosystem services. Glob Environ Change 26:152–158

Curtis AJ, Helmig D, Baroch C, Daly R, Davis S (2014) Biogenic volatile organic compound emissions from nine tree species used in an urban tree-planting program. Atmos Environ 95:634–643

Dahmus ME, Nelson KC (2014) Yard stories: examining residents' conceptions of their yards as part of the urban ecosystem in Minnesota. Urban Ecosyst 17(1):173–194

de Groot RS, Alkemade R, Braat L, Hein L, Willemen L (2010) Challenges in integrating the concept of ecosystem services and values in landscape planning, management and decision making. Ecol Complex 7:260–272

Dobbs C, Kendal D, Nitschke CR (2014) Multiple ecosystem services and disservices of urban forest establishing their connections with landscape structure and sociodemographics. Ecol Ind 43:4–55

Doick KJ, Peace A, Hutchings TR (2014) The role of one large greenspace in mitigating London's nocturnal urban heat island. Sci Total Environ 493:662–671

Donovan GH, Butry DT (2009) The value of shade: estimating the effect of urban trees on summertime electricity use. Energy Build 41:662–668

Dos Santos AR, da Rocha CFD, Bergallo HG (2010) Native and exotic species in the urban landscape of the city of Rio de Janeiro, Brazil: density, richness, and arboreal deficit. Urban Ecosyst 13(2):209–222

Duarte DHS, Shinzato P, dos Santos Gusson C, Abrahão Alves C (2015) The impact of vegetation on urban microclimate to counterbalance built density in a subtropical changing climate. Urban Clim 14:224–239
Dunnett N, Qasim M (2000) Perceived benefits to human well-being of urban gardens. HortTechnology 10(1):40–45
Dvorak B, Volder A (2010) Green roof vegetation for North American ecoregions: a literature review. Landscape Urban Plann 96(4):197–213
Escobedo FJ, Kroeger T, Wagner JE (2011) Urban forests and pollution mitigation: analyzing ecosystem services and disservices. Environ Pollut 159(8):2078–2087
Everard M, Moggtidge HL (2012) Rediscovering the value of urban rivers. Urban Ecosyst 15:293–314
Fallmann J, Forkel R, Emeis S (2016) Secondary effects of urban heat island mitigation measures on air quality. Atmos Environ 125:199–211
Folke C, Jansson Å, Rockström J, Olsson P, Carpenter SR, Chapin FS III, Crépin A-S, Daily G, Danell K, Ebbesson J, Elmqvist T, Galaz V, Moberg F, Nilsson M, Österblom H, Ostrom E, Persson Å, Peterson G, Polasky S, Steffen W, Walker B, Westley F (2011) Reconnecting to the biosphere. Ambio 40(7):719–738
Garden JG, McAlpine CA, Peterson A, Jones D, Possingham H (2006) Review of the ecology of Australian urban fauna: a focus on spatially explicit processes. Austral Ecol 31:126–148
Godefroid S (2001) Temporal analysis of the Brussels flora as indicator for changing environmental quality. Landscape Urban Plann 52:203–224
Godschalk DR (2003) Urban hazard mitigation: creating resilient cities. Nat Hazards Rev 4 (3):136–143
Gómez-Baggethun E, Barton DN (2013) Classifying and valuing ecosystem services for urban planning. Ecol Econ 86:235–245
Gonzalea-Redin J, Luque S, Poggio L, Smith R, Gimona A (2016) Spatial Bayesian belief networks as a planning decision tool for mapping ecosystem services trade-offs on forested landscapes. Environ Res 144:15–26
Gibb H, Hochuli DF (2002) Habitat fragmentation in an urban environment: large and small fragments support different arthropod assemblages. Biol Conserv 106:91–100
Haase D (2015) Reflections about blue ecosystem services in cities. Sustain Water Qual Ecol 5:77–83
Hedquist BC, Brazel AJ (2014) Seasonal variability of temperatures and outdoor human comfort in Phoenix, Arizona, USA. Build Environ 72:377–388
Holt AR, Mears M, Maltby L, Warren P (2015) Understanding spatial patterns in the production of multiple urban ecosystem services. Ecosyst Serv 16:33–46
Inkiläinen ENM, McHale MR, Blank GB, James AL, Nikinmaa E (2013) The role of the residential urban forest in regulating throughfall: a case study in Raleigh, North Carolina, USA. Landscape Urban Plann 119:91–103
Irga PJ, Burchett MD, Torpy FR (2015) Does urban forestry have a quantitative effect on ambient air quality in an urban environment? Atmos Environ 120:173–181
Jansson Å (2013) Reaching for a sustainable, resilient urban future using the lens of ecosystem services. Ecol Econ 86:285–291
Jim CY, Chen WY (2009) Ecosystem services and valuation of urban forests in China. Cities 26 (4):187–194
Jim CY, Chen WY (2011) Bioreceptivity of buildings for spontaneous arboreal flora in compact city environment. Urban For Urban Greening 10(1):19–28
Kaźmierczak A (2013) The contribution of local parks to neighbourhood social ties. Landscape Urban Plann 109:31–44
Kowarik I (2011) Novel urban ecosystems, biodiversity and conservation. Environ Pollut 159:1974–1983
Kremer P, Andersson E, McPhearson T, Elmqvist T (2015) Advancing the frontier of urban ecosystem services research. Ecosyst Serv 12:149–151

La Rosa D, Spyra M, Inostroza L (2016) Indicators of cultural ecosystem services for urban planning: a review. Ecol Ind 61:74–89

Laćan I, McBride JR (2008) Pest vulnerability matrix (PVM): a graphic model for assessing the interaction between tree species diversity and urban forest susceptibility to insects and diseases. Urban For Urban Greening 7(4):291–300

Larondelle N, Haase D (2013) Urban ecosystem services assessment along a rural–urban gradient: a cross-analysis of European cities. Ecol Ind 29:179–190

Lauf S, Haase D, Kleinschmit B (2014) Linkages between ecosystem services provisioning, urban growth and shrinkage—a modeling approach assessing ecosystem service trade-offs. Ecol Ind 42:73–94

Lin BB, Philpott SM, Jha S (2015) The future of urban agriculture and biodiversity-ecosystem services: challenges and next steps. Basic Appl Ecol 16:189–201

Liu W, Chen W, Peng C (2015) Influences of setting sizes and combination of green infrastructures on community's stormwater runoff reduction. Ecol Model 318:236–244

Livesley SJ, Baudinette B, Glover D (2014) Rainfall interception and stem flow by eucalypt street trees-the impacts of canopy density and bark type. Urban For Urban Greening 13:192–197

Lovell ST (2010) Multifunctional urban agriculture for sustainable land use planning in the United States. Sustainability 2(8):2499–2522

Lowenstein DM, Matteson KC, Minor ES (2015) Diversity of wild bees supports pollination services in an urbanized landscape. Oecologia 179:811–821

Luederitz C, Brink E, Gralla F, Hermelingmeier V, Meyer M, Niven L, Panzer L, Partelow S, Rau A-L, Sasaki R, Abson DJ, Lang DJ, Wamsler C, von Wehrden H (2015) A review of urban ecosystem services: six key challenges for future research. Ecosyst Serv 14:98–112

Lundholm JT, Marlin A (2006) Habitat origins and microhabitat preferences of urban plant species. Urban Ecosyst 9(3):139–159

Lundholm JT, Richardson PJ (2010) Habitat analogues for reconciliation ecology in urban and industrial environments. J Appl Ecol 47:966–975

Lundy L, Wade R (2011) Integrating sciences to sustain urban ecosystem services. Prog Phys Geogr 35(5):653–669

Lyytimäki J (2015) Ecosystem disservices: embrace the catchword. Ecosyst Serv 12:136

Maraja R, Jan B, Teja T (2016) Perceptions of cultural ecosystem services from urban green. Ecosyst Serv 17:33–39

McIntyre NE (2000) Ecology of urban arthropods: a review and a call to action. Ann Entomol Soc Am 93(4):825–835

McKinney ML (2006) Urbanization as a major cause of biotic homogenization. Biol Conserv 127:247–260

McKinney ML (2008) Effects of urbanization on species richness: a review of plants and animals. Urban Ecosyst 11:161–176

McLain R, Poe M, Hurley PT, Lecompte-Mastenbrook J, Emery MR (2012) Producing edible landscapes in Seattle's urban forest. Urban For Urban Greening 11:187–194

McPhearson T, Kremer P, Hamstead ZA (2013) Mapping ecosystem services in New York City: applying a social-ecological approach in urban vacant land. Ecosyst Serv 5:e11–e26

MEA—Millennium Ecosystem Assessment (2005) Ecosystems and human well-being synthesis. Island Press, Washington, DC

Meffert PJ, Dziock F (2013) The influence of urbanisation on diversity and trait composition of birds. Landscape Ecol 28(5):943–957

Morgenroth J, Östberg J, van den Bosch CK, Nielsen AB, Hauer R, Sjöman H, Chen W, Jansson M (2016) Urban tree diversity-taking stock and looking ahead. Urban For Urban Greening 15:1–5

Mücke H-G, Wagener S, Werchan M, Bergmann K-C (2014) Measurements of particulate matter and pollen in the city of Berlin. Urban Climate 10:621–629

Mullaney J, Lucke T, Trueman SJ (2015) A review of benefits and challenges in growing street trees in paved urban environments. Landscape Urban Plan 134:157–166

Niinemets Ü, Peñuelas J (2008) Gardening and urban landscaping: significant players in global change. Trends Plant Sci 13(2):60–65
Norton BA, Coutts AM, Livesley SJ, Harris RJ, Hunter AM, Williams NS (2015) Planning for cooler cities: a framework to prioritise green infrastructure to mitigate high temperatures in urban landscapes. Landscape Urban Plann 134:127–138
Nowak DJ, Crane DE, Stevens JC (2006) Air pollution removal by urban trees and shrubs in the United States. Urban For Urban Greening 4:115–123
Nowak DJ, Greenfield EJ, Hoehn RE, Lapoint E (2013) Carbon storage and sequestration by trees in urban and community areas of the United States. Environ Pollut 178:229–236
Nowak DJ, Hirabayashi S, Bodine A, Greenfield E (2014) Tree and forest effects on air quality and human health in the United States. Environ Pollut 193:119–129
O'Farrell PJ, Anderson PM, Le Maitre DC, Holmes PM (2012) Insights and opportunities offered by a rapid ecosystem service assessment in promoting a conservation agenda in an urban biodiversity hotspot. Ecol Soc 17(3):27
Ortega-Álvarez R, MacGregor-Fors I (2009) Living in the big city: effects of urban land-use on bird community structure, diversity, and composition. Landscape Urban Plann 90(3):189–195
Ossola A, Hahs AK, Livesley SJ (2015) Habitat complexity influences fine scale hydrological processes and the incidence of stromwater runoff in managed urban ecosystems. J Environ Manage 159:1–10
Panagopoulos T, González Duque JA, Bostenaru Dan M (2016) Urban planning with respect to environmental quality and human well-being. Environ Pollut 208:137–144
Pataki DE, Carreiro MM, Cherrier J, Grulke NE, Jennings V, Pincetl S, Pouyat RV, Whitlow TH, Zipperer WC (2011) Coupling biogeochemical cycles in urban environments: ecosystem services, green solutions, and misconceptions. Front Ecol Environ 9:27–36
Pincetl S (2012) Nature, urban development and sustainability-what new elements are needed for a more comprehensive understanding? Cities 29:S32–S37
Potter A, LeBuhn G (2015) Pollination service to urban agriculture in San Francisco, CA. Urban Ecosyst 18:885–893
Radford KG, James P (2013) Changes in the value of ecosystem services along a rural–urban gradient: a case study of Greater Manchester, UK. Landscape Urban Plann 109(1):117–127
Raupp MJ, Shrewsbury PM, Herms DA (2010) Ecology of herbivorous arthropods in urban landscapes. Annu Rev Entomol 55:19–38
Renaud V, Rebetez M (2009) Comparison between open-site and below-canopy climatic conditions in Switzerland during the exceptionally hot summer of 2003. Agric For Meteorol 149(5):873–880
Richards PJ, Farrell C, Tom M, Williams NSG, Fletcher TD (2015) Vegetable raingardens can produce food and reduce stromwater runoff. Urban For Urban Greening 14:646–654
Salmond JA, Williams DE, Laing G, Kingham S, Dirks K, Longley I, Henshaw GS (2013) The influence of vegetation on the horizontal and vertical distribution of pollutants in a street canyon. Sci Total Environ 443:287–298
Sandström UG, Angelstam P, Mikusiński G (2006) Ecological diversity of birds in relation to the structure of urban green space. Landscape Urban Plann 77:39–53
Savard J-PL, Clergeau P, Mennechez G (2000) Biodiversity concepts and urban ecosystems. Landscape Urban Plann 48:131–142
Schaefer V (2009) Alien invasions, ecological restoration in cities and the loss of ecological memory. Restor Ecol 17:171–176
Schubert S, Grossman-Clarke S (2013) The influence of green areas and roof albedos on air temperatures during extreme heat events in Berlin, Germany. Moteorogosche Zeitschrift 22:131–143
Schulman A, Peters CA (2008) GIS analysis of urban schoolyard landcover in three US cities. Urban Ecosyst 11(1):65–80
Schuyt KD (2005) Economic consequences of wetland degradation for local populations in Africa. Ecol Econ 53:177–190

Setälä H, Viippola V, Rantalainen AL, Pennanen A, Yli-Pelkonen V (2013) Does urban vegetation mitigate air pollution in northern conditions? Environ Pollut 183:104–112

Speak AF, Rothwell JJ, Lindley SJ, Smith CL (2013) Rainwater runoff retention on an aged intensive green roof. Sci Total Environ 461:28–38

Speak AF, Mizgajski A, Borysiak J (2015) Allotment gardens and parks: provision of ecosystem services with an emphasis on biodiversity. Urban For Urban Greening 14:772–781

Sung CY, Li M-H, Rogers GO, Volder A, Wang Z (2011) Investigating alien plant invasion in urban riparian forests in a hot and semi-arid region. Landscape Urban Plann 100:278–286

Tait CJ, Daniels CB, Hill RS (2005) Changes in species assemblages within the Adelaide metropolitan area, Australia, 1836–2002. Ecol Appl 15:346–359

TEEB (2011) TEEB manual for cities: ecosystem services in urban management. www.teebweb.org

Tobias S (2013) Preserving ecosystem services in urban regions: challenges for planning and best practice examples from Switzerland. Integr Environ Assess Manage 9:243–251

Tomlinson I, Potter C, Bayliss H (2015) Managing tree pests and diseases in urban settings: the case of Oak Processionary Moth in London, 2006–2012. Urban For Urban Greening 14 (2):286–292

Uslu A, Bariş E, Erdoğan E (2009) Ecological concerns over cemeteries. Afr J Agric Res 4 (13):1505–1511

Vanacker V, Bellin N, Molina A, Kubik PW (2014) Erosion regulation as a function of human disturbances to vegetation cover: a conceptual model. Landscape Ecol 29(2):293–309

Venn SJ, Niemelä JK (2004) Ecology in a multidisciplinary study of urban green space: the URGE project. Boreal Environ Res 9:479–489

Verboven HAF, Aertsen W, Brys R, Hermy M (2014) Pollination and seed set of an obligatory outcrossing plant in an urban-peri-urban gradient. Perspect Plant Ecol Evol Syst 16:121–131

Vilisics F, Hornung E (2009) Urban areas as hot-spots for introduced and shelters for native isopod species. Urban Ecosyst 12(3):333–345

Völker S, Kistemann T (2015) Developing the urban blue: comparative health responses to blue and green urban green spaces in Germany. Health Place 35:196–205

von Döhren P, Haase D (2015) Ecosystem disservices research: a review of the state of the art with a focus on cities. Ecol Ind 52:490–497

Wang T (2014) Study on agriculture as pastoral complex in urban residential area. J Qingdao Technol Univ 35(3):76–79

Wang Z-H, Zhao X, Yang J, Song J (2016) Cooling and energy saving potentials of shade trees and urban lawns in a desert city. Appl Energy 161:437–444

Weissert LF, Salmond JA, Schwendenmann L (2014) A review of the current progress in quantifying the potential of urban forests to mitigate urban CO_2 emissions. Urban Clim 8:100–125

Xian X-D, Feng Y-K, Willison JHM, Ai LJ, Wang P, Wu Z-N (2015) Restoring ecosystem services to littoral zones of rivers in the urban core of Chongqing, China. Environ Sci Pollut Res 22:12576–12584

Yang L, Zhang L, Li Y, Wu S (2015) Water-related ecosystem services provided by urban green space: a case study in Yixing city (China). Landscape Urban Plann 136:40–51

Yao L, Chen L, Wei W, Sun R (2015) Potential reduction in urban runoff by green spaces in Beijing: a scenario analysis. Urban Forest Urban Green 14:300–308

Part III
The Forms of Urban Green Spaces

Chapter 10
Blue-Green Infrastructure: New Frontier for Sustainable Urban Stormwater Management

Kuei-Hsien Liao, Shinuo Deng and Puay Yok Tan

Abstract Blue-green infrastructure (BGI) has been recognized as an important tool for sustainable urban stormwater management. BGI is ecosystem-based, relying on biophysical processes, such as detention, storage, infiltration, and biological uptake of pollutants, to manage stormwater quantity and quality. Rain gardens, bioswales, constructed wetlands, retention and detention basins, and green roofs are most commonly used BGI systems. Unlike the single-functioned grey infrastructure, which is the conventional urban drainage system, these landscape systems collectively provide multiple ecosystem services, including flood risk mitigation, water quality treatment, thermal reduction, and urban biodiversity enhancement. In recent years, BGI is increasingly embraced through different initiatives around the world, driven by the urgency to tackle different local challenges, such as water quality standards, water security, increased flood risk, and aquatic ecosystem degradation. Whereas BGI is a relatively new term, the idea and practice are not new. In this chapter, we also showcase four cities—Portland, New York City, Singapore, and Zhenjiang—that are active and progressive in implementing BGI. Although BGI receives increasing attention, mainstreaming BGI remains a challenge today. To promote widespread BGI implementation, future research should focus on case studies on practical BGI experiences to inform strategies for overcoming the barriers to mainstreaming BGI in different cities.

K.-H. Liao (✉) · P.Y. Tan
National University of Singapore, Singapore, Singapore
e-mail: kueihsienl@gmail.com; akiliao@nus.edu.sg

P.Y. Tan
e-mail: akitpy@nus.edu.sg

S. Deng
AECOM Asia Company Limited, Sha Tin, Hong Kong
e-mail: shinuo.deng@gmail.com

P.Y. Tan and C.Y. Jim (eds.), *Greening Cities: Forms and Functions*,
Advances in 21st Century Human Settlements,
DOI 10.1007/978-981-10-4113-6_10

Keywords Blue-green infrastructure · Sustainable stormwater management · Ecosystem-based solutions · Ecosystem services · Low-impact development · Urban drainage

10.1 Introduction

Managing urban stormwater is increasingly challenging as the world sees increasing urbanized areas and extreme storms. Stormwater runoff from the ever expanding impervious surfaces not only exacerbates flood risk but also further degrades the aquatic ecosystem that receives it. The problems and limitations of the conventional management approach to urban stormwater are well recognized. This has led to the emergence of 'green infrastructure' in recent years as a supplement and even an alternative to the existing 'grey infrastructure', which typically consists of roadside drains and sewers.

The term 'green infrastructure' commonly refers to a connected network of multi-functional green and open spaces that provide ecosystem services (Benedict and McMahon 2002). At the local scale, green infrastructure often refers specifically to sustainable stormwater management features that utilize natural processes, e.g., rain gardens, bioswales, constructed wetlands (Rouse and Bunster-Ossa 2013). The term 'blue-green infrastructure' (BGI) is also used, albeit less frequently. In this chapter, we discuss BGI as a particularly type of green infrastructure and define it as a network of landscape systems, which often combines both natural and artificial materials and is purposefully designed and managed to provide stormwater-related ecosystem services.

The essence of BGI as an approach to stormwater management is that it is ecosystem-based, relying on natural processes as opposed to engineering structures. It utilizes biophysical processes, such as detention, storage, infiltration, and biological uptake of pollutants, to manage stormwater quantity and quality. By including both blue and green components, the notion of BGI explicitly emphasizes the fact that aquatic and terrestrial ecosystems are interconneted, and so are water, vegetation, and soil.

While BGI is a relatively new term, the idea and practice are hardly new. It has been promoted and implemented under various terms or programmes, such as Stormwater Best Management Practices (BMPs) and Low Impact Development (LID) in the US, Sustainable Urban Drainage System (SUDS) in UK, Water Sensitive Urban Design (WSUD) in Australia, Low Impact Urban Design and Development (LIUDD) in New Zealand, ABC (Active, Beautiful and Clean) Waters Programme in Singapore, and more recently the Sponge City initiative in China. The reader can refer to Fletcher et al. (2015) for a review of some of these ideas. Here, the landscape systems associated with these programmes are all considered BGI.

In the remainder of this chapter, several common BGI systems are first introduced. We then outline the multiple ecosystem services provided by BGI and review the scientific evidence of these claimed benefits. Next we discuss the drivers behind current BGI implementation and introduce several cities that are in the forefront. Finally, we identify further BGI research agenda to better promote sustainable urban stormwater management.

10.2 Common BGI Systems

In contrast to the conventional urban drainage system, which focuses on efficient removal of stormwater runoff, it has been widely agreed upon that the more sustainable management approach is to tackle stormwater at source. BGI involves a variety of ecosystem-based landscape systems, which are designed to mimic natural hydrology and are implemented individually or in combination to manage the quantity and quality of stormwater runoff on site. It should be noted that what is perceived as sustainable stormwater management also includes measures such as permeable paving, rainwater harvesting cisterns, and underground flood storage tanks; however, here they are not considered BGI systems because they are not ecosystem-based.

10.2.1 Rain Gardens

A rain garden is a vegetated, shallow depression designed to collect and treat stormwater runoff from nearby impervious surfaces. It is also called bioretention basin. Stormwater runoff is treated through filtration, sedimentation, adsorption, and plant and microbial uptake (Lucke and Nichols 2015).

The rain garden mainly consists of the filter media and vegetation, although the detailed design varies in different bio-climatic contexts. The filter media—soil—functions not only to support plant growth but also for water quality treatment. The vegetation also functions for water quality treatment through the biofilms on the roots that adsorb pollutants. It also prevents soil erosion and keeps it porous to avoid clogging (CSIRO 2005). Native species are often used because they adapt well to the local weather and may serve as habitat for local wildlife.

Although a rain garden is often modest in size, three different planting zones should be considered: wet, moist, and dry (Hinman 2007). At the bottom is the wet zone, where the vegetation must tolerate fluctuating water levels and periodic standing water. Along the slope is the moist zone, subject to less frequent water fluctuation, and the vegetation should function for erosion control and tolerate slightly drier soil condition. At the highest elevation is the dry zone, where the soil should be well-drained and hence the vegetation must tolerate extremely dry condition.

To prevent overflow, sometimes underneath the filter media there would also be an under-drainage layer consisting of coarse sand or fine gravel, where perforated drain pipes are embedded. The under-drainage layer is designed to convey treated flows into the perforated drain pipes. However, it is also argued that the under-drainage layer could result in less filtration and poor nitrogen removal because the layer causes the filter media to become mostly aerobic, such that organic nitrogen and ammonium in the stormwater runoff are transformed into nitrate and released in the effluent (Brown and Hunt 2011).

10.2.2 Bioswales

Also a bioretention system, a bioswale is a shallow, vegetated channel used to convey stormwater runoff and treat it prior to its entry into the receiving water body. But in stormwater management a swale is used mainly for conveyance. While the bioswale is capable of treatment (Stagge et al. 2012), it is often not designed to treat the runoff to the degree to meet water quality standards, but only to filter out coarse sediment (CSIRO 2005). The bioswale is often used along the street and in the parking lot.

Similar to the rain garden, the bioswale consists of vegetation, soil, and in some cases also the under-drainage layer. What makes the bioswale different from the rain garden is its long and linear shape. It often has a parabolic or trapezoidal cross-section, with mild side slopes.

10.2.3 Constructed Wetlands

A constructed wetland is designed mainly for water quality treatment in an environmentally more controlled fashion, compared to the natural wetland. In stormwater management it can also help to slow down the flow of runoff to dampen its peak flow with its dense vegetation and relatively flat gradient.

There are two major types of constructed wetlands. One is the free water surface wetland, which consists of a series of vegetated basins, through which water flows at relatively shallow depth and low velocity; the other is the subsurface flow wetland, which is a gravel and sand-filled basin planted with vegetation, and the water level is designed to remain below the surface (Naja and Volesky 2011). The former type can be used to treat stormwater runoff, whereas the later for domestic, municipal, and industrial wastewater (USEPA 1993). As runoff enters the wetland and makes its way to the outlet, pollutants are removed through several mechanisms, including plant uptake, microbial biodegradation by biofilms, chemical adsorption, physico-chemical adsorption, mechanical filtration, and sedimentation (Naja and Volesky 2011).

10.2.4 Retention and Detention Basins

Manmade ponds have been used to intercept runoff to attenuate peak flow during an extreme storm event. There are two types of such ponds, namely detention and retention basins, or dry and wet ponds. The fundamental difference is whether there is a permanent pool of water. The detention basin has an outlet pipe at the bottom to drain the water completely after the storm, and it stays dry in between the events, hence also referred to as the dry pond; whereas the outlet pipe of the retention basin, also referred to as the wet pond, is at a higher elevation to retain some water. Traditionally, both types of basins are designed solely for stormwater management purposes. Often fenced-off and aesthetically unattractive, they are purely hydraulic structures—grey infrastructure. However, if designed creatively to be multifunctional, they can be BGI.

The retention basin, with permanent ponding, can be designed as an aesthetically pleasing water feature for the community, such as the case in the High Point redevelopment project in Seattle. Since the water is not drained quickly after the storm, the retention basin can also become a constructed wetland for water quality treatment. The detention basin, when dry, can be designed as a playground, picnic area, sports field, parking lot, etc. (Park et al. 2014). For example, the City of Elk Grove in California recently retrofitted a 2.5-ha detention basin to serve as a neighbourhood park with enhanced wildlife habitat.

10.2.5 Green Roofs

A green roof, or vegetated roof, typically consists of several layers, including vegetation, growing media, drainage layer, root barrier, and waterproofing membrane. It can absorb some rainfall falling on the roof, thereby reducing the runoff going into the downspouts.

There are two types of green roofs, extensive and intensive (Bliss et al. 2009). The extensive green roof has a thin soil layer as shallow as a few cm, often installed on a building with low load-bearing roof slab. The intensive green roof has a deeper soil layer of 15 cm or more and can support a variety of plants, including shrubs and even trees (Locatelli et al. 2014). Because of its thick substrate layer, the intensive green roof can absorb more water and hence has higher rainfall retention capacity (Mentens et al. 2006).

10.3 Multiple Ecosystem Services of BGI

Contrary to the conventional urban drainage system—grey infrastructure—that functions solely for preventing pluvial flooding, the major advantage of BGI is its multi-functionality. BGI can provide multiple benefits, or ecosystem services, including flood hazard mitigation, water quality treatment, thermal reduction, and urban biodiversity enhancement. While we focus on regulating and supporting services here, it should be noted that BGI can also deliver cultural services such as recreation, education, and aesthetic appreciation.

10.3.1 Flood Hazard Mitigation

As the city continues to expand and densify, it is increasingly difficult—space- and finance-wise—to upgrade the conventional drainage system to handle increasing runoff. As such, BGI is increasingly used to supplement the existing drainage system. BGI can intercept, retain, absorb, and evapotranspire stormwater locally to reduce the runoff going to the storm drains to overwhelm the drainage network, thereby also delaying the peak flow to mitigate downstream flood risk.

Selective research findings on the hydraulic performance of rain gardens, constructed wetlands, and green roofs are listed in Tables 10.1, 10.2, and 10.3 respectively. Most studies show that BGI is effective in flood hazard mitigation. In particular, there are extensive evidences for green roofs. While green roofs are shown to retain more rainfall than conventional roofs do, the rainfall retention rate can vary substantially (Table 10.3). It decreases as the storm progresses and the soil reaches saturation (Bliss et al. 2009). The retention rate is also affected by soil type and thickness, vegetation, slope, and age of the roof (Buccola and Spolek 2011; VanWoert et al. 2005; Getter et al. 2007).

Table 10.1 Selected research findings of the hydraulic performance of rain gardens[a]

Study	Location	Runoff reduction (%)	Peak flow reduction (%)
Dietz and Clausen (2005)	Haddam, CT, USA	98.8	–
Hirschman and Collins (2008)	Virginia, USA (computer modelling)	40–80	–
Hunt et al. (2008)	Charlotte, NC, USA	–	99
Chapman and Horner (2010)	Seattle, WA, USA	48	–
Hathaway et al. (2011)	Wilmington, NC, USA	61–63	–

[a]Rain gardens here refer to all bioretention systems, which are not necessarily called rain gardens in the studies included here

Table 10.2 Selected research findings of the hydraulic performance of constructed wetlands

Study	Location	Runoff reduction (%)	Peak flow reduction (%)
Cohen and Brown (2006)	Dade County, FL, USA	31	–
Hirschman and Collins (2008)	Virginia, USA (computer modelling)	0	–
Al-rubaei et al. (2014)	Vaxjo, Sweden	–	72
Javaheri and Babbar-Sebens (2014)	Indianapolis, IN, USA (computer modelling)	–	20–41

Table 10.3 Selected research findings of the hydraulic performance of green roofs

Study	Location	Runoff reduction (%)	Peak flow reduction
Kumar and Kaushik (2005)	Yamuna Nagar, India	–	–
VanWoert et al. (2005)	Michigan, USA	60.6	–
Berndtsson et al. (2005)	Augustenborg, Sweden	51	–
Carter and Rasmussen (2006)	Georgia, USA	50–90	18 min
Getter et al. (2007)	Michigan, USA	80.8	
Hathaway et al. (2008)	North Carolina, USA	64	>75
Hirschman and Collins (2008)	Virginia, USA (computer modelling)	45–60	–
Bliss et al. (2009)	Pennsylvania, USA	5–69	5–70
Fioretti et al. (2010)	Italy	68	89
Susca et al. (2011)	New York City, USA	–	–
Buccola and Spolek (2011)	(laboratory at Portland State University, USA)	20–65	4–8 min
Morau et al.(2012)	Reunion Island, Indian Ocean	–	–
Stovin et al. (2012)	Sheffield, UK	50.2	60
Kok et al. (2013)	Kuala Lumpur, Malaysia	–	24
Qin et al. (2013)	Singapore	11.4	65
Locatelli et al. (2014)	Denmark	43–68	0–40 min

10.3.2 Water Quality Treatment

Stormwater runoff is considered non-point source (or diffuse) pollution. As it flows through different parts of the catchment it carries with it various types of pollutants (Table 10.4). Mobile vehicles produce a considerable amount of pollutants, such as oil, grease, and tire and brake wear; and industrial activities contribute to substances eroded from open stacks of raw and finished products (Liu et al. 2015). Runoff is often of higher temperature and could increase the temperature of the receiving water body; particularly, urban water bodies are often devoid of riparian vegetation, which worsens the impact (Natarajan and Davis 2010; Susca et al. 2011). Warmer temperature can harm the aquatic organisms, especially cold-water species such as trout and salmon (Long and Dymond 2014).

Most BGI systems, except green roofs, are found effective in removing toxic chemicals, filtering sediments, breaking down bacteria, and neutralizing acidic waters. Selected research findings of rain gardens, constructed wetlands, retention and detention basins, and green roofs are listed in Tables 10.5, 10.6, 10.7, and 10.8 respectively.

Rain gardens are effective in removing most pollutants except total phosphorus (TP) and bacteria (Table 10.5). However, different rain gardens also exhibit divergent performances, probably due to different soil depths and hydraulic loadings (Hathaway et al. 2011). Constructed wetlands (Table 10.6) and retention and detention basins (Table 10.7) also exhibit various pollution reduction rates, and it can be attributed to the antecedent weather condition, rainfall intensity, and design parameters (e.g., geometry, size) (Herb et al. 2009; Al-Rubaei et al. 2014). For retention and detention basins, it also depends on residence duration of the water (Wang et al. 2004). A retention basin could have higher pollutant reduction rate if the water could have longer contact with the vegetation and sediments rich in organic matter (Mallin et al. 2002).

The pollutant removal capacity of green roofs, nevertheless, is still highly uncertain, as there are contradictory research results. In some cases, higher concentrations of total nitrogen (TN), TP, and metals are even found in the green roof outflow (Vijayaraghvan et al. 2012; Hathaway et al. 2008). In others, the green roof

Table 10.4 Common types of stormwater pollutants

Pollutant	Sources
Sediment	Soil erosion, construction sites, building weathering
Nutrients	Fertilizer, animal waste, septic system overflow Example: total nitrogen, total phosphorus
Heavy metals	Automobile exhausts, tires, fuel combustion Example: copper, iron, lead, and zinc
Bacteria	Animal waste, septic system overflow Example: *E. Coli*, faecal coliform
Toxic contaminants	Pesticides, herbicides, oil and gas leakage from vehicles

Source Bakri et al. (2008), Scholz (2015)

Table 10.5 Selected research findings on pollution reduction rates of rain gardens (the numbers are rounded up; unit in %)*

Study	Location	Heavy metal				TN	TP	Bacteria (*E. coli*)	BOD	COD	TSS
		Cd	Cu	Zn	Pb						
Dietz and Clausen (2005)						32	−111				
Hunt et al. (2006)	North Carolina, USA		99	98	81	40	−240 to 65				
USEPA (2006)	Villanova, PA, USA			74		46	28				99
Davis (2007)	Maryland, USA		57	62	83		76				47
Hunt et al. (2008)	Charlotte, NC		54	77	31	32	31	71	63		60
Hirschman and Collins (2008)	Virginia, USA, (computer modelling)					40–60	25–50				
Chapman and Horner (2010)	Seattle, WA, USA		80	80	86	63	67				87
Battiata et al. (2010)	Mechanicsville, VA, USA (computer modelling)					40–60	25–50				
Hathaway et al. (2011)	Wilmington, NC, USA							−119 to 70			
Mei and Yang (2011)	Beijing, China					31–50	47–58				
Wang et al. (2015)	(laboratory at Beijing University, China)	>90	>90		>90						

TN Total nitrogen; *TP* total phosphorus; *BOD* biological oxygen demand; *COD* chemical oxygen demand; *TSS* total dissolved solid
*Negative value represents an increase in concentration.

Table 10.6 Summary of selected research findings on pollution reduction rates of constructed wetlands (the numbers are rounded up; unit in %)

Study	Location	Heavy metal					TN	TP	Herbicide		BOD	COD	SS
		Cd	Cr	Cu	Zn	Pb			Diuron	Simazine			
Walker and Hurl (2002)	Adelaide, Australia		0	48	57	71							
Poe et al. (2003)	North Carolina, USA						30						
Cohen and Brown (2006)	Dade County, FL, USA							27					36
Hirschman and Collins (2008)	Virginia, USA, (computer modelling)						25–55	50–75					
Lu et al. (2009)	Kunming City, China						61						
Battiata et al. (2010)	Mechanicsville, VA, USA (computer modelling)						25–55	50–75					
Page et al. (2010)	Adelaide, Australia								33–51	20–60			
Scholz and Hedmark (2010)	Scotland, UK						15–98				53–93	38–71	
Zhang et al. (2011)	(laboratory at Nanchang University, China)						29–48	65–91					
Li et al. (2011)	North China						98	90				91	
Ye and Li (2009)	Ningbo, China						83	64				85	89
Babatunde et al. (2011)	Dublin, Ireland						15–76	52–100			18–88	18–84	16–93
Rai et al. (2013)	Shantikunj, Haridwar, India		35	95	55	92					90		65
Al-rubaei et al. (2014)	Vaxjo, Sweden	90	89	91	90	96	61	86					96
Beutel et al. (2014)	Yakima, WA, USA							40–60					
Lynch et al. (2014)	Virginia Beach, VA, USA						25–40	4–48					

Table 10.7 Selected research findings on pollution reduction rates of retention/detention basins (the numbers are rounded up; unit in %)

Study	Location	Heavy metal				TN	TP	Bacteria (*E. coli*)	COD	TSS
		Cd	Cu	Zn	Pb					
Middleton and Barrett (2008)	Austin, Texas, USA		55	62	69	58	52		64	91
Hirschman and Collins (2008)	Virginia, USA, (computer modelling)					25–35	20–40			
Battiata et al. (2010)	Mechanicsville, VA, USA (computer modelling)					30–40	50–75			
Rosenzweig et al. (2011)	Princeton, NJ, USA					68				
Vezzaro et al. (2012)	Stockholm, Sweden and Melbourne, Australia (computer modelling)		90	88						
Beaudry et al. (2014)	Grand Fork, ND, USA					40	73	76		76
Stanley (2015)	Greenville, NC, USA	54	26	26	55					71

Table 10.8 Selected research findings on pollution reduction rates of green roofs

Study	Location	TN	TP
Berndtsson et al. (2005)	Augustenborg, Sweden	58%	Increase
Hathaway et al. (2008)	North Carolina, USA	Increase	Increase
Hirschman and Collins (2008)	Computer modelling	0%	0%
Bliss et al. (2009)	Pennsylvania, USA	0%	Increase
Battiata et al. 2010)	Mechanicsville, VA, USA (computer modelling)	0%	0%
Stovin et al. (2012)	Sheffield, UK	–	–

behaves as a sink of TN and heavy metals (Berndtsson et al. 2009). What is slightly more certain is green roof's ability to mitigate mild acid rain through rapid neutralization of the acid deposition (Bliss et al. 2009; Viyayaraghavan et al. 2012). In any case, most researchers stress the importance of soil media composition and proper maintenance in water quality treatment (Berndtsson et al. 2005).

10.3.3 Thermal Reduction

Urban areas typically suffer from the urban heat island effect. Theoretically, BGI can cool the air temperature through evapotranspiration and shading by vegetation and the moisture-containing soil (Zhang et al. 2012). Runoff flowing through the paved surface of high thermal capacity is often warmed by conduction (Natarajan and Davis 2010). But it can be cooled during the infiltration process and as it mixes with the shallow groundwater (Erickson et al. 2013).

Green roofs are the most studied BGI systems with regard to thermal reduction performance, and they are found effective in lowering the air temperature above and below the roof, and therefore can reduce energy consumption of air conditioning (Parizotto and Lamberts 2011; Morau et al. 2012). Selected research findings are listed in Table 10.9.

There is little research on thermal reduction performance of rain gardens and constructed wetlands. It is however pointed out that the soil depth of the rain garden plays an important role to affect the temperature of the outflowing water (Jones 2008). A well-vegetated wetland might reduce thermal loads by substantial shading of the water surface (Herb et al. 2007). Retention basins can be a source of thermal pollution. This is because most water surface is exposed to direct sunlight during hot days, and as new runoff enters the pond, the previous heated water is displaced and discharged, thereby raising the temperature of the receiving water body (Herb et al. 2009; Erickson et al. 2013). However, since the retention basin reduces runoff directly discharging into the water body, it can reduce the temperature of the receiving water body at peak flow (Erickson et al. 2013).

10.3.4 Urban Biodiversity Enhancement

As changes in hydrology and biodiversity in urban areas share a common driver—land use and land cover changes, it is logical to assume that more sustainable approaches to urban stromwater management should have complementary benefits on urban biodiversity. BGI may support biodiversity by providing wildlife habitat and temporary refuges, as well as by enhancing landscape connectivity (Chester and Robson 2013; Hassall 2014).

Table 10.9 Selected research findings on thermal reduction of green roofs

Study	Location	Thermal reduction
Kumar and Kaushik (2005)	Yamuna Nagar, India	5.1 °C (indoor air)
Susca et al. (2011)	New York City, USA	2 °C (indoor air)
Morau et al. (2012)	Reunion Island, Indian Ocean	6.7 °C (roof surface)
Kok et al. (2013)	Kuala Lumpur, Malaysia	1.5 °C (indoor air)
Qin et al. (2013)	Singapore	7.3 °C (roof surface)

Tan and Ng (2015) reviewed more than fifty papers conducted from 1980 to 2015, exploring the ability of BGI to enhance urban biodiversity. Most studies report that BGI has led to increases in species of different flora and fauna groups. The strongest evidence is for aquatic faunal groups, i.e., macroinvertebrates, anuran, fish, and odonates, but there are limited studies on terrestrial faunal groups.

However, the extent of biodiversity enhancement of BGI is highly variable across different studies (Tan and Ng 2015). Moreover, few studies evaluate whether biodiversity supported locally contributes to long-term survival of metapopulations across larger geographic regions. This is possibly because of the variation of the surrounding land uses. Furthermore, different design parameters of a BGI system, such as depth of the water body, shoreline complexity, proportion of macrophytes, composition of macrophytes, and size, also exert influence on abundance and composition of biodiversity (Hamer et al. 2012; Scheffers and Paszkowski 2013). For amphibians, the combined effects of the proximity to upland supporting habitats, the characteristics of stormwater runoff received by BGI (e.g., quantity, periodicity, pollutant load), pond age, and the amount of other ponds in a larger area have been shown to be important (Birx-Raybuck et al. 2010; Holzer 2014). The potential of BGI to enhance urban biodiversity, therefore, is affected by complex interacting factors.

10.4 Drivers Behind BGI Implementation

In recent years, BGI is increasingly embraced through different initiatives in different nations and cities. The implementation of BGI has been driven by the urgency to solve different local problems or challenges. These drivers mainly include water quality standards, water security, increased flood risk, and aquatic ecosystem degradation.

10.4.1 Water Quality Standards

Where industrial and domestic wastewater discharges have been largely treated, stormwater runoff has become a major source of pollution. In US, Europe, Australia, Singapore, etc., controlling stormwater pollution has been a major challenge. Some cities adopt the combined sewer system, where stormwater runoff, along with other streams of wastewater, is delivered to the treatment plant before discharged into the receiving water body. Nevertheless, heavy precipitation events often overwhelm the system to cause combined sewage overflows (CSOs), where untreated sewage is discharged directly into the water body.

The emergence of BMPs and LID in the US was driven by the Clean Water Act, specifically, the associated National Pollutant Discharge Elimination System (NPDES) stormwater permit programme, administered by the US Environmental Protection Agency (USEPA) (Keeley et al. 2013). The permit requires stormwater runoff discharge to meet certain water quality standards, and USEPA requires the use of BMPs to meet those standards. Therefore, almost every jurisdiction in the US has adopted BMPs in the stormwater design manual by the early 1990s (Fletcher et al. 2015).

In the European Union (EU) , the Water Framework Directive (WFD) of 2000 sets the goal of attaining 'good status' for Europe's water bodies by 2015. Explicitly addressing stormwater pollution, WFD sets standards for the qualities of both runoff discharge and the receiving water body. For each water body, WFD also requires an integrated river basin management plan for achieving the 'good status', which prompted the use of SUDS by many EU members (Nickel et al. 2014).

10.4.2 Water Security

The need to control stormwater pollution can be closely linked with the issue of water supply, especially in water scarce nations. For example, Singapore's ABC Waters Programme is related to water security, which is a top priority in Singapore because of a history of water shortage (Tan et al. 2009). Surface water is a major source of water supply and is collected through a network of rivers, canals, drains, as well as 17 reservoirs across the nation. Two-thirds of this densely populated city-state function as water catchments, including built-up areas. Recognizing stormwater pollution as a threat to Singapore's water security, the ABC Waters Programme places an emphasis on using BGI systems for water quality control.

Having experienced extended droughts in recent years, Australia has shifted the focus of stormwater management from aquatic ecosystem protection to long-term water security (Morison and Brown 2011), and stormwater is considered as a source of water (Wong 2006). WSUD has evolved from a management approach to stormwater quality and quantity to a framework that integrates urban design with three 'urban water streams', i.e., potable water, wastewater, and stormwater (Wong 2006). To address water security, WSUD involves rainwater harvesting, as well as storing locally treated stormwater runoff in the aquifer.

China is also challenged by water shortage. Managing for water security is one of the major objectives of China's Sponge City initiative, which only recently started in 2014. The Sponge City initiative aims to make the city metaphorically like a sponge to be able to absorb, store, infiltrate, and purify stormwater and also be able to release water when it is needed.

10.4.3 Increased Flood Risk

While flood hazard mitigation is one of BGI's multiple functions, it is not a focus in BGI implementations in US, Europe, Australia and Singapore, mainly because the basic drainage infrastructure is already in place. However, in UK the problem of pluvial flooding plays a bigger role in its adoption of SUDS because of the concern that the existing drainage system will be increasingly inadequate in the face of climate change (Ellis 2013). Furthermore, the Flood and Water Management Act that was introduced in 2010 requires the implementation of SUDS in both new development and redevelopment projects (Ashley et al. 2013).

Increasing flood risk is a major driver behind China's Sponge City initiative because the country is still undergoing rapid urbanization. As a result of massive increases of impervious surfaces and inadequate or non-existent drainage infrastructure, in recent years numerous Chinese cities, such as Beijing, Shanghai, and Shenzhen, just to name a few, suffer frequently from severe pluvial flooding.

10.4.4 Aquatic Ecosystem Degradation

Stormwater runoff has been understood as a major threat to urban aquatic ecosystems in the developed nations, such as US and Australia. Untreated stormwater runoff not only pollutes the water body but also imposes other harms to the aquatic ecosystem. It is particularly detrimental to smaller streams because the hydrologic regime can be dramatically altered to become flashy as the conventional drainage system quickly sends runoff to the stream (Walsh et al. 2005). Although aquatic ecosystems in the urban area are subject to multiple stressors, the alteration of the hydrological regime is considered a major cause of ecological degradation (Booth 2005).

However, while most existing BGI programmes address stormwater pollution, few explicitly emphasize the wider ecological implications. An exception is Australia's WSUD, as its emergence is partly a response to the ecological degradation associated with stormwater runoff (Wong 2006). For example, Melbourne has implemented WSUD with an explicitly stated goal of 'protection of the environment, with a specific emphasis on the aquatic ecosystem including rivers, riparian zones and wetlands' (City of Melbourne 2016: 26).

Although BMPs or LID in the US generally focuses on meeting water quality standards, salmon restoration also serves as a powerful driver in the Northwest of the US, where five species of Pacific salmon are listed under the Federal Endangered Species Act. For example, Seattle's 'Green Stormwater Infrastructure' programme prioritizes basins with salmon-bearing waterways. The Seattle government and local NGOs also explicitly communicate to the general public that salmon restoration is a reason for alternative stormwater management. To promote salmon-friendly land management practices, an NGO in Portland has developed the 'Salmon-Safe' certification programme to acknowledge practices that keep the watershed clean enough for native salmon to spawn and thrive.

10.5 Examples of BGI Implementation in Cities

In this section we showcase four cities that are relatively more active and progressive in implementing BGI. We include Portland and New York City from the West but note that many other western cities (e.g., Melbourne, Copenhagen) also have notable achievements. Two cities from the East, Singapore and Zhenjiang, that are relatively less known for BGI, are also included to better reflect the current extent of BGI implementation across the world.

10.5.1 Portland, Oregon, USA

Striving to tackle the problem of CSO, Portland is a pioneer of BGI in the US. Its major policy is the Green Streets programme, which turns conventional streets into 'green streets' by installing 'stormwater street planters'—a form of rain gardens—in the sidewalks, curb extensions, roundabouts, and traffic islands. These planters are located close to the storm drains to intercept, slow, cleanse, and infiltrate runoff to keep it out of the combined sewer system. The first green street was completed in 2003.

Portland also promotes 'ecoroofs', that is, green roofs. Since 1999, when the ecoroof was officially recognized as a stormwater management tool in Portland, over 560 ecoroofs have been installed, covering 15.4 ha. Any city-owned building is required to install ecoroofs to cover at least 70% of the total roof area. Incentives are also available to encourage ecoroofs on private buildings, including FAR (floor area ratio) bonus and monetary refund (US$5 for each square foot of ecoroof built).

The performances of existing BGI systems have been monitored through the Sustainable Stormwater Management Program to quantify benefits, improve design, and lower maintenance cost. According to Portland's Bureau of Environmental Services, both ecoroofs and green streets have shown positive results in runoff reduction.

There is also the Green Street Steward Program to encourage community members to volunteer in the care and maintenance of BGI systems. To further promote BGI, the city government has partnered with local schools to install BGI systems in schoolyards for education on sustainable stormwater management.

10.5.2 New York City, USA

Also plagued by CSO, the New York City (NYC) has carried out the Green Infrastructure Plan since 2010 to reduce CSO through retrofitting streets, sidewalks, and public and private properties. The objectives are '[r]educing CSO volume by an additional 3.8 billion gallons (11.4 million m^3) per year; capturing the first 2.5 cm of rainfall from 10% of the impervious area in watersheds with combined sewers through green infrastructures; and providing substantial, quantifiable sustainability benefits, such as cooling the city, reducing energy use, increasing property values, and cleaning the air' (NYCDEP 2010).

To make BGI implementation cost-effective, the Green Infrastructure Plan identifies 'priority areas', which are drainage basins with frequent CSO incidents or high CSO volume. A major BGI system installed in the priority areas is the 'right-of-way bioswale', which tackles runoff from the public right of way. Other BGI systems include right-of-way rain gardens, stormwater green streets, green roofs, and other types of rain gardens and bioswales, all of which are considered 'green infrastructure assets'. The target of the Green Infrastructure Plan is a total of 5905 such assets, and by 2015 NYC has established 3830 assets to manage 179 ha or 0.6% of the impervious area within combined sewer tributary (NYCDEP 2015).

The GIS-based Project Tracking and Asset Management System have been established to monitor the operation and maintenance of the green infrastructure assets. There is also the NYC Green Infrastructure Co-benefits Calculator. This open access online tool allows a designer or planner to specify any type of green infrastructure asset and its parameters to estimate the cost and environmental, social and economic benefits. In addition to facilitating the process of planning and designing a green infrastructure asset, the quantification of the benefits can also facilitate stakeholder buy-ins and public outreach.

10.5.3 Singapore

Singapore has promoted BGI through the ABC Waters Programme since 2006. The objective of the programme is to transform the utilitarian drains, canals and reservoirs throughout Singapore into 'beautiful and clean streams, rivers and lakes with postcard-pretty community spaces for all to enjoy' (PUB 2014). Because almost all waterways in Singapore have been heavily channelized, managed solely for drainage efficiency, they are largely external to the everyday life of people. The ABC Waters Programme is to better integrate these waterways and other water bodies with the rest of the urban landscape to foster a sense of ownership, through improving the quality of water, physical appearance, and recreational value of the water body.

The programme addresses water quality using BGI systems, referred to as 'ABC Waters features', including vegetated swales, bio-retention swales, bio-retention basins, sedimentation basins, constructed wetlands, and cleansing biotopes. Specific and achievable water quality targets were set, subject to change over time based on monitoring results. However, an overall monitoring programme does not exist.

The ABC Waters Programme is administered by Singapore's national water authority, the Public Utilities Board (PUB). To increase the adoption of ABC Waters features throughout the nation, PUB also launched the ABC Waters Certification scheme in 2010 to encourage other public and private sectors to incorporate ABC Waters features in their development projects.

10.5.4 Zhenjiang, Jiangsu Province, China

Like other rapidly developing Chinese cities, Zhenjiang is challenged by inadequate drainage infrastructure and stormwater pollution (Sheng et al. 2011). In 2007, the Zhenjiang government began to study the idea of LID and its local implementation. Since 2010, LID measures have been incorporated into the Guantang New Town—a new urban district built from scratch, as well as the redevelopment of the old inner city area. As of 2015, Zhenjiang has built a total of 16 km of bioswales, 350,000 m^2 of green roofs, 350,000 m^2 of rain gardens, 40 km of road with pervious paving, and 1 million m^3 of rainwater storage facilities (Zhenjiang Housing and Development Bureau 2015).

In April 2015, the prior experiences on LID resulted in Zhenjiang being selected as one of the 16 'pilot sponge cities' for China's Sponge City initiative. The Sponge City initiative is backed by a strong political will top-down from President Xi Jinping with a substantial budget. The Zhenjiang Sponge City project involves a total area of 22 km^2 and 302 different sub-projects, with a total investment of RMB 8 billion (US$1.2 billion), and the ultimate goal is to tackle 75% of the total annual volume of stormwater runoff for flood safety against the 30-year storm, and to reduce non-point pollution by 60% (Zhenjiang Housing and Development Bureau 2015). It is unclear what exact BGI systems are to be built to achieve these goals. Since the Sponge City initiative is relatively new, its actual implementation and effects remain to be seen.

10.6 Future Research Agenda and Concluding Remarks

The conventional drainage system is single-functioned infrastructure that solves one problem while creating many others. BGI has been increasingly recognized as a desirable alternative, which taps into local natural processes to manage urban stormwater in a more sustainable fashion. While BGI is widely discussed in academia and increasingly practised, it is far from mainstream. Around the world, grey infrastructure continues to dominate stormwater management in existing cities, as well as new development and redevelopment projects. The major challenge today, therefore, is to mainstream BGI. Addressing the barriers to mainstreaming BGI in different cities should be an important future BGI research agenda.

However, identifying the barriers is a daunting task itself, as it depends on different socioeconomic and environmental contexts. Although different cities face different challenges of stormwater management, valuable lessons could still be drawn from existing BGI implementations, be they successful or not. Therefore, we stress the importance of conducting case studies on practical BGI experiences. In this chapter, we have touched upon several BGI programmes and reviewed various important drivers behind them. These cases, along with those not mentioned here, deserve to be studied further, particularly on the process from conception to implementation to dissemination, with a focus on the challenges and associated solutions. A good example is the study of Brown and Clarke (2007) on Melbourne's implementation of WSUD.

Furthermore, the case study should further explore the drivers behind the programme. Understanding the drivers is important because a paradigm shift often only takes place when there is a major crisis that provides the opportunity for a change. This emphasis, however, does not imply that it is impossible to mainstream BGI without some crisis. The understanding of the drivers could lead to strategies for using an existing or potential water-related problem as a leverage to promote BGI. For example, since climate change is likely to affect most cities around the world, every city would need to re-examine the resilience or robustness of its water sector to identify the current and future weak points. This could serve as an opportunity for introducing BGI to the public.

While we place an emphasis on the research on practical implementation, we note that strengthening the science behind BGI is no less important. We have reviewed the scientific evidences for the claimed benefits or ecosystem services of BGI. There remain many evident gaps in such research. While water quality treatment and flood hazard mitigation are much better researched, thermal reduction and urban biodiversity enhancement are far less understood, not to mention cultural services associated with BGI. We also do not know the relationship between these different ecosystem services. For example, would strengthening the function of water quality treatment influence the function of flood hazard mitigation? In other words, can BGI simultaneously provide multiple services equally well? Or is there tradeoff? The more we understand the actual effects of BGI, the better we know how to design BGI systems properly to make them more effective, and the better we can promote BGI with convincing evidences.

Finally, despite our focus on stormwater management, we stress that BGI should concern all water sectors, as demonstrated in the conceptualization of WSUD as a holistic urban water management approach, from water supply to wastewater treatment to stormwater management (Wong 2006). BGI can potentially be an alternative to other single-functioned water infrastructure and a framework for integrating those conventionally isolated and independent water sectors because of the multi-functionality embedded in its design. The idea of integrated urban water management is nothing new, but BGI as an integrated urban water management approach should be further explored. We believe that pushing the research frontier through identifying divers and barriers and developing integrated solutions could further contribute to urban sustainability and resilience.

References

Al-Rubaei A, Engstrom M, Viklander M, Blecken G (2014) Long-term treatment efficiency of a constructed stormwater wetland: preliminary results. In: 13th International Conference on Urban Drainage, Sarawak, Malaysia

Ashley R, Lundy L, Ward S, Shaffer P, Walker AL, Morgan C, Saul A, Wong T, Moore S (2013) Water-sensitive urban design: opportunities for the UK. Municipal Eng 166:65–76

Babatunde AO, Zhao YQ, Doyle RJ, Rackard SM, Kumar JLG, Hu YS (2011) Performance evaluation and prediction for a pilot two-stage on-site constructed wetland system employing dewatered alum sludge as main substrate. Bioresour Technol 102:5645–5652

Bakri DA, Rahman S, Bowling L (2008) Sources and management of urban stormwater pollution in rural catchments, Australia. J Hydrol 356:299–311

Battiata J, Collins K, Hirschman D, Hoffmann G (2010) The runoff reduction method. J Contemp Water Res Educ 146:11–21

Beaudry A, Lim Y, Gullicks H, Moretti C (2014) Removal efficiency of water quality pollutants in a wet detention basin in Grand Forks, ND. Thesis. Bachelor of Science, University of North Dakota

Benedict MA, Mcmahon ET (2002) Green infrastructure: smart conservation for the 21st century. Renew Res J 20(3):12–17

Berndtsson JC, Emilsson T, Bengtsson L (2005) The influence of extensive vegetaed roofs on runoff water quality. Sci Total Environ 355:48–63

Berndtsson JC, Bengtsson L, Jinno K (2009) Runoff water quality from intensive and extensive vegetaed roofs. Ecol Eng 35:369–380

Beutel MW, Morgan MR, Erlenmeyer JJ, Brouillard ES (2014) Phosphorus removal in a surface-flow constructed wetland treating agricultural runoff. J Environ Q 43:1071–1080

Birx-Raybuck DA, Price SJ, Dorcas ME (2010) Pond age and riparian zone proximity influence anuran occupancy of urban retention ponds. Urban Ecosyst 13:181–190

Bliss DJ, Neufeld RD, Ries RJ (2009) Storm water runoff mitigation using a green roof. Environ Eng Sci 26:407–417

Booth DB (2005) Challenges and prospects for restoring urban streams: a perspective from the Pacific Northwest of North America. J North Am Benthological Soc 24(3):724–737

Brown R, Clarke J (2007) Transition to water sensitive urban design: the story of Melbourne, Australia. Report No. 07/1, Facility for Advancing Water Biofiltration, Monash University, June 2007

Brown RA, Hunt WF (2011) Underdrain configuration to enhance bioretention exfiltration to reduce pollutant loads. J Environ Eng 1082–1091

Buccola N, Spolek G (2011) A pilot-scale evaluation of greenroof runoff retention, detention, and quality. Water Air Soil Pollution 216:83–92

Carter TL, Rasmussen TC (2006) Hydrologic behavior of vegetated roofs. J Am Water Resour Assoc 42:1261–1274

Chapman C, Horner R (2010) Performance assessment of a street-drainage bioretention system. Water Environ Res 109

Chester ET, Robson BJ (2013) Anthropogenic refuges for freshwater biodiversity: their ecological characteristics and management. Biol Conserv 166:64–75

City of Melbourne WSUD Guidelines: Applying the Model WSUD Guidelines: An initiative of the Inner Melbourne Action Plan. https://www.melbourne.vic.gov.au/SiteCollectionDocuments/wsud-full-guidelines.pdf. Accessed 9 Sept 2016

Cohen WJ, Brown MT (2006) A model examining hierarchical wetland networks for watershed stormwater management. Ecol Model 201:179–193

CSIRO [Commonwealth Scientific and Industrial Research Organization] (2005) WSUD engineering procedures: stormwater [e-book]. CSIRO Publishing, Collingwood, Vic. Available from: eBook Collection (EBSCOhost), Ipswich, MA. Accessed 22 July 2016

Davis AP (2007) Field performance of biorention: water quality. Environ Eng Sci 24:8

Dietz ME, Clausen JC (2005) A field evaluation of rain garden flow and pollutant treatment. Water Air Soil Pollut 167:123–138

Ellis JB (2013) Sustainable surface water management and green infrastructure in UK urban catchment planning. J Environ Planning Manage 56:24–41

Erickson AJ, Weiss PT, Gulliver JS (2013) Optimizing stormwater treatment practices. Springer, New York

Fioretti R, Palla A, Lanza LG, Principi P (2010) Green roof energy and water related performance in the Mediterranean climate. Build Environ 45:1890–1904

Fletcher TD, Shuster W, Hunt WF, Ashley R, Butler D, Arthur S, Trowsdale S, Barraud S, Semadeni-Davis A, Bertrand-Krajewski JL, Mikkelsen PS, Rivard G, Uhl M, Dagenais D, Viklander (2015) SUDS, LID, BMPs, WSUD and more—the evolution and application of terminology surrounding urban drainage. Urban Water J 12(7):525–542

Getter KL, Rowe DB, Andresen JA (2007) Quantifying the effect of slope on extensive green roof stormwater retention. Ecol Eng 31:225–231

Hamer AJ, Smith PJ, McDonnell MJ (2012) The importance of habitat design and aquatic connectivity in amphibian use of urban stormwater retention ponds. Urban Ecosystems 15:451–471

Hassall C (2014) The ecology and biodiversity of urban ponds. Wiley Interdisc Rev Water 1:187–206

Hathaway AM, Hunt WF, Jennings GD (2008) A field study of green roof hydrologic and water quality performance. Am Soc Agric Biol Eng 51:37–44

Hathaway JM, Hunt WF, Graves AK, Wright JD (2011) Field evaluation of bioretention indicator bacteria sequestration in Wilmington, North Carolina. J Environ Eng 137:1103–1113

Herb W, Mohseni O, Stefan H (2007) A model for mitigation of surface runoff temperatures by a wetland basin and a wetland complex. Vermillion River Watershed Joint Powers Organization, Dakota County

Herb WR, Mohseni O, Stefan HG (2009) Simulation of temperature mitigation by a stormwater detention pond. J Am Water Resour Assoc 45(5):1164–1178

Hinman C (2007) Rain garden handbook for western Washington homeowners. Washington State University Extension, Tacoma

Hirschman D, Collins K (2008) Technical memorandum: the runoff reduction method. Center for Watershed Protection and Chesapeake Stormwater Network

Holzer KA (2014) Amphibian use of constructed and remnant wetlands in an urban landscape. Urban Ecosyst 17:955–968

Hunt WF, Jarrett AR, Smith JT, Sharkey LJ (2006) Evaluating bioretention hydrology and nutrient removal at three field sites in North Carolina. J Irrig Drain Eng 600–608

Hunt WF, Smith JT, Jadlocki SJ, Hathaway JM, Eubanks PR (2008) Pollutant removal and peak flow mitigation by a bioretention cell in urban Charlotte, NC. J Environ Eng 403–408

Javaheri A, Babbar-Sebens M (2014) On comparison of peak flow reductions, flood inundation maps, and velocity maps in evaluating effects of restored wetlands on channel flooding. Ecol Eng 73:132–145

Jones MP(2008) Effect of urban stormwater BMPs on runoff temperature in trout sensitive regions. PhD thesis, North Carolina State University, Raleigh, NC

Keeley M, Koburger A, Dolowitz DP, Medearis D, Nickel D, Shuster W (2013) Perspectives on the use of green infrastructure for stormwater management in Cleveland and Milwaukee. Environ Manage 51:1093–1108

Kok KH, Sidek LM, Abidin MRZ, Basri H, Muda ZC, Beddu S (2013) Evaluation of green roof as green technology for urban stormwater quantity and quality controls. IOP Conference Series: Earth and Envrionmental. Science 16:012045

Kumar R, Kaushik SC (2005) Performance evaluation of green roof and shading for thermal protection of buildings. Build Environ 50:1505–1511

Li Y, Zhang Y, Zhang X (2011) Heat preservation of subsurface flow constructed wetland in cold area in winter and its operation effect. Procedia Environ Sci 10:2182–2188

Liu Y, Bralts VF, Engel BA (2015) Evaluating the effectiveness of management practices on hydrology and water quality at watershed scale with a rainfall-runoff model. Sci Total Environ 511:298–308

Locatelli L, Mark O, Mikkelsen PS, Arnbjerg-Nielsen K, Jensen MB, Binning PJ (2014) Modelling of green roof hydrological performance for urban drainage applications. J Hydrol 519:3237–3248

Long DL, Dymond RL (2014) Thermal pollution mitigation in cold water stream watersheds using bioretention. J Am Water Resour Assoc 50(4):977–987

Lu S, Zhang P, Jin X, Xiang C, Gui M, Zhang J, Li F (2009) Nitrogen removal from agricultural runoff by full-scale constructed wetland in China. Hydrobiologia 621:115–126

Lucke T, Nichols PWB (2015) The pollution removal and stormwater reduction performance of street-side bioretention basins after ten years in operation. Sci Total Environ 536:784–792

Lynch J, Fox L, Owen JS Jr, Sample D (2014) Evaluation of commercial floating treatment wetland technologies for nutrient remediation of stormwater. Ecol Eng 75:61–69

Mallin MA, Ensign SH, Wheeler TL, Mayes DB (2002) Pollutant removal efficacy of three wet detention ponds. J Environ Qual 31:654–660

Mei Y, Yang X (2011) The effect of nutrients removal for bio-retention system in rainwater runoff. J Environ Eng 520–526

Mentens J, Raes D, Hermy M (2006) Green roofs as a tool for solving the rainwater runoff problem in the urbanized 21st century? Landscape Urban Plann 77:217–226

Morau D, Libelle T, Garde F (2012) Performance evaluation of green roof for thermal protection of buildings in Reunion Island. Energy Procedia 14:1008–1016

Morison PJ, Brown RR (2011) Understanding the nature of publics and local policy commitment to water sensitive urban design. Landscape and Urban Planning 99:83–92

Naja GM, Volesky B (2011) Constructed wetlands for water treatment. Compr Biotechnol 353–369

Natarajan P, Davis AP (2010) Thermal reduction by an underground storm-water detention system. J Environ Eng 520–526

Nickel D, Schoenfelder W, Medearis D, Dolowitz DP, Keeley M, Shuster W (2014) German experience in managing stormwater with green infrastructure. J Environ Plann Manage 57:403–423

NYCDEP [New York City Department of Environmental Protection] (2010) NYC green infrastructure plan

NYCDEP [New York City Department of Environmental Protection] (2015) NYC green infrastructure 2015 annual report

Page D, Dillon P, Mueller J, Bartkow M (2010) Quantification of herbicide removal in a constructed wetland using passive samples and composite water quality monitoring. Chemosphere 81:394–399

Parizotto S, Lamberts R (2011) Investigation of green roof thermal performance in temperate climate: a case study of an experimental building in Florianopolis city, southern Brazil. Energy Build 43:1712–1722

Park D, Jang S, Roesner L (2014) Evaluation of multi-use stormwater detention basins for improved urban watershed management. Hydrol Process 28:1104–1113

Poe AC, Piehler MF, Thompson SP, Paerl HW (2003) Denitrification in a constructed wetland receiving agricultural runoff. Wetlands 23(4):817–826

PUB (2014) Active, beautiful, clean waters design guidelines. PUB, Singapore

Qin X, Wu X, Chiew Y, Li Y (2013) A green roof test bed for stormwater management and reduction of urban heat island effect in Singapore. Br J Environ Clim Change 2:410–420

Rai UN, Tripathi RD, Singh NK, Upadhyay AK, Dwivedi S, Shukla MK, Mallick S, Singh SN (2013) Constructed wetland as an ecotechnological tool for pollution treatment for conservation of Ganga river. Bioresour Technol 148:535–541

Rosenzweig BR, Smith JA, Baeck ML, Jaffe PR (2011) Monitoring nitrogen loading and retention in an urban stormwater detention pond. J Environ Qual 40:598–609

Rouse DC, Bunster-Ossa I (2013) Green infrastructure: a landscape approach. American Planning Association, Chicago

Scheffers BR, Paszkowski CA (2013) Amphibian use of urban stormwater wetlands: the role of natural habitat features. Landscape Urban Plann 113:139–149

Scholz M (2015) Wetlands for water pollution control [e-book]. Elsevier Science. Accessed 22 July 2016

Scholz M, Hedmark A (2010) Constructed wetlands treating runoff contaminated with nutrients. Water Air Soil Pollut 205:323–332

Sheng JG, Zeng P, Zhang C, Yan Q (2011) Characteristics of pollution in precipitation runoff in Zhenjiang urban area. J Jiangsu Univ Sci Technol (Natural Science Edition) 25(5):496–499

Stagge JH, Davis AP, Jamil E, Kim H (2012) Performance of grass swales for improving water quality from highway runoff. Water Res 46:6731–6742

Stanley DW (2015) Pollutant removal by a stormwater dry detention pond. Water Environ Res 68 (6):1076–1083

Stovin V, Vesuviano G, Kasmin H (2012) The hydrological performance of a green roof test bed under UK climatic conditions. J Hydrol 414–415:148–161

Susca T, Gaffin SR, Dell'Osso GR (2011) Positive effects of vegetation: urban heat island and green roofs. Environ Pollut 159:2119–2126

Tan PY, Ng C (2015) Biodiversity enhancement as a goal of blue-green infrastructure development in cities. A report for research project "enhancing blue green and social performance in high density urban environments". National University of Singapore, p 76

Tan YS, Lee TJ, Tan J (2009) Clean, green and blue: Singapore's journey towards environmental and water sustainability. ISEAS Publishing, Singapore

USEPA (United States Environmental Protection Agency) (1993) Guidance manual for developing best management practices (BMP). EPA, Washington, DC

VanWoert ND, Rowe DR, Andresen JA, Rugh CL, Thomas R, Xiao L (2005) Green roof stormwater retention: effects of roof surface, slope, and media depth. J Environ Qual 34:1036–1044

Vezzaro L, Eriksson E, Ledin A, Mikkelsen PS (2012) Quantification of uncertainty in modelled partitioning and removal of heavy metals (Cu, Zn) in a stormwater retention ponds and a biofilter. Water Res 46:6891–6903

Vijayaraghvan K, Joshi UM, Balasubramanian R (2012) A field study to evaluate runoff quality from green roofs. Water Res 46:1337–1345

Walker DJ, Hurl S (2002) The reduction of heavy metals in a stormwater wetland. Ecol Eng 18:407–414

Walsh CJ, Roy AH, Feminella JW, Cottingham PD, Groffman PM, Morgan RP II (2005) The urban stream syndrome: current knowledge and the search for a cure. J North Am Benthological Soc 24(3):706–723

Wang G, Chen S, Barber ME, Yonge D (2004) Modeling flow and pollutant removal of wet detention pond treating stormwater runoff. J Environ Eng 1315–1321

Wang J, Zhang P, Yang L, Huang T (2015) Adsorption characteristics of construction waste for heavy metals from urban stormwater runoff. Chin J Chem Eng 23:1542–1550

Wong THF (2006) Water sensitive urban design—the journey thus far. Australian J Water Resour

Ye F, Li Y (2009) Enhancement of nitrogen removal in towery hybrid constructed wetland to treat domestic wastewater for small rural communities. Ecol Eng 35:1043–1050

Zhang R, Wan Y, Zhou W (2011) Simulation constructed wetland system for treating stormwater runoff in southern China. In: International conference on remote sensing, environment and transportation engineering (RSETE), 2011, Nanjing, pp 2483–2486

Zhang Y, Chen Y, Qing D, Jiang P (2012) Study on urban heat island effect based on normalized difference vegetated index: a case study of Wuhan, China. Procedia Environ Sci 13:574–581

Zhenjiang Housing and Development Bureau (2015) Developing the sponge city: Zhenjiang as a pilot city. http://jsj.zhenjiang.gov.cn/xwdt/cjxw/201503/t20150310_1440058.htm. Accessed 9 Sept 2016

Chapter 11
Highrise Greenery: Ancient Invention with New Lease of Life

Chi Yung Jim

Abstract Many cities especially compact ones are beset by urban heat island effect compounded by climate change and poor environmental quality. Urban green infrastructure can provide promising relief, but its implementation in dense cities is constrained by inadequate solution space. Departing from conventional thinking, greenroofs offer an innovative alternative of converting the negative amenity of barren roofs to pleasant greenery plus handsome bonus of multiple ecosystem services. The ancient origin of greenroofs is traced to the pragmatic need to build primitive shelters in harsh climate. Gradual refinement of the precursor has allowed development of a cultural invention. Despite continued installation in rural areas, its adoption in cities remained scanty in historical times. The notable classical exemplars in pre-industrial and industrial periods are assessed as pioneers. With fortuitous combination of factors, the idea was revived in Germany from the 1960s, spearheaded by scientific research and technological innovations. The new materials and designs, in conjunction with enabling public policies, have pump-primed the modern greenroof movement which subsequently spread to other European countries and then worldwide. The critical technological advances and the directions for further improvements are critically evaluated. The need to deepen understanding and enhance the key functions of cooling, warming and stormwater management is highlighted. Some inspiring recent projects are surveyed with respect to their outstanding innovative elements. Future developments could focus on tailor-made, cost-effective and environmentally-friendly dimensions.

Keywords Greenroof design · Greenroof technology · Greenroof benefit · Vegetated roof · Skyrise greenery · Urban green infrastructure

---

C.Y. Jim (✉)
University of Hong Kong, Pokfulam, Hong Kong
e-mail: hragjcy@hku.hk

P.Y. Tan and C.Y. Jim (eds.), *Greening Cities: Forms and Functions*,
Advances in 21st Century Human Settlements,
DOI 10.1007/978-981-10-4113-6_11

11.1 Introduction

Many cities around the world, especially compact ones experiencing fast growth, are beset by an inadequate supply of urban green spaces (UGS). Buildings and roads often usurp most of the land at the street level, creating an exceptionally cramped and impermeable landscape. The attendant harsh environmental conditions for people and wildlife, with impacts on the quality of life and urban sustainability, demand smart solutions.

Established built-up areas with rather fossilized town plans may attempt to relax the development density by converting building sites back to open space. However, such macro-scale surgical operations, often encountering tenacious institutional barriers and being invariably time-consuming and very costly, are seldom effective or satisfactory. Thus UGS deficit tend to remain a chronic and vexing problem with little solution in sight (Jim and Chan 2016).

The reinstatement of nature in tightly-packed city areas could fruitfully move to an innovative mode. Instead of focusing on the conventional ground-level UGS, they can be supplied above the ground level. Urban greening can literally be freed from the ground-hugging bondage and extend to the highrise domain. The numerous bare rooftops and facades of buildings offer feasible alternatives to install greenery and restore nature. In recent decades, greenroof (also known as eco-roof or vegetated roof) and the associated greenwall movement has been assiduously promoted in some cities. Greenwall (vertical greening) is an integral component of skyrise or highrise greening, but its detailed assessment lies outside the scope of this study.

The term *skyrise greenery* was initially used in Singapore (National Parks Board 2002; Tan 2013) and adopted recently by Hong Kong and other places. It refers to a wide range of vegetated sites on the envelope of building structures. Situated above the ground on mainly level or sloping surfaces, they include rooftop, podium, terrace, balcony, flyover edge, footbridge edge and top, and decking above road, railway and transport station. For underground structures such as carpark or transport station, the greened rooftop could be at grade (at ground level). On vertical surfaces, they encompass building façade, free-standing wall and frame, pole and cable. Regardless of the site location, they share a common pool of materials and techniques for installation and maintenance.

This chapter explores the utilization of a hitherto somewhat neglected resource, the large reservoir of above-ground space, with ample potentials to compensate for nature-deficit in cities. A historical survey traced the ancient root of the conception with the preserved tradition expressed and enhanced by modern transformation. The adoption of the idea was underpinned by some opportune, enabling and motivating factors. Recent researches have clarified the multiple functions and benefits, and provided pointers for future developments.

11.2 Precursors to Modern Revamp

11.2.1 From Tradition to Necessity

The Scandinavian vernacular tradition of sod roof or turf architecture (Osmundson 1999) has been kept alive for millennia in different parts of the countryside and occasionally in towns. The harsh high-latitude weather conditions necessitated human adaption for survival. It began as a form of primitive shelter made from local natural materials. Putting earth and vegetation on the roofs and walls of the early dwellings could create a less trying indoor space. With widespread adoption of modern greenroof technology in different parts of the world, it may be worthwhile to revisit the ancient skills bequeathed by our ancestors. They laid the foundation in terms of fundamental principles and designs for modern greenroofs.

Sod roofs are installed on pitched roofs using age-old methods with the following natural material layers laid in sequence: wooden roof frame providing structural support, continuous wooden planks base, several layers of birch bark harvested in the form of broad sheets, and two layers of natural sod cut from meadows. The first sod layer is placed upside down so that the dead grass provides internal space to facilitate drainage at the grass-bark interface. The second sod layer is laid so that its roots will enter the first to bond them together (van Hoof and van Dijken 2008; Ignatieva and Bubnova 2014). The high natural biodiversity of the suspended meadow could be inherited by the sod roof. The dry load amounts to 250 kg/m^2, and saturated load up to 400–500 kg/m^2. The raw materials are widely available and construction is conducted usually by family members with the help of neighbours. Other cultures have developed and preserved similar natural roofs on their dwellings.

Some recent natural events would trigger human response to the vicissitudes of nature and the need to maintain resilience. Climate change has increased the frequency and intensity of extreme weather events, one of which is heat wave with sombre consequences on human health and mortality in both developed and developing countries (Centers for Disease Control and Prevention 2016; World Health Organization 2016). The more alarming cases include the Chicago heat wave of 1995 killing 735 residents, which was one of the worst in the history of USA. The extensive heat wave in Europe in August 2003 took a staggering 35,000 lives, with 14,000 in France alone. Subsequent detailed study of the catastrophe estimated over 70,000 excessive deaths attributed to the extreme weather event (Robine et al. 2008). In 2015, the India heat wave took 2500 lives. The world needs solutions to the climate-change challenges which have been compounded by relentless urbanization inducing the urban heat island (UHI) effect and other environmental problems.

The urban green infrastructure has been promoted as a nature-friendly and cost-effective way to tackle high-temperature stresses. The provision of amenity and recreational spaces and various ecological-environmental services would further justify installation. For many compact cities, deficiency in ground-level space could be resolved by moving upwards to rooftops and walls. The ancient greenroof concept has recently been given a new lease of life due to a thorough and successful technological revamp. The efforts of individuals and companies would continually improve the greenroof idea and techniques. Some prominent historical examples using old methods and materials deserve to be enlisted.

11.2.2 Prominent Classical Pioneers

The quaint medieval Guinigi Tower was built by the rich merchant family in Palazzo Guinigi in 1384 in Lucca city, Tuscany, Italy. The imposing 38 m-tall fortified red-brick structure, built for defence purpose and as manifestation of power and wealth, has a small square garden perching on the top. The owner bestowed a signature feature on the tower by planting seven Holm Oak trees (*Quercus ilex*) in strip planters to refine and soften the look and to connote rebirth and renewal. It is believed to be one of the oldest existing greenroofs in the world established in the pre-industrial era.

As the Mediterranean tree par excellence, the long-lived native evergreen species, strong and adaptable, can reach a final height of 12–20 m and a crown spread of 15–18 m. The trees on the tower have been confined by the limited soil volume, hence its biological potential size has been curtailed. The original oaks are still dwelling on the prominent site which stands proudly above the surrounding houses. Both the structure and the trees are faring well after some 600 years. For its rich cultural heritage and monuments, the historic centre of Lucca has been put in 2006 on a tentative list of the UNESCO World Heritage Site (UNESCO 2016).

From 1933 to 1936, the Rockefeller Center at the Fifth Avenue of New York was constructed despite the prevailing economic depression, embellished with five greenroofs. They denote the oldest greenroofs on existing commercial buildings in North America (Greenroof.com 2016). The secluded gardens were installed on the top of low-rise seven-storey blocks, and on the eleventh-floor setback portions of high-rise blocks (Fig. 11.1). Different sections have adopted unique but largely formal-geometric garden styles, including neatly-clipped low hedges and topiaries, flower beds, small trees in containers, lawns and a pond (Dailey 2014). In total, the mainly semi-intensive greenroofs occupy about 7000 m^2 which have remained private to in situ office workers. The property has been declared a National Historic Landmark in 1987.

The Derry and Toms Department Store was opened in 1933 at High Street Kensington in London. Its Roof Garden (Fig. 11.2), completed in 1938, has remained operational and preserved most of the original designs after nearly 80 years (Anon 2007; London Parks and Gardens Trust 2016). The 6000 m^2 site on the large rooftop is composed of three connected gardens with distinctive motifs, namely English-Woodland, Moorish-Spanish, and Tudor-Walled. The eclectic spectrum of styles covers intensive to semi-intensive, formal to informal and manicured to naturalistic. The elaborate gardens are richly endowed with medium

Fig. 11.1 One of the greenroofs with formal design on the podium of the Rockefeller Center in New York. It is composed of extensive and semi-intensive types and a shallow pond. Opened in 1936, it is the oldest remaining greenroof on a commercial building in North America (*Photo credit* Tishman Speyer)

and small trees, shrubs, flowers, lawns, courtyards, walls and pergolas with climbers, streams and ponds with waterfowls including four resident pink flamingos, ducks and mandarins, fish, waterfalls, fountains, and wooden and stone bridges. It is now occupied by a private club which allows occasional public access. The gardens and then the building itself were listed as Grade II protected heritage respectively in 1978 and 1981.

11.2.3 New Approaches to Ancient Practice

The modern greenroof revival was initiated in the 1960s in Germany. In response to continued environmental degradation, environmental awareness was lifted with popular demand to protect the environment and live in harmony with nature. The community was eager to adopt environmental solutions to arrest degradation and to bring rehabilitation and improvement. The immediate push was the oil crisis and the urge to find ways to reduce energy consumption. There was a fortuitous spatial-temporal convergence of a basket of factors at an opportune time in Germany. They were conducive to triggering and sustaining the development of modern greenroof technology.

Fig. 11.2 Opened in 1933 as the earliest intensive greenroof in London, this Spanish-Moorish garden is one of the three elaborate roof gardens installed on the then Derry and Toms department store. It provides a fine exemplar for high-quality, complicated yet durable roof greening that has remained continuously in use for nearly eight decades (*Photo credit* C.Y. Jim)

The historical backdrop to the new advances provided the basis to usher greenroof improvements. In nineteenth century Berlin, with rapid urbanization and the need to house the burgeoning rank of workers, many inexpensive housing blocks called *Mietskasernen* (rental barracks) were built. Their flat roofs used tar, a relatively inexpensive material, as the waterproofing barrier. Unfortunately, the exposed material is highly flammable and posed a grave fire hazard. From around 1880, an alternative simple and economic method was invented by a roofer called H. Koch. He shielded the tar with a layer of sand-gravel mixture to reduce the ignition risk. Many German cities henceforth adopted the tar-paper-gravel (TPG) roof cover which could furnish additional thermal insulation (Thuring and Dunnett 2014).

The sand-gravel surface layer with some water-holding capacity (WHC) and nutrient supply was conducive to spontaneous vegetation colonization. The natural seed rain brought propagules by wind and animal agents, whereupon they germinated and grew on the inadvertent growing medium. *Spontaneous green roofs* soon sprang up extensively on such nature-conducive rooftops to provide interesting landscape and practical functions. In the 1960s, some remaining Koch greenroofs were rediscovered and studied (Köhler and Keely 2005). The species composition would be largely determined by the sand-gravel mixing ratio, particle-size distribution, and layer thickness. The TPG roofs with nature's greenery gift could range from simple xeric to mixed-herb communities. Such unintentional urban ecology provided the stimulus and impetus to emulate and improve greenroofs for other buildings.

Research on modern greenroof in Germany generated a series of inventions to revamp and promote the traditional idea. They include light-weight plastic drainage layer, filter, root barrier, waterproof membrane, growth media, and plant-species performance assessment. They aimed at revolutionizing the greenroof know-how to align with new technology and materials. Importantly, their light weight and easy installation method permitted retrofitting on many existing buildings. The scientific achievements initiated and then pump-primed the greenroof movement and market in Germany. Promoted and incentivized by enabling policies (Buehler et al. 2011), millions of square metres of skyrise greenery were soon installed in different cities. The revamped technology and products were quickly embraced by the country with spillover impacts on other European countries.

Reinhard Bornkamm at the Free University of Berlin has been widely regarded as the father of modern greenroof movement. He initiated the study of the Koch greenroofs and spearheaded scientific research from 1961. The 3000 m^2 greenroof on the Geno Haus Bank in Stuttgart, one of the pioneering projects, was built in 1969 under his inspiration. Other researchers such as Gerda Gollwitzer and Werner Wirsing reported their findings on modern greenroofs in their seminal book published in 1971 entitled *Roof areas inhabited, viable, and covered by vegetation*.

The German Landscape Research, Development and Construction Society (Forschungsgesellschaft Landschaftsentwicklung Landschaftsbau) (FLL 2008), was founded in 1975. Encompassing representatives of the greenroof industry, it operates as an independent non-profit and non-government professional-scientific organization. It developed detailed technical greenroof specifications and practical instructions which have been regarded as the state-of-art blue-ribbon standard widely adopted in Germany and other countries around the world. The first issue of the guideline has since been revised and an English edition was published in 2002 to cater to the earnest international demand.

11.3 Enhanced Designs and Prospects

11.3.1 Key Technological Advances

The development of modern greenroof materials in Germany and other European countries was based firmly on basic and applied research. The findings in a basket of cognate sciences subsumed under urban horticulture and urban forestry spearheaded the revamp. A group of related disciplines contributed synergistically to the technological advances, including botany, ecology, soil science, climatology, hydrology, and material science. To transform scientific findings to commercial merchandises required engineering and manufacturing capabilities. Enabling public policies provided the catalyst and impetus.

A series of closely-related new products to be laid in sequence as a package was developed. Each material layer performs a specific function. Different companies generated somewhat different designs using similar materials. However, there is a convergence in general principles and practical functions. Comparing with traditional greenroofs which rely on natural ingredients, the new products mainly use synthetic substances aiming at performance and durability. Massive commercialization and mass production have brought standardization of materials and techniques, and homogenization of greenroof technology down to a small number of variants. Overall, they are characterized by a multiple-layer structure presented as a system, with regular and orderly organization and intimate associations and interactions between components.

The new products were generated from inventions that are usually protected by patents, intellectual-property legislations and conventions. From ideas to marketable goods, the processes involved notable investments. Thus the improvements have brought better but more expensive products, pushing the cost of greenroof installation to a rather high level. Whereas the elevated expenses could be afforded in developed economies, they constitute a barrier to adoption in less developed countries. Since the germinal stage, manufacturing was dominated by a few companies in Europe. Thereafter, the newcomers established outside Europe are concentrated in developed countries in North America, Australia and Japan. The products emerging recently in some developing countries tend to imitate established products and are beset by low quality.

With traditional greenroof serving as the prototype, the following types of products have been used widely in modern installations, to begin from the bottom upwards: (a) root barrier; (b) separation layer (optional); (c) water-absorption felt (optional); (d) drainage layer; (e) filter sheet; (f) water-storage board (optional); (g) growing medium; and (h) vegetation. Waterproofing and thermal insulation materials, as integral parts of the conventional building roof structure, have not been included in the setup. Strictly speaking, they are not greenroof components. Whether a greenroof is installed or not, they have to be included in the roof slab.

The popular adoption of the rather simple Sedum greenroof using analogous methods has led to widespread installation of similar greenroofs in terms of appearance, function and ecology (Fig. 11.3). Due to the limited number of Sedum species in nature, and that not all are suitable for nursery production and greenroof application, the species in use is limited to a small subset.

11.3.2 Boosted Functions of Modern Greenroof

Research on modern greenroof products and designs aims at optimizing the key properties. They allow greenroofs to perform their functions more efficiently, and simplify the installation and maintenance processes.

Fig. 11.3 A Sedum greenroof planted on the top of a one-storey kindergarten building in Bonn. This is an extensive type which requires a shallow substrate with limited water holding capacity to suit the drought-tolerant physiology of the succulent plant. It is light-weight, durable and does not demand frequent maintenance (*Photo credit* Gwendolyn Wong)

(a) Roof live load and greenroof installation

Reducing the greenroof weight can allow more existing buildings with relatively low load bearing capacity to receive greenroofs (weight per unit area, such as kg/m^2). It allows wider adoption of the innovation by retrofitting older buildings with marginal load-supporting design. In greenroof practice, it is essential to ascertain the live load of the roof slab to see whether it can safely take the weight of the added greenroof. Live load is denoted by engineers in SI pressure unit of kPa. For conversion, 1 kPa is equivalent to slightly more than 100 kg/m^2. For practical applications, the round-off value can be used for weight calculations. For new buildings, a light greenroof can reduce the cost of constructing the roof slab.

Different places may have different minimum live load requirement for roofs. The common threshold is about 1.5 kPa, which is meant for access only by maintenance workers. Usually, such roofs are accessible by a vertical ladder rather than a normal staircase. Other roofs may take a higher live load to permit different kinds of use, such as recreation, greening, or to accommodate building-service machines. Before greenroof installation, the site should be inspected to see whether new material layers or machines have been placed after occupation of the building. If so, such added appurtenances would have shifted a portion of the live load to dead load. In judging whether the roof can receive the greenroof, such extra weight should not be neglected.

As different types of greenroofs have different weights, a suitable one can be selected to match the structure. If the live load is too low even for the lightest greenroof, installation should never be recommended, as taking the risk of structural failure is not an option. It is theoretically possible to add structural reinforcement to the roof slab to raise the live load, but it is too expensive and hence not cost-effective to be contemplated. Once a greenroof is installed, it should be taken as an added component of dead load rather than live load. The remaining reduced live load can reckon the number of people and installations that the greened roof can receive. If there is little remnant live load, access should be limited only to maintenance workers and it cannot allow visitor use.

(b) Light-weight drainage and substrate layers

The drainage and substrate layers are the heaviest components of a conventional greenroof. Modern greenroof research accorded priority to developing a light-weight drainage material to substitute the old gravel drainage layer which has to be rather thick to be effective, and hence it is heavy. It has been replaced by a light and thin plastic drainage layer with a characteristic dimpled configuration which shoulders dual roles, namely drainage and water storage. It has indentations or cups to hold some water percolated from the substrate. The stored water can be returned to the substrate by capillary rise or evaporation and condensation. In calculating the drainage-layer weight in greenroof design, the worst-case saturated scenario should be used, assuming that water may not be shed at a sufficiently fast rate.

The substrate weight can be reduced by minimizing its depth to match the need of different plant growth forms. In sequence, Sedum and other succulents need a thin substrate of about 5 cm. Herb and turfgrass can perform reasonably well at about 10 cm. Shrub can do with about 20 cm, and tree can adapt to around 80 cm. These thickness thresholds can be trimmed if a water-storage board is provided. Light-weight minerals (such as vermiculite or perlite) or organic materials (such as peat moss or coconut coir dust) can be blended into the soil to partly replace the heavy mineral soil (Jim 1996; Noguera et al. 2003). Various synthetic organic polymers can be added as hydrogels to the substrate to increase WHC and reduce overall weight (Farrell et al. 2013).

The optional water-absorption felt below the drainage layer or water-storage board below the substrate provide a supplementary source of water for plant growth. They can replace part of the substrate's water-supply function, and hence its thickness and weight could be correspondingly depressed. Regarding the vegetation layer, planting Sedum and other drought-tolerant succulents can notably lower the weight. The use of a particularly light substrate, such as a thin layer of crushed bricks or a mat composed of weaved organic substances, can support Sedum growth quite well (Grant 2006).

(c) Strong root-barrier and drainage layers

Two layers demand mechanically strong materials to fulfil their functions well in the long term. The root barrier at the bottom of the greenroof should withstand puncturing or breakage by roots (tensile strength). It is made of strong plastic such as high-density polyethylene. The gap between adjacent strips could be breached by penetrating roots. Two installation methods have been proposed by manufacturers to tackle this weakness, either by a minimum overlap of 20 cm or by bonding. A continuous band of water-resistant adhesives can glue the strips together. The more durable and effective option is to bond by heat-welding.

The drainage layer has a general dimpled shape that varies with manufacturers, made of plastic such as high-impact polystyrene. The design provides ample internal space to facilitate reception of water from the substrate and lateral shedding of water. The presence of ample space requires a design to use a small amount of materials to resist the pressure imposed by the substrate and other material layers as well as people walking on the greenroof. The plastic structure must not deform, or in the worst case, collapse due to the applied pressure. It must have sufficient compressive strength (normally expected to be >300 kN/m^2) to prevent material failure and squashing. In humid areas with high-intensity rainfall, the drainage capacity requirement demands a thick layer (>25 mm). A thicker layer requires stronger material and design to meet the compressive strength specifications.

(d) Cooling effect and energy saving in the warm season

The thermal behaviour of greenroofs has been studied intensively in different geographic zones. They provide different degree of cooling benefits in the warm season despite climatic differences in the study areas. Two kinds of cooling effect could be differentiated, namely above and below the greenroof (Jim 2014a). Most studies investigate the upward ambient air cooling near the greenroof surface to mitigate the UHI effect. Some studies assess the downward indoor cooling to reduce air-conditioning energy consumption and improve human comfort. Upstream environmental benefits at the power plants include reduction of greenhouse-gas and air-pollutant emissions.

For upward cooling, the evapotranspiration (ET) rate is regulated by inherent factors such as plant species, vegetation cover, and soil moisture availability, and external factors such as solar radiation, temperature, relative humidity and wind speed. Greenroof design and maintenance can regulate the inherent factors to enhance the ET cooling effect due to absorption of latent heat of vaporization. The substrate is more effectively cooled on sunny days than cloudy and rainy days (Jim and Peng 2012). Provision of irrigation can sustain ET and cooling during dry periods. Thicker substrate, and substrate with a good soil structure and hence higher porosity, furnish higher WHC. Adding the water-absorption felt and water-storage board can increase WHC. The water storage function of the drainage layer supplements the substrate WHC. These designs allow more water to be held in the greenroof during rainfall or irrigation events to sustain ET between them.

Additional cooling functions can be attributed to physical shading of the roof surface from direct insolation, and the relatively high albedo of the vegetation. In conjunction with ET, the greenroof temperature on warm days in summer would often register a temperature lower than the indoor. A thermal gradient would be generated to create an upwards heat flux. Indoor heat can be literally drawn upwards to dissipate in the ambient air to result in cooling. Without the greenroof, the bare roof can be heated to a high temperature vis-à-vis the cooler indoor space to generate a reverse thermal gradient to push heat into indoor space.

The different material layers can offer good thermal insulation (R-value, in m^2 K/W). Moreover, the effectiveness of the building's own thermal insulation can influence the efficacy of indoor cooling. The greenroof furnishes an additional thermal barrier to trim heat flux into indoor space. However, under humid-tropical summer conditions, the greenroof substrate often maintains a high moisture content. The high specific heat of water can store a considerable amount of heat derived from intense solar irradiance. The continuity of water in the pores of substrate can raise thermal conductivity and reduce the R-value. Under these circumstances, the warmed greenroof can push heat into the indoor space. A thick substrate with poor internal drainage could keep the soil moist for a long time to accentuate this indoor warming effect, which can increase the cooling load and raise energy consumption (Jim 2014b, 2015). This adverse warming rather than cooling effect of greenroofs has escaped attention until recently.

The study of thermal performance of intensive greenroofs in the tropics verifies that a substrate over 10 cm can effectively suppress downward heat flux into indoor space on hot summer days (Jim and Tsang 2011a, b). This finding could be applied to greenroof practice by designing for a relatively thick substrate which can support semi-intensive or intensive greenroofs. More complex biomass structure can drive higher ET and additional intra-vegetation thermal insulation to achieve better cooling (He and Jim 2010; Jim 2012).

(e) Indoor heat conservation in the cold season

In the cold season, greenroofs are known to form a blanket-like thermal barrier to retard the loss of indoor heat upward through the roof. For artificially-heated indoor space, it can significantly reduce energy consumption. The thermal insulation of greenroof materials in conjunction with the building's own insulation layer can jointly regulate the efficacy of thermal resistance. It can allow saving in indoor heating energy.

The same greenroof R-value that resists ingress of heat to indoor space in the warm season hopefully can work well in the opposite direction to contain leakage of indoor heat. For thin extensive greenroof coupled with building roof slab with poor insulation, heat retention could be ineffective to allow escape of indoor heat through the roof. With high moisture content, the substrate could behave as a thermal conductor rather than a thermal insulator, in which case heat outflow will be accelerated (Coma et al. 2016).

In subtropical areas where artificial indoor heating is seldom practised in winter, the elderly, young and weak could suffer from low indoor temperature during cold spells. Installing an irrigated greenroof can allow absorption and retention of solar energy as sensible heat. The greenroof serves as a thermal mass to store heat in daytime and gradually release it downwards to warm indoor space in both daytime and nighttime (Jim 2014c). This warming effect in the cold season in the subtropics could be actively incorporated into greenroof custom-designed for the climatic zone.

(f) Stormwater management and hydrological restoration

Increasingly, greenroofs are expected to contribute to stormwater management in cities. Urbanization has imposed a high proportion of impervious surface cover with attendant environmental problems. The hydrological behaviour of urbanized watersheds has been substantially altered, resulting in early arrival of runoff peak, higher peak, and shorter duration of discharge. The net result is a temporally-compressed and accentuated discharge regime with high flood risk, demanding heavy investments in constructing and maintaining a large stormwater drainage system.

The sustainable urban drainage system (SUDS), and cognate measures such as Best Management Practices (BMPs), Low Impact Development (LID) and Water Sensitive Urban Design (WSUD), offers stormwater management solutions (Chap. 10 provides detailed analysis of their functions). They include engineering measures such as stormwater ponds, retention basins, vegetated swales, treatment wetland, underground storage tank, and porous paving (Berndtsson 2010). However, such structural techniques require extensive undeveloped lands which are rare in built-up areas especially in compact cities. Greenroof as a non-structural and at-source measure can partly rehabilitate the urban watershed by increasing permeable and evaporating green areas and restoring natural hydrological conditions. The clear advantage is using existing above-ground building tops that would otherwise contribute to hydrological problems (Berghage et al. 2009).

Greenroofs can supply a package of related stormwater benefits (Chap. 10 explains the flood mitigation and other benefits). Similar to natural vegetation sites, different parts of the system can retain water, a portion of which is released to the atmosphere through ET rather than feeding stormwater drains. The vegetation, substrate and drainage layers can jointly participate in retention. Some of the retained water is released at a reduced rate as runoff, resulting in runoff-peak shaving, runoff detention, delayed occurrence of runoff peak, and runoff-duration extension to stop well after the end of the rainfall event (Carter and Rasmussen 2006). These changes can reduce the urban flooding risk which could bring costly and devastating damages. The capital and recurrent cost of the stormwater drainage system can be considerably scaled down. The savings could well be ploughed back to subsidize greenroof installation.

The greenroof can be designed to optimize the stormwater benefits. A thicker substrate with a good soil structure and sufficient porosity of the right size has higher WHC. The moisture bracket that can be filled and released actively should be identified. It is the pore range labelled macro-pores at >60 μm, lying above the field capacity moisture constant at which micro-pores and meso-pores respectively at <0.2 μm and 0.2–60 μm are filled with water (Jim and Peng 2012). The soil composition and structure can be adjusted to ensure a balanced distribution of these three pore-size classes so as to enhance hydrological functions whilst maintaining the quality of a growing medium.

The water-absorption felt or the water-storage board can be added to enhance retention and detention capacities. The intensive greenroof can retain more water than the extensive type. Older greenroofs with more organic matter accumulation and better soil structure can accommodate more WHC. A gentler slope down to a minimum of about 2% can hold more water for a longer time. In general, precipitation depth is inversely proportion to retention (VanWoert et al. 2005). A long duration of the antecedent dry weather days (ADWD) can contribute considerably to regeneration of retention capacity (Stovin et al. 2012).

In the tropical region with frequent and intense rainfalls, the greenroof hydrological responses may differ from findings derived from temperate-latitude studies. With a relatively thin substrate of 4–8 cm, retention capacity is limited, but peak reduction and delay remain notable (Wong and Jim 2014). Even after depletion of retention capacity, peak shaving and runoff detention continue to operate. Adding a rockwool water-storage board can enhance both retention and detention. In the context of the tropical rainfall regime and flood risk, detention is more important than retention which has a limited capacity vis-à-vis the rainfall.

The filter sheet placed between the substrate and drainage layers can block the entry of soil particles into the drainage layer. It is a durable material, usually a non-woven geotextile fabric made of polypropylene. Its permeability is a function of the density and diameter of the punctured micro-holes. By reducing small particles from being carried with the runoff, it helps to improve the quality of the discharged water. In the long run, the micro-holes could be choked by fine sediments to reduce drainage efficiency. This issue needs to be investigated to understand the process and find solutions.

Some greenroofs can contribute to water pollution due to leaching of pollutants from its substrate to induce a disservice (Buffam et al. 2016). Nutrient-rich substrate materials are more vulnerable. Besides nutrient ions, heavy metals and harmful organic compounds from agrochemicals and other sources could contaminate the runoff. Greenroof design could reduce the amount and type of organic matter in the substrate. FLL (2008) recommends a weight threshold of 4–8% organic content, allowing more for substrate with lower bulk density. The source material of substrates, including recycled materials, could be assessed by laboratory tests to ascertain their pollution potential. New greenroofs tend to release more undesirable materials in solution as well as suspended form, whereas aged ones are usually cleaner.

11.4 Inspiring Recent Projects

11.4.1 European Exemplars

Advances in greenroof technology has encouraged installation of many greenroofs. Whereas most sites follow the routine, some are unique and inspiring. In Europe, the Austrian artist Friedensreich Hundertwasser designed the iconic residential block called Hundertwasserhaus in Vienna in 1986 (Kraftl 2009). The fundamental tenet is to build a house in harmony with nature and with the nature of humans with the notable help of tree tenants (Zaraś-Januszkiewicz et al. 2015). It is pioneering due to the thick soil layer that supports a dense and complex woodland on the rooftop, and some pockets of the building devoted to trees. Another Hundertwasser's work, much bigger and more elaborate also with woodland cover, was the Waldspirale (the Forest Spiral) completed in 2000 in Darmstadt in Germany (Silvestre 2007). The name refers to the gradual spiral climb of the greenroof from the ground level to a high point. It could be interpreted as an enhanced and expanded variant of his concept of amalgamating nature with people and houses with roof afforestation.

The Promenade Plantée (French for tree-lined walkway) in Paris, also called Coulée verte (French for green course), is an elevated greenway built on an obsolete railway viaduct (Campaña 2002). It represents the world's first public green space built on an elevated viaduct, as a greenroof installed on the narrow train bed (Fig. 11.4). The old viaduct built by bricks in 1859, abandoned since 1969, could have been demolished. Fortunately, it has been given a new lease of life, completed in 1994, to serve an innovative adaptive use. Without major civil engineering works, it is not as costly as other projects. The disused facility has been repurposed and remade to become an elevated linear park about 9 m above the surroundings. A narrow strip varying from 9 to 30 m wide, it offers an attractive ribbon garden of 6.5 ha.

The revamped site provides varied habitats for diverse plant life and spontaneous wildlife. It features attractive ponds with aquatic organisms. The project has a successful commercial component by adding some 7000 m^2 of shop and 18,500 m^2 of office space under its arches. It serves as an economic catalyst to regenerate the old district. New property developments were triggered by restoring old residences and constructing new ones. The surrounding property value has been lifted by the new amenity. This French novelty has inspired the analogous High Line in New York, which was completed in phases from 2009 to 2014. It should offer insights and a role model for adaptive reuse of old buildings and structures especially in post-industrial societies (Heathcott 2013).

The Petuelpark project in Munich denotes an innovative way to transform the transport land use of a major highway into a linear park (Baureferat 2004). Initially, the project aimed at concealing the noisy and polluting trunk road with a concrete

Fig. 11.4 La promenade plantée, French for tree-lined walkway (aka Coulée verte, French for green course), is a linear greenroof park situated on the old train bed of an abandoned railway viaduct in Paris. Opened in 1993, it denotes the world's first attempt to revitalize an otherwise derelict transport wasteland into a verdant greenway to bring wholesome ecosystem services and salubrious outdoor recreational opportunities to the neighbourhood (*Photo credit* C.Y. Jim)

cover, which was completed in 2002. The road has been literally buried underground by sinking and covering with a concrete roof which measures 650 m by 60 m. The extensive lid has successful abated the nuisance due to the traffic volume of some 120,000 vehicles per day. It returns good air quality and tranquility to adjacent residents.

Subsequent work to install the Petuelpark on the roof of the concrete deck has transformed the site into a nature-in-city haven for people and wildlife (Fig. 11.5). It also re-unites the neighbourhoods which were cut off previously by the road barrier. Instead of environmental blight, the road was converted into a large park covering 7.4 ha which was opened in 2004. It provides pleasant at-grade linkage to nearby existing green spaces. As the roof deck is 2.6 m above the general ground level, and a sufficient soil layer has to be added to support tree growth, the finished level of the park is 3.5 m. The green space is endowed with hundreds of trees which include birch, cherry, pine and plane, thousands of shrubs and herbs and extensive lawns. The construction cost runs to 10.5 million Euros which was considered as rather high.

The analogous Bahndeckel project in Quartier Theresienhöhe of Munich, opened in 2010, signifies another innovative project to use an old reinforced concrete cover

Fig. 11.5 The Petuelpark in Munich presents a pioneering attempt to cover a major highway with a concrete deck to abate its chronic environmental nuisance. A greenroof has been installed on the top of the huge deck to bestow considerable landscape and ecological value on the neighbourhood. Opened in 2002, it denotes an innovative solution to urban blight caused by major transport infrastructure in cities (*Photo credit* C.Y. Jim)

on existing railway tracks. A garden was installed on the large elongated deck (Baureferat 2010). The adjacent residential areas have benefitted from the tranquil and verdant public open space at their doorsteps. It offers a friendly and safe passage for residents living on the two sides of the strip. The concrete slab measures 50 m wide and 300 m long, and provided a spacious garden covering 16,800 m^2 for the neighbourhood. Several thousand flowering herbs, shrubs, small trees, lawns, artificial turf, sand and gravel beds, low profile (3 m) knolls with styrofoam core and rubber veneer, imitation dunes and innovative play apparatus were installed. Due to the limited load bearing capacity of the old deck built in the 1980s, thick soil and large trees cannot be supported. Light-weight materials and designs were liberally used to circumvent the loading constraint (Folkerts 2011). The project cost was 5.4 million Euros. As cities have considerable lands in transport use, decking of railway tracks and highways can offer valuable solution space for green infrastructure installation. Compact built-up areas lacking open space can be inspired by the literal creation of new lands.

11.4.2 Asian Exemplars

Many outstanding greenroofs were built in Japanese cities from the 1980s, and the prominent exemplars could be evaluated. The first greenroof in the traditional Japanese garden style was constructed on the top of the Mitsukoshi department store in Tokyo, which was destroyed in the Kanto earthquake in 1923 and rebuilt (Mikami 2005). The Mitsui Sumitomo headquarters building in the Kanda-Surugadai area Tokyo, completed in 1984, features an attractive dense woodland planted on the low-rise podium block (Kajima Corporation 2016). The 2600-m^2 site has a portion devoted to allotment gardens opened to the public who can obtain the right to cultivate free-of-charge for a year by drawing lots. The site offers a pleasant oasis in the heart of the metropolis, providing ecosystem services such as enrichment of urban biodiversity and cooling to ameliorate the UHI effect.

The iconic ACROS (Asian Crossroads Over the Sea) Fukuoka Prefectural International Hall was completed in 1995. The building occupies a sizeable portion of the precious Tenjin Central Park, and the design was required to literally return land and nature to the people. The 60-m tall structure has 15 levels of terraces covered by lush greenery composed of diverse assemblage of trees, shrubs and herbs (Velazquez 2011). The 50,000 plants supported by 30–60 cm soil depth are derived from 120 species, including voluntary invasions brought by natural seed rain (Greenroof.com 2016). From a distance, the elevated intensive greenroofs of the stepped garden, covering 97,500 m^2, resemble a forested hill. It maintains the proud record as the building with the largest coverage and biomass of vegetation in the world.

The Osaka Municipal Central Gymnasium in Yahataya Park, constructed in 1993–1996, presents another unique and pioneering Japanese project. The main arena with 10,000 seats is shielded by a greenroof covering the partly subterranean stadium structure. The sprawling facility has an unusually wide maximum span of 110 m, maximum height of 26 m, and a wooden floor area of 3580 m^2. The adjacent smaller sub-arena, also covered by a greenroof, has 52 m diameter and 1540 m^2 floor area (Osaka City 2016a, b). The greenroof maintains a world record of the widest roof span. The two extensive low-profile dome-like rooftops have been filled with soil, vegetated and returned to public use as a notable portion of the urban park. At the ground level, they look like two low rolling hills; from the air, they are completely concealed by greenery. Without this ground-breaking multiple-level land use, as much as one-third of the Park would not be usable as green space. The above three Japanese greenroofs have been resoundingly resourceful and successful in resolving the conflicting demands of buildings and urban green spaces.

11.4.3 North American Exemplars

The United States began large-scale roof greening using modern technology from around 2000. The European experience especially regarding species choice, mix and culture may not be directly transferable to North America. In any case, there is the earnest desire to respect native ecology and use indigenous species. The first major public building with a greenroof was the Chicago City Hall installed in 2001 (Dvorak and Carroll 2008). It was conceived as a pilot project to assess feasibility in terms of material, method, maintenance, performance and sustainability for widespread applications in the city. The old building with 11 storeys completed in 1911 has an extensive flat roof covering 3800 m^2. After structural assessment of load bearing capacity, 1100 m^2 were selected for extensive and only 10 m^2 above some interior columns for intensive greenroofs.

The project began with the intention to emulate the high diversity of natural temperate grassland vegetation. As many as 120 plant species in different floristic combinations were assigned to zones to evaluate the factors of solar radiation regime and greenroof system design. Planting was spread from 2000 to 2001, and initial establishment was reasonably promising. Subsequent interactions and competitions modified the species mix with domination by some aggressive tall-grass species. Some portions lost a considerable amount of plant cover. In response, the management revised the maintenance strategy aiming at a prairie garden to echo the pre-urbanization dominant natural vegetation. A continual re-seeding programme gradually modified the species composition. Proactive removal of undesirable species was attempted to regulate species composition. The planting performance highlights the unpredictability and capriciousness of nature in juxtaposing many species on the roof plots. The formal horticulture-oriented objectives of creating regimented patterns may not be achieved despite assiduous maintenance inputs. Perhaps the naturalistic approach could give more berths for nature to run its own course and adjustments.

The Millennium Park in Chicago completed in 2004 is a huge at-grade greenroof sitting on the top of a large underground parking structure and the rebuilt Millennium Station (Gilfoyle 2006). The large Millennium Park Garage has 9.3 ha of parking space for 2126 cars. Measuring nearly 10 ha, the Park probably represents the largest greenroof in the world. Its at-grade level may have defied the impact in comparison with a huge elevated garden. The project denotes a successful urban renewal scheme by adaptive re-use of a large and unsightly brown site previously occupied by disused railway tracks and parking lots. The land was originally earmarked as a portion of the extensive Grant Park, subsequently allocated to railway development, and then reverted to a high-order community amenity after railway use became obsolete.

The huge greenroof differs from others in four ways. It was designed as an urban park sitting on the top of a sprawling underground reinforced concrete structure. It is a multiple-use urban green infrastructure with some substantial, elaborate and expensive facilities and artworks. Its funding mode echoes the fruitful synergy of a public-private partnership with generous donations from individuals, foundations and corporations. The project required US$484 million. It is managed by a private not-for-profit Millennium Park Foundation rather than the municipal government.

11.5 Conclusion

The greenroof idea is deeply rooted in ancient times, believed to be developed at least several thousand years ago initially in a serendipitous manner. The harsh high-latitude climatic conditions probably mobilized human ingenuity to find ways to seek effective shelters from the elements. From primitive conception, the skill has been progressively honed and enhanced to become a well-established cultural endowment. Subsequent conscious and formal installations on domestic structures in pre-historical and historical times were largely confined to rural areas in regions with tough weather conditions. The advent of urbanization has seldom adopted this invention which could have provided a tool for climate adaptation and human comfort in our indoor cocoon.

Revival of the idea in Europe in the mid-twentieth century triggered scientific research and technological innovations to revolutionize the materials, methods and designs of greenroof. The movement dovetailed with the rising environmental awareness and societal desire to re-establish a cordial relationship with nature. Despite the major technical revamps, the fundamental principles and functions of roof greening have persisted with little modifications. After half a century of modernization, greenroofs have taken root in different parts of the world, extending from the European locus to many cities. As a well-known innovation with many far-reaching benefits, it has been earnestly embraced by most municipal authorities and citizens.

Thus far, the new materials and designs demonstrate remarkable convergence. The real-world applications in diverse climatic zones and varied building conditions have generated in general similar benefits. Their performance, however, tends to differ quite markedly due to local climate and building and site circumstances. In some places, the expected benefits have been feebly expressed or even contradicted. The present materials, designs and applications have room for further improvement to cater to the increasingly diverse as well as specific needs. A detailed stocktaking of the past and present experiences offers hints for future improvements. It is hoped that continued research can steer the innovation to a more tailor-made, cost-effective and environmentally-friendly mode.

Acknowledgements The research grant supports kindly provided by the Matching Fund of the University Grants Committee, Dr. Stanley Ho Alumni Challenge Fund and MTR Corporation are gratefully acknowledged.

References

Anon (2007) Derry and Toms: a history of the building and roof gardens. http://www.ralphhancock.com/theroofgardensatderry%26toms. Accessed 10 June 2016

Baureferat (2004) Petuelpark information. Landeshauptstadt, München

Baureferat (2010) Quartiersplatz Theresienhöhe. Landeshauptstadt, München

Berghage RD, Beattie D, Jarrett AR, Thuring C, Razaei F, O'Connor TP (2009) Green roofs for stormwater runoff control. Environmental Protection Agency, Washington, DC

Berndtsson JC (2010) Green roof performance toward management of runoff water quantity and quality: a review. Ecol Eng 36:351–360

Buehler R, Jungjohann A, Keeley M, Mehling M (2011) How Germany became Europe's green leader: a look at four decades of sustainable policymaking. Solutions 2(5). http://www.thesolutionsjournal.com/node/981. Accessed 18 July 2016

Buffam I, Mitchell ME, Durtsche RD (2016) Environmental drivers of seasonal variation in green roof runoff water quality. Ecol Eng 91:506–514

Campaña JG (2002) From Promenade Plantée to the New York High Line. Research Paper, Hixon Center for Urban Ecology, Yale University. http://hixon.yale.edu/sites/default/files/files/fellows/paper/gonzalex-campana_javier_2002_report.pdf. Accessed 8 May 2016

Carter TL, Rasmussen TC (2006) Hydrologic behavior of vegetated roofs. J Am Water Res Assoc 42:1261–1274

Centers for Disease Control and Prevention (2016) Climate change and extreme heat events. Washington, DC

Coma J, Pérez G, Solé C, Castell A, Cabeza LF (2016) Thermal assessment of extensive green roofs as passive tool for energy savings in buildings. Renew Energy 85:1106–1115

Dailey J (2014) Rockefeller Center's rooftop gardens are a hidden urban treasure. Inhabitat New York City. http://inhabitat.com/nyc/the-rockefeller-centers-rooftop-gardens-are-a-hidden-urban-treasure/. Accessed 20 July 2016

Dvorak B, Carroll K (2008) Chicago City Hall green roof: its evolving form and care. Greening rooftops for sustainable communities, Baltimore, MD, 30 Apr–2 May 2008

Farrell C, Ang XQ, Rayner JP (2013) Water-retention additives increase plant available water in green roof substrates. Ecol Eng 52:112–118

FLL (2008) Guideline for the planning, execution and upkeep of green-roof sites. Forschungsgesellschaft Landschaftsentwicklung Landschaftsbau, Bonn

Folkerts T (2011) (ed) Topotek 1 Rosemarie Trockel: a landscape sculpture for Munich. Birkhäuser, Basel

Gilfoyle TJ (2006) Millennium Park: creating a Chicago landmark. University of Chicago Press, Chicago

Grant G (2006) Extensive green roofs in London. Urban Habitats 4(1):51–65

Greenroof.com (2016) Rockefeller Center roof gardens. http://www.greenroofs.com/projects/pview.php?id=666. Accessed 20 July 2016

He HM, Jim CY (2010) Simulation of thermodynamic transmission in green roof ecosystem. Ecol Model 221:2949–2958

Heathcott J (2013) The Promenade Plantée: politics, planning, and urban design in postindustrial Paris. J Plann Educ Res 33:280–291

Ignatieva M, Bubnova A (2014) The new is well forgotten old: Scandinavian vernacular experience on biodiverse green roofs. The Nature of Cities. http://www.thenatureofcities.com/2014/08/28/the-new-is-well-forgotten-old-scandinavian-vernacular-experience-on-biodiverse-green-roofs/. Accessed 20 July 2016

Jim CY (1996) Edaphic properties and horticultural applications of some common growing media. Commun Soil Sci Plant Anal 27:2049–2064

Jim CY (2012) Effect of vegetation biomass structure on thermal performance of tropical green roof. Landscape Ecol Eng 8:173–187

Jim CY (2014a) Green roof energetics: systematic organization and integration of knowledge. In: Britz J, Koehler M, de Felipe I (eds) Green cities in the world: progression, innovation, organization. Editorial Agricola Espanola, Ciudad Universitaria, Madrid, pp 251–270
Jim CY (2014b) Heat-sink effect and indoor warming imposed by tropical extensive green roof. Ecol Eng 62:1–12
Jim CY (2014c) Passive warming of indoor space induced by tropical green roof in winter. Energy 68:272–282
Jim CY (2015) Assessing climate-adaptation effect of extensive tropical green roofs in cities. Landscape Urban Plann 138:54–70
Jim CY, Chan MWH (2016) Urban greenspace delivery in Hong Kong: spatial-institutional limitations and solutions. Urban Forest Urban Greening 18:65–85
Jim CY, Peng LLH (2012) Substrate moisture effect on water balance and thermal regime of a tropical extensive green roof. Ecol Eng 47:9–23
Jim CY, Tsang SW (2011a) Ecological energetics of tropical intensive green roof. Energy Build 43:2696–2704
Jim CY, Tsang SW (2011b) Biophysical properties and thermal performance of an intensive green roof. Build Environ 46:1263–1274
Kajima Corporation (2016) Nature in an urban setting: a green network at Mitsui Sumitomo Insurance. http://www.kajima.com/news_events/special_features/vol3/vol3-4.html. Accessed 12 July 2016
Köhler M, Keely M (2005) Berlin: green roof technology and policy development. Green roofs: ecological design and construction. Schiffer, Atglen, pp 108–112
Kraftl P (2009) Architectural movements, utopian moments: (in)coherent renderings of the Hundertwasser-Haus, Vienna. Geogr Ann Ser B Hum Geogr 92:327–345
London Parks and Gardens Trust (2016) The roof gardens: formerly Derry and Toms Roof Garden. http://www.londongardensonline.org.uk/gardens-online-record.asp?ID=KAC055. Accessed 12 June 2016
Mikami T (2005) Tokyo: cooling rooftop gardens. Green roofs: ecological design and construction. Schiffer, Atglen, pp 113–116
National Parks Board (2002) Handbook on skyrise greening in Singapore. National Parks Board and Centre for Total Building Performance, Department of Building, School of Design and Environment, National University of Singapore
Noguera P, Abad M, Puchades R, Maquieira A, Noguera V (2003) Influence of particle size on physical and chemical properties of coconut coir dust as container medium. Commun Soil Sci Plant Anal 34:593–605
Osaka City (2016a) Osaka Municipal Central Gymnasium. https://www.yahataya-park.jp/wp/wp-content/themes/yahataya/common2/dl/brochure_arena_jp.pdf. Accessed 12 June 2016
Osaka City (2016b) How to build the OMC gymnasium (in Japanese). http://www.yahataya-park.jp/osaka_arena/analysis/index.html. Accessed 12 June 2016
Osmundson T (1999) Roof gardens: history, design, and construction. Norton, New York
Robine JM, Cheung SL, Le Roy S, Van Oyen H, Griffiths C, Michel JP, Hermann FR (2008) Death toll exceeded 70,000 in Europe during the summer of 2003. CR Biol 331(2):171–178
Silvestre FL (2007) Del dachgärtenal green roofscape (1970–2005) aportación a la historia reciente del paisaje urbano. Saitabi 57:169–203
Stovin V, Vesuviano G, Kasmin H (2012) The hydrological performance of a green roof test bed under UK climatic conditions. J Hydrol 414–415:148–161
Tan PY (2013) A vertical garden city. Straits Times Press, Singapore
Thuring CE, Dunnett N (2014) Vegetation composition of old extensive green roofs (from 1980s Germany). Ecol Process 3(4):1–11
UNESCO (2016) Historic centre of Lucca. World Heritage Conservation. http://whc.unesco.org/en/tentativelists/340/. Accessed 21 July 2016
van Hoof J, van Dijken F (2008) The historical turf farms of Iceland: architecture, building technology and the indoor environment. Build Environ 43:1023–1030

VanWoert ND, Rowe DB, Andresen JA, Rugh CL, Fernandez RT, Xiao L (2005) Green roofs stormwater retention: effects of roof surface, slope, and media depth. J Environ Qual 34:1036–1044

Velazquez L (2011) ACROS Fukuoka Prefectural International Hall. Sky Gardens Blog. http://www.greenroofs.com/blog/2011/08/12/gpw-acros-fukuokaprefectural-international-hall/. Accessed 20 July 2016

Wong GKL, Jim CY (2014) Identifying keystone meteorological factors of green-roof stormwater retention to inform design and planning. Landscape Urban Plann 143:173–182

World Health Organization (2016) Climate change and health. Geneva, Switzerland

Zaraś-Januszkiewicz E, Fornal-Pieniak B, Żarska B (2015) Tree in Teodor Talowski's and Friedensreich Hundertwasser's organic architecture. Plants in urban areas and landscape. Slovak University of Agriculture in Nitra, Faculty of Horticulture and Landscape Engineering, pp 167–171

Chapter 12
Urban Ecological Networks for Biodiversity Conservation in Cities

Abdul Rahim Hamid and Puay Yok Tan

Abstract Triggered by concerns of global biodiversity loss, cities are increasingly called upon to play an increased role in biodiversity conservation, leading to a surge in interest in urban biodiversity conservation. In playing this role, greening of cities needs to move beyond mere provision of amenities or ecosystem services to one of providing habitats for native biodiversity. This chapter describes one of the approaches for enhancing urban biodiversity conservation through the ecological network approach. The concept of ecological networks is not new in the field of ecology. However, its application to cities, both in conceptual and operational forms, is highly limited. As a high-rise, high-density city in which biodiversity conservation is threatened by other competing land uses, Singapore is used as an example to illustrate the development and application of the ecological network approach. The ecological network is built on the concept of *network*, *spatial* and *landscape* cohesions. Using methods in landscape ecology, remote sensing, biodiversity conservation and the Analytic Hierarchy Process, this chapter describes how a toolkit for ecological network can be developed, as well as the efficacy of its use for biodiversity management. The toolkit is categorized into *monitoring tools*, *mapping tools*, and *communication and decision making tools*. The learning outcomes gleaned from the research are presented as the *5-multis*: *multispecies*, *multiscalar*, *multilevel*, *multifunctionality*, and *multidisciplinarity*.

Keywords Ecological network · Urban biodiversity conservation · Sustainable landscape · Patchiness index · Analytic hierarchy process · Landscape planning

---

A.R. Hamid (✉) · P.Y. Tan
National University of Singapore, Singapore, Singapore
e-mail: akiarah@nus.edu.sg

P.Y. Tan
e-mail: akitpy@nus.edu.sg

P.Y. Tan and C.Y. Jim (eds.), *Greening Cities: Forms and Functions*,
Advances in 21st Century Human Settlements,
DOI 10.1007/978-981-10-4113-6_12

12.1 Importance of Urban Biodiversity

For long periods in the urban development history of human settlements, cities are traditionally perceived as places where development and economic growth take place, and where the natural environment, ecological processes, and biodiversity or wildlife have limited roles. Planned green spaces, and those that act as buffers or fillers in urban developments such as commercial and residential estates, are usually not designed as potential habitats for biodiversity, let alone as part of a city's plan for biodiversity conservation. Instead, parks and green open spaces found dotted in many cities often cater only to the recreational needs of urban dwellers.

With the realisation that extinction of species is taking place globally at an unprecedented pace (Sodhi et al. 2007; Seto et al. 2012; Pimm et al. 2014), international efforts such as the Convention on Biological Diversity and the Millennium Ecosystem Assessment, have begun to foster more global, national and local efforts to tackle biodiversity loss (Secretariat of the Convention on Biological Diversity 2006, 2012), including efforts directed at urban areas. Such endeavours are also driven by the increasing amount of scientific evidence pointing to immense consequences of biodiversity loss. The impacts of biodiversity loss can be summarized as: (1) it reduces ecosystem functions, such as resource capture, biomass production, and nutrient recycling (Cardinale et al. 2012; Naeem et al. 2012; Balvanera et al. 2006); (2) it reduces resilience of ecosystems to shocks or stress (Cardinale et al. 2012; Steudel et al. 2012); (3) it now ranks alongside climate change as a key environmental driver (Cardinale et al. 2012); and (4) it has negative impacts on human well-being, including direct impacts on our physical and mental health (Hough 2014; Horwitz et al. 2015).

Given that the majority of humans now live in cities, which exert immense influence on all ecosystems within and outside their boundaries, it has also been argued that cities therefore have a role to play in biodiversity conservation (Chan and Djoghlaf 2009), as well as an ethical duty to conserve biodiversity (Dearborn and Kark 2010). The role of cities can be direct, for instance, through integration of biodiversity conservation objectives, strategies, and targets into their sustainability and land use plans (Chou 2010; Pickett and Zhou 2015; Secretariat of the Convention on Biological Diversity 2012), and the conservation of remnant native vegetation within cities and in their hinterlands. The role can also be indirect, by shaping environmentally responsible behaviour towards the preservation of natural ecosystems through exposing urban dwellers to natural environment in cities (Savard et al. 2000; Miller 2006). Indeed, given the important role played by cities, urban biodiversity conservation and enhancement as a worthy cause and necessary research topic is increasingly voiced by different scholars (Farinha-Marques et al. 2011; Brown and Grant 2005; Blaustein 2013). However, what are suitable approaches for urban biodiversity conservation and enhancement in cities?

One common approach is the use of nature reserves or other protected areas. In many cities and their immediate regions, remnant habitats for biodiversity are often preserved within protected areas. However, based on the premise that there is not enough land for all biodiversity to be encapsulated within such reserves, habitat loss is inevitable when further land expansion and exploitation of natural ecosystems occur. Protecting biodiversity through protected land alone is considered inadequate as certain species may have limited mobility (Kostyack et al. 2011), and reserves are also affected by ecological conditions of land around them (Laurance et al. 2012). Biodiversity protection in reserves is thus necessary, but inadequate.

Other approaches have also been used, including incorporating biodiversity conservation objectives in the design and management of urban green infrastructure. 'Green infrastructure', defined as 'an interconnected network of natural areas and other open spaces that conserves natural ecosystem values and functions, sustains clean air and water, and provides a wide array of benefits to people and wildlife' (Benedict and McMahon 2006), can be designed to provide habitat and facilitate biodiversity movement. Often conceived as part of the green infrastructure are 'ecological corridors' formed through contiguous linear vegetated areas, and which incorporate overpasses or underpasses to facilitate faunal movement. However, there does not appear to be a consensus on the efficacy of corridors for biodiversity conservation. For instance, Hilty et al. (2006) cautioned the assumption that corridors will inevitably improve connectivity among habitat fragments. This is because a corridor to humans may not be judged the same by target species, and that the results of species dispersal via corridors may depend on the spatial and temporal scales of the observations. Rosenberg et al. (1997) also reviewed the functional efficacy of corridors for conservation and suggested alternatives to corridors. The lack of information on the effectiveness of such corridors for urban landscapes is even more pronounced. Given that the field of ecological connectivity research in urban areas is still in its infancy (LaPoint et al. 2015), much remains to be explored and understood, including translating conceptual understanding from conservation science to urban areas, methodologies and approaches, and collection of empirical evidence. In Chap. 7, Werner and Kelcey also described the approach of managing urban biodiversity through the focus on actions at three spatial scales.

This chapter describes the ecological network approach to biodiversity conservation, in which a cohesive network of ecosystem types is interlinked with the greenery found in the urban matrix, forming an ecological network. The following sections describe the conceptual foundation of the approach, methodology of deriving the ecological network applied to Singapore as a high-density compact city, and the advantages of the methodology. In addition, given that the application of the concept to cities is still uncommon, we also highlight additional areas for further research to develop the concept to its full potential.

12.2 Ecological Networks for Urban Biodiversity Conservation

12.2.1 Conceptual Bases of Ecological Network

Many cities have incorporated biodiversity conservation objectives into their land use policies to varying extents. There are, however, few realized examples in which urban biodiversity programmes are implemented based on principles from conservation science. In fact, we suggest that even though a city may be physically green, the mere presence of greenery does not necessarily support a wide range of biodiversity, especially those of high conservation importance. This is because different biodiversity groups have specific requirements in terms of access to resources, be they water, food and shelter, as well as a sufficient habitat size for foraging and reproduction. Habitat size, home range, and barriers to dispersal are also ecologically scaled, and thus the conservation of specific species requires a species-based approach for habitats to be provided within the green spaces in the city. An ecological network approach attempts to incorporate knowledge from conservation science, including landscape ecology which provides the tools and methods that relate spatial patterns to ecological processes, as well as urban ecology which provides an understanding of not just the ecological, but also the social determinants of cities as habitats.

The ecological network concept has been proposed more than two decades ago by Cook and van Lier (1994). Definitions of 'ecological network' emphasize the 'coherence' of natural areas or ecosystems, achieved via linking them together into a 'system', as well as interacting with the surrounding landscape matrix or the urban greenery found in the urban matrix (Jongman 2004; Opdam et al. 2006). Such networks can be characterized by four structural features, namely 'total network area', 'network density', 'network quality', and 'permeability of the matrix' (Opdam et al. 2006). The total network area and density are dependent on the amount and spatial distribution of vegetated patches. The closer the fragmented patches are, the higher the network density will be, even if the total area of all the patches is the same as another group of patches that are further apart. Network quality refers to a number of aspects: the suitability of the habitats found in the network local population's growth and extinction rate, and the intensity of the dispersal stream of species across the landscape (Opdam et al. 2003). Thus, even with local extinctions at individual patches, as long as there are recolonizations through species dispersal, metapopulations of various species over the entire region can remain above the minimum threshold for each species. In addition, the suitability of the habitats is dependent on specific plant species and growth requirements for every stage of the organism's life cycle, as well as the multi-layered canopy structure of a tropical biome.

Opdam et al. (2003) proposed a comprehensive framework that integrates the requirements of individual species into multispecies indicators at the landscape scale. It builds upon the more fundamental concept of 'habitat networks', which is a

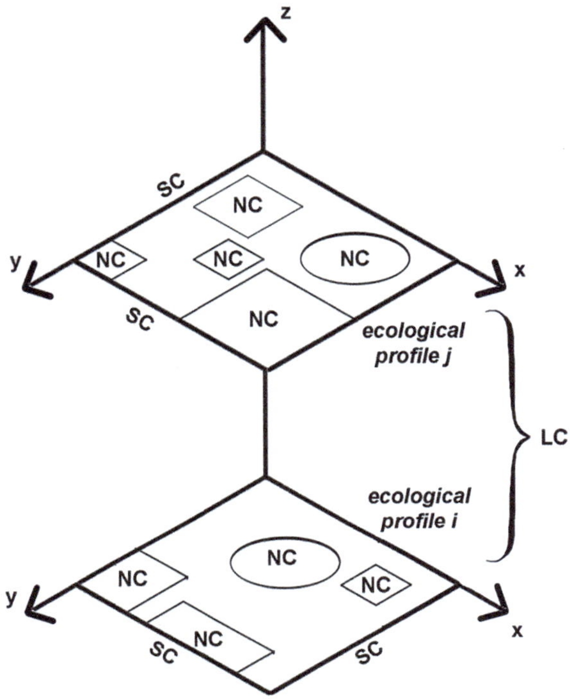

Fig. 12.1 A diagrammatic representation of the ecological network framework by Opdam et al. (2003, 2006). The layers represent the ecological profiles under consideration. Within each layer is a *large square* representing the study area or region, which constitutes the 'spatial cohesion (SC). The *smaller shapes* are notional representations of the individual habitat networks for each ecological profile at various locations within the study area. Each habitat network has its own 'network cohesion' index (NC). 'Landscape cohesion' (LC) represents a multi-species index for the whole study area

single network for a single species within a specified spatial scale (Opdam 2002). The framework includes a 'system of ecological profiles', an index for 'network cohesion' (NC) for a single habitat network, an index for 'spatial cohesion' (SC) which integrates the entire habitat networks for a region, and a multi-species index for 'landscape cohesion' (LC) (Fig. 12.1).

The ecological network as suggested by Opdam et al. (2006) is also meant to fulfil the objective of landscape sustainability via three conditions. Firstly, the concept supports the condition that landscape structure needs to reinforce the necessary ecological, social and economic processes so that goods and services in the form of biodiversity as its prime 'resource' can be delivered to the current and future generations. Only a large-scale coherent structure of ecosystems that supports ecological processes at various scales can attain this objective, underpinned by empirical scientific knowledge and theory.

Secondly, as habitats fragment, the network population transforms into a 'metapopulation' once the fragmentation threshold is passed (Verboom et al. 2001). The concept however allows for changes to the area, location, and configuration of the components of the networks depending on land use needs, as long as the 'persistence probability' of the target metapopulations does not fall below a certain threshold in the long run (Opdam et al. 2003). The flexibility conferred by the metapopulation concept enables different configurations of the ecological network to be planned as strategies for 'ecologically sustainable landscapes', yet having the same aim of biodiversity conservation, thus conferring a 'sustainable trajectory' to landscapes and ecosystems rather than a 'steady state' (Haines-Young 2000). This 'adaptive approach' to biodiversity conservation is fundamentally different from the protected areas approach adopted by some cities (Bouwma et al. 2003).

Thirdly, the ecological network concept emphasizes the importance of communication between multiple stakeholders for effective decision making. This is also the stage in which knowledge transfer between ecology and planning happens. However, there is still a big gap between ecology and planning (Moss 2000; Opdam 2002). In addition, due to human preferences coming into play at this stage and the complex processes involved, the communication process should be geared towards setting an 'ambition level', which is a measure of socio-economic feasibility, as an aspiration towards biodiversity conservation goals (Opdam et al. 2006). Moreover, a delineated ecological network can facilitate the decision of identifying and prioritizing areas of high biodiversity potential or key corridors which need to be left intact, as compared to areas which can be compromised, or integrated with other land uses as multifunctional landscapes or mixed-use developments.

Despite the concept having been proposed more than twenty years ago, the development of ecological networks is still limited, particularly for application to cities. Some examples of studies include the Netherlands (Bruinderink et al. 2003), Bologna and Modena in Italy (Bolck et al. 2004), two north-eastern cities of the Seoul Metropolitan Area, Namyangju and Guri (Oh et al. 2011), and Florida (Hoctor et al. 2004). In most studies, the focus is also limited to a single taxonomic group, such as mammals for Amsterdam (Bruinderink et al. 2003), Kuala Lumpur (Reza et al. 2013), and Namyangju and Guri (South Korea) (Oh et al. 2011). A limitation of such studies is that due to ecological scaling, different species interact differently with the landscape. Thus, the structure of the habitat network for each species and their associated processes are at different scales for the same landscape. Therefore, some of the important components of a complete ecological network, such as the consideration of multiple taxonomic groups, may be inadvertently excluded from planning considerations.

As for developing Southeast Asian cities in a tropical biome, a limited number of studies focus on the location and delineation of protected areas, urban green space planning, and changes in landscape spatial patterns (Phua and Minowa 2005; Abdullah and Nakagoshi 2006). There are also scientific studies on ecological networks that approach from a species level, but which has yet to be translated into city planning (Dover and Settele 2008; Dennis et al. 2013).

12.2.2 The Challenge for Cities

Cities face the challenge of attempting to connect urban green spaces with habitats because, unlike the well-defined ecological corridors found in the regional landscapes external to the city, the boundary between patches and the urban matrix is increasingly blurred in a compact city. Moreover, the distinction between interior species, edge species, and those which have adapted and prefer urbanized areas imbues an increased complexity to the task at hand. Furthermore, the least-cost path analysis and the dispersal models used to derive these ecological networks have usually been applied to large regional areas rather than to highly dense and compact cities.

This scenario thus poses a challenge as to how a tapestry of habitats can be embedded within the highly urbanized and dense city fabric or vice versa. It remains to be seen as to how an ecological network approach could be applied to highly urbanized and dense cities, whereby the ecological network serves primarily as a network for the metapopulation of multiple species, especially when multiple barriers exist in urban areas. Another gap is the lack of studies or application of the ecological network concept to cities and regions in a tropical biome, especially in cities undergoing rapid urbanization in which native species are replaced by exotic species in urban areas. There are no precedents to guide this approach.

The next section shows how the ecological network framework described above by Opdam et al. (2003, 2006) serves as a springboard for determining ecological networks for urbanised areas and cities, firstly by incorporating its existing green infrastructure, secondly by capitalising on the ability of small, disturbed remnants of primary forest to retain a large proportion of their non-vertebrate diversity after a long-time isolation (Corlett 1992), and thirdly by enhancing the role of parks, reserves and fragments for biodiversity conservation (Koh and Sodhi 2004).

12.3 Tools for Developing and Monitoring Ecological Networks

Since Opdam et al. (2003, 2006) did not provide a mechanism for the operationalization of their framework, the rest of this chapter describes how this can be achieved via a synthesis of various methods and techniques employed by other researchers, including those for studies of landscape fragmentation, vegetation composition and structure, biodiversity conservation, habitat and dispersal characteristics of various fauna, and decision making processes. The methodology is underpinned by the concepts of ‘land mosaics’ (Forman 1995), ‘ecological scaling’ (Vos et al. 2001), and metapopulation under the ‘non-equilibrium paradigm’ of ecology (Fiedler et al. 1997; Wu and Loucks 1995; Wiens 2002).

12.3.1 Methodological Framework

There are three broad stages to characterize the ecological network for a city (Fig. 12.2). The first stage is to derive a vegetation cover map from a remotely sensed image, followed by characterizing its spatial structure, which consists of the composition and configuration of vegetation patches in relation to one another. The objective is to develop a detailed understanding of the spatial distribution and structure of green spaces as a starting point to map out the components of the ecological network. Depending on the resolution and type of satellite image obtained, various analytical and modelling processes can be applied. The key steps involve the use of 'landscape metrics', which are 'indices that quantify specific spatial characteristics of patches, class of patches, or entire landscape mosaics' (McGarigal and Marks 1995), as well as statistical analysis to select only the key metrics out of numerous possibilities.

The second stage corresponds to a species-based approach as a link between the spatial structure of vegetation and ecological processes. The objective of this stage is to apply the ecological network framework by Opdam et al. (2003, 2006) to a

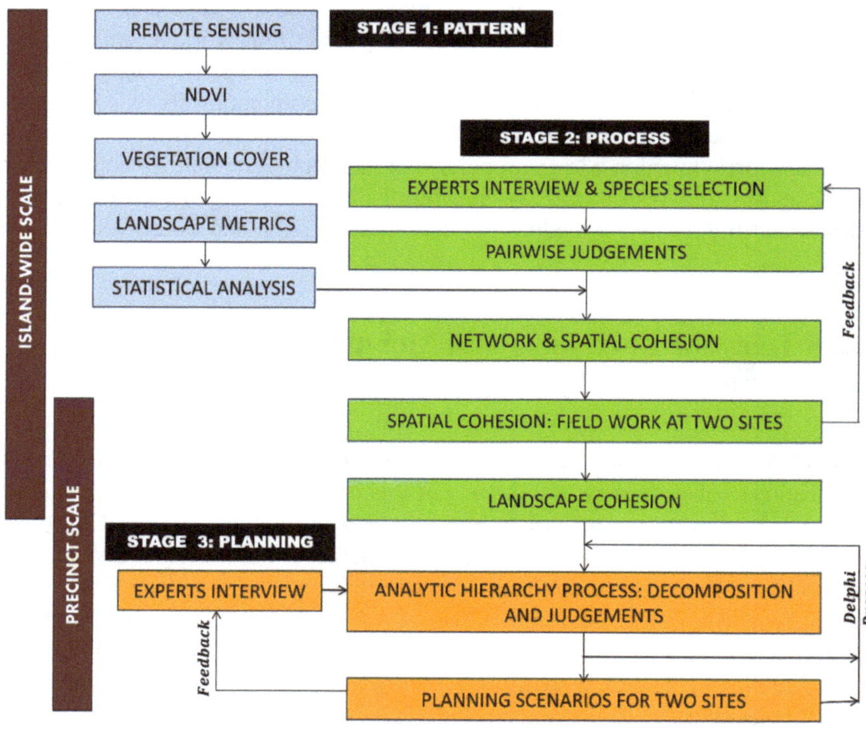

Fig. 12.2 Overall methodology to determine ecological networks for urbanized areas. *Stages 1*, *2* and *3* correspond to the *blue*, *green* and *orange boxes* respectively

highly urbanized area in the assessment of focal species-landscape structure interactions. The concept of ecological scaling is incorporated, which is the species-specific perception of the landscape, ranging from those with small home ranges to larger ones (Vos et al. 2001; Verboom and Pouwels 2004). It involves habitat suitability modelling and determining the least cost corridors for multiple species at an island-wide scale for a suite of 'focal species' (Lambeck 1997; Coppolillo et al. 2004).

This is followed by a further in situ vegetation assessment from patch-corridor-patch, as well as ground truthing of the maps derived earlier. In particular, it verifies the degree of vertical stratification of vegetation, the amount of canopy cover, the presence of plant resources for each species, and the confirmation of barriers. This is done at sites deemed as priority in terms of delineating areas for conservation versus those which can be compromised for development.

The third stage involves the application of the defined ecological networks into planning by employing both the 'decomposition' and 'judgement' steps of the Analytic Hierarchy Process (AHP) developed by Saaty (1980, 2001, 2005, 2008) and Saaty and Vargas (2001). The objective is to facilitate communication and decision-making between stakeholders, while highlighting the trade-offs needed when reconciling biodiversity conservation with land use needs. In addition, several scenarios are possible by assigning a 'power' to the weights assigned by each corresponding actor in turn, and one where all actors have the same 'power' (Saaty 2008).

The AHP is applied at the planning unit scale. Employing this scale at this stage can facilitate the negotiations for an agreed 'ambition level' while allowing for human preferences to come into play. In addition, the planning unit scale matches the 'chorological' scale at which humans perceive landscape scenery (Zonneveld 1994), thus enabling a more effective communication pathway with the public in the effort to integrate ecology with planning, rather than at the regional or an island-wide scale for the whole of Singapore. Three dimensional visualisation techniques can also be incorporated in the generation of scenarios to complement stakeholder or community engagement in planning for biodiversity conservation within the planning unit.

The results of the AHP can then be presented to the stakeholders individually according to a Delphi process, after which they can decide if they would like to amend their judgements. The Delphi process aims to seek consensus amongst several experts, or to extract key expert-based opinions via a 'structured communication process', which may involve the use of questionnaires, or systematic analyses of statements or key phrases (Linstone and Turoff 1975). A modified Delphi approach allows for a group discussion after one round of individual interviews or questionnaires in order to open up new insights into the issues, have honest discussions, and achieve a final consensus (James et al. 2009). This opens up channels of communication between stakeholders with the aid of maps and diagrams, as well as facilitates the decision of the preferred scenario for each site, followed by the accompanying mitigation or compensation measures to ensure that the overall ecological network remains coherent.

The methodological framework is both comprehensive and efficient in terms of characterizing the ecological network for urbanized areas. Both top-down and bottom-up approaches are used, including matching the species to the landscape scales of biodiversity. The use of computational tools such as Geographic Information Systems (GIS) and FRAGSTATS enables fast and efficient analyses of vegetation and other data needed to derive the ecological network. The scale of the analysis covers an island-wide or regional spatial scale to the more detailed planning unit scale, which facilitates the fine-grain analysis of habitat quality. This is usually limited by a lack of GIS data for vegetation composition and other environmental variables.

The flexible and modular nature of GIS which organises data into layers allow for any desired number of criteria and experts as required for both the elucidation of the ecological network, as well as for its application into the planning process (Haines-Young et al. 1993). Taking a layered approach also implies that the analysis could incorporate a range of fauna, while most studies focus only on one taxon or a particular species. The breadth of analysis is limited only by the extent of species and criteria that the planner wishes to incorporate. Only with the derivation of the separate habitat networks for each species prior to their integration into an ecological network would the concept of 'spatial cohesion' followed by 'landscape cohesion' make sense (Opdam et al. 2003, 2006). This approach is also a methodical way of incorporating ecological scaling into the analysis. The ecological network can then be built upon over time as research in the various species progresses.

Moreover, all four components of the ecological network framework by Opdam et al. (2003) are embedded within the methodology, which is important if the concept of ecological networks is to be effectively addressed within the planning framework. Unlike most studies on species distribution, this study includes the dimension of habitat quality, which covers both specific vegetation characteristics as well as the multi-layered vertical stratification of tropical vegetation. Matrix permeability is also taken into account, based on the ideal vegetation and environmental characteristics of dispersal corridors denoted as least-cost corridors or paths, rather than Euclidean distance. The definition of the matrix is also dependent on the type of species under consideration.

The third stage demonstrates how the concept of ecological networks can be incorporated into a communication and decision-making tool while taking into account both the ecological and social aspects of landscape provision. The results are presented as maps and charts using GIS and mapping technology which facilitates communication and understanding even for the layperson, and enables choices to be made in terms of what is important for the community. It is an iterative process whereby the outcomes may feedback into the formulation of new objectives, conservation goals, and planning outcomes, or refining existing ones. As much feedback can be sought from all stakeholders, while scientific findings can be communicated back to them. Therefore, this tool has the potential to provide an opportunity for community participation early in the planning process to determine the ecological network for a city.

12.3.2 Monitoring Tools: Locations of Cohesive Networks for Biodiversity and Use of Indicators

An updated vegetation cover map is essential in the determination of ecological networks. In our study, it was derived from a multispectral SPOT (*Satellite pour l'Observation de la Terre*) satellite image of Singapore for June 2012 using the Normalised Difference Vegetation Index (NDVI) on the GIS software, ArcMap version 10. Since the pixel resolution of 10 m by 10 m is rather low, Gutman and Ignatov (1998) introduced the concept of 'green vegetation fraction', which is a function of the NDVI, to estimate the amount of vegetation found within each pixel. This led to a map of the vegetation cover in five percentage vegetation cover classes, namely 0, 25, 50, 76, and 100% (Fig. 12.3). The green vegetation cover was found to be 323.5 km^2 or 44% of the total area of Singapore.

Due to the plethora of landscape metrics available for use and the potential redundancy between them, the study by Tian et al. (2011) was adapted to limit the number of metrics to only the key ones that are relevant for deriving the ecological network. Using a grid composed of hexagonal units of 120 ha, pixels with 100% vegetation cover within each hexagon were clipped and exported into FRAGSTATS version 4.1 (McGarigal and Marks 1995), which was used to compute the values for an initial set of fourteen landscape metrics for each hexagon.

The Pearson product-moment correlation coefficient and a partial correlation analysis were carried out to eliminate correlated pairs of landscape metrics which led to a reduced set of six. A Principal Components Analysis (PCA) was then conducted, in which the Fragmentation Index (FI) proposed by Tian et al. (2011) was assembled using the results of the PCA. A further multiple regression analysis was performed with the FI as the response variable and the top loading metrics for each component, the 'effective mesh size' (MESH) and the 'mean Euclidean nearest neighbour distance' (ENN_MN), as the predictor variables. A regression model was used to construct a Patchiness Index (PI) (Eq. 12.1), where α, β, and γ are the B values for MESH, ENN_MN and the constant respectively.

$$PI = \alpha \, \mathrm{MESH} + \beta \, \mathrm{ENN\_MN} + \gamma \tag{12.1}$$

The regression model derived is PI = 0.191 MESH − 0.032 ENN_MN − 1.577. The values for each hexagon was computed, reclassified using the 'quantile' algorithm, and visualised on ArcMap (Fig. 12.3). The two metrics making up the PI, the 'effective mesh size' and the 'mean Euclidean nearest neighbour distance', act as proxies to the physical components of ecological networks, namely 'total network area' and 'network density' respectively. A higher 'effective mesh size' implies a larger area of the largest unfragmented patch in the landscape. Thus, two animals placed at any point in the landscape will have a higher probability of finding each other (Jaeger 2000). As for the 'mean Euclidean nearest neighbour distance', as its value increases, the inter-patch distance of fragmented patches increases, reducing the probability of inter-patch dispersal, and hence for two

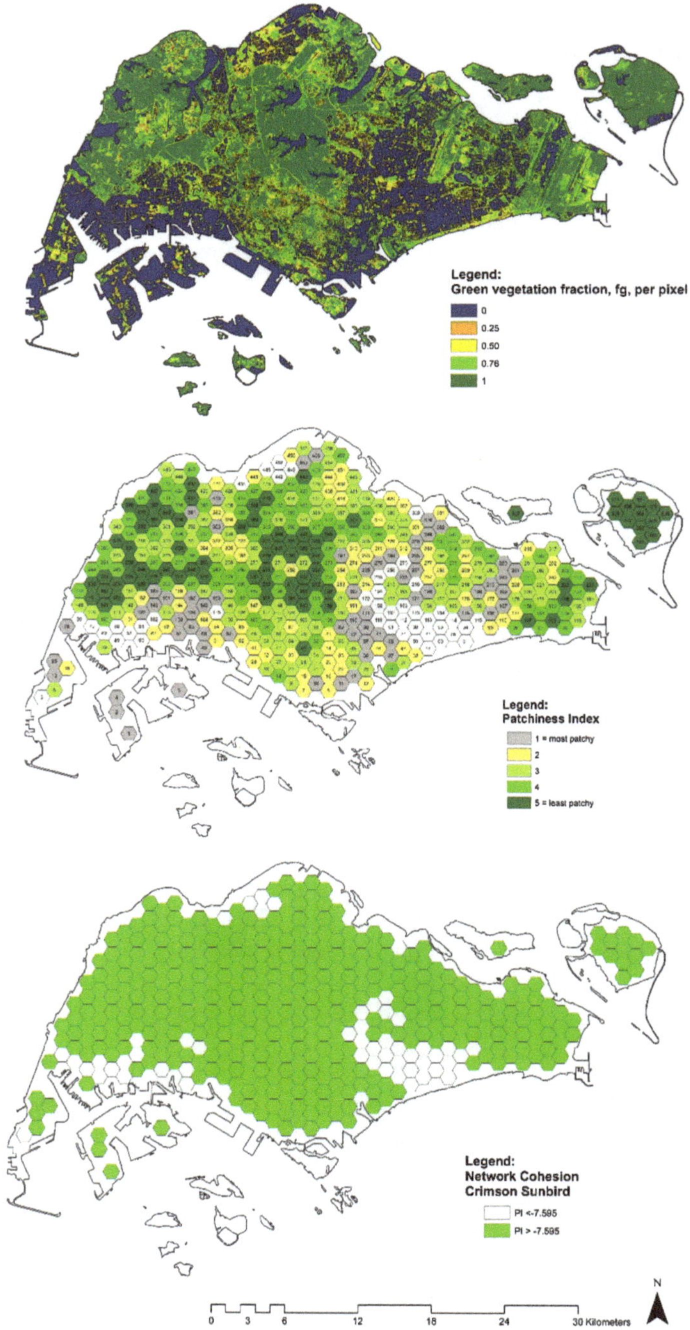

Fig. 12.3 *Top* Green vegetation cover map of Singapore for June 2012. *Middle* Distribution of patchiness indices of pixels with full vegetation cover. *Bottom* Network cohesion map for the Crimson Sunbird (*Aethopyga siparaja*)

animals to find each other if each is placed in different patches. Furthermore, as inter-patch distances increase, the patch density within a particular area decreases as a result. The two metrics are thus sufficient and critical for the characterization of the spatial distribution and configuration of ecological networks at the 120-ha scale.

The third component is to link the Patchiness Index to a species-based approach at a 120-ha scale. Nine experts from the field of biodiversity conservation and ecology were consulted in order to understand how the chosen species interact with their habitats, landscapes and urban elements. Ten 'focal species' were suggested, two each from five taxonomic classes namely, butterflies, birds, amphibians, reptiles, and mammals. The experts were also asked to state the minimum habitat size required for a mating pair of each species to reproduce assuming the presence of plant resources, and the maximum Euclidean distance that the species can traverse between patches without the required resources. The two key metrics, namely the 'effective mesh size' and 'mean Euclidean nearest neighbour distance' may also act as proxies for the two requirements, respectively. This enables the computation of the Patchiness Index values for each species by using the PI equation derived in Stage 1, which can then be visualised in ArcMap such as the one for the Crimson Sunbird (*Aethopyga siparaja*) (Fig. 12.3). The resulting network cohesion map for each species represents the hexagons containing habitat patches above the minimum PI threshold level, thus having a suitable level of 'network cohesion'.

A composite map of all the ten focal species can be derived using the *map calculator* tool in ArcMap, which highlights hexagonal areas that contain cohesive habitat networks for greater biodiversity over others (Fig. 12.4). Thus, a 'bridge' has been made between generic landscape indices to the 'persistence probability of species populations', linking the species level right up to the landscape level of biodiversity (Opdam et al. 2003). This understanding can be translated into a set of indicators for the long term monitoring of potentially cohesive habitat networks for biodiversity (Table 12.1), which could contribute to studies in patch dynamics, and studies in land use and land cover changes. Moreover, Opdam et al. (2003) stressed that political decisions do involve, and are affected by such indicators or indices, particularly on the choice of target species and ambition level for conservation.

12.3.3 Mapping Tools: Location of Potential Biodiversity and Dispersal Corridors for Biodiversity

As part of the Analytic Hierarchy Process (AHP), the nine experts were asked to assign 'judgements' to pairwise comparisons of factors using the 'Fundamental Scale for Pairwise Comparisons'. The purpose of the AHP is to derive the weights to be applied to the factors in habitat suitability and cost map modelling. The latter derives a map of the 'resistance' or 'cost' incurred by each species as it moves from pixel to pixel across the whole island. The factors needed to derive habitat potential maps are *canopy cover*, *vertical stratification*, presence of *water bodies*,

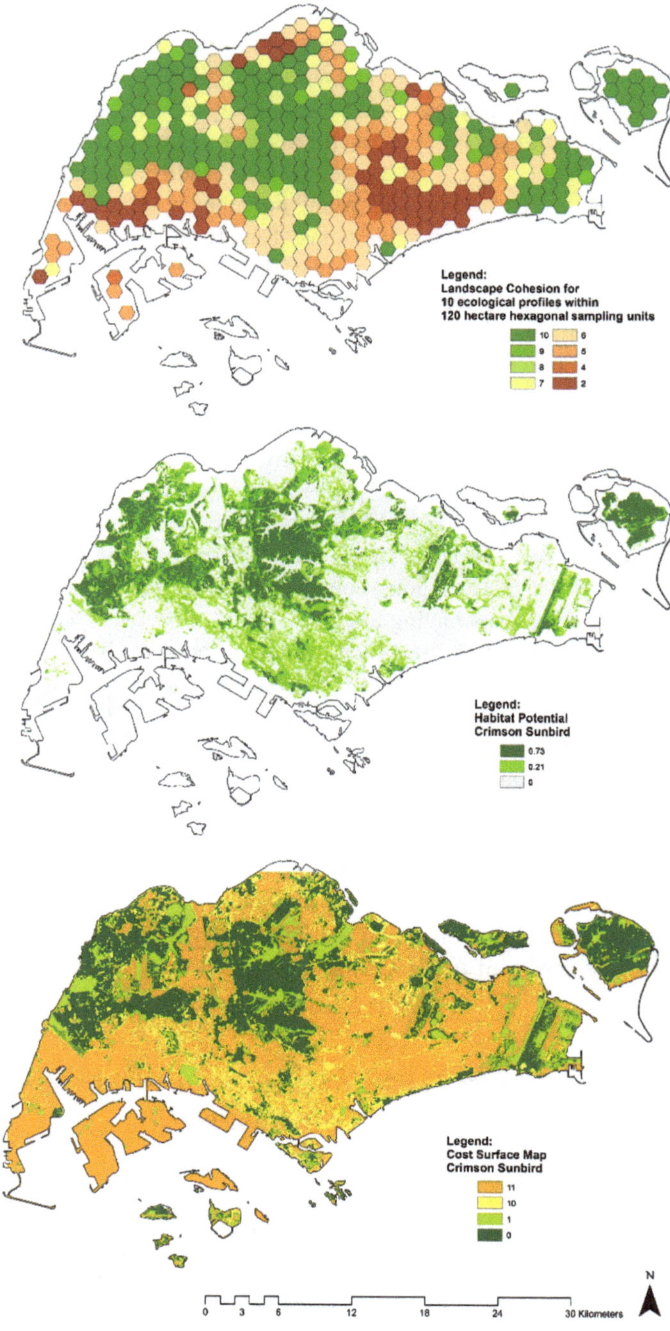

Fig. 12.4 *Top* Map of cohesive habitat networks for biodiversity employing the landscape cohesion concept by incorporating 10 ecological profiles, based on the aggregation of all network cohesion maps. *Middle* Habitat Potential Map for the Crimson Sunbird (*Aethopyga siparaja*). *Bottom* Cost Surface for the Crimson Sunbird (*Aethopyga siparaja*)

environmental factors, and *urbanised areas*; while those for the cost surface relates to their ability to facilitate the dispersal of the species through the urban matrix, such as *canopy cover* and presence of *water bodies*, as well as for their properties as barriers to dispersal, such as *roads*, *canals*, and *development intensity*. Incorporating specific criteria for habitat requirements and AHP into the formulation of habitat patches enables the creation of a map of 'effective patch areas' weighted by habitat quality, and becoming more ecologically relevant (Schooley and Branch 2011). The weights were then fed into an overlay analysis in ArcMap while incorporating relevant digitized or derived base maps to represent these factors, followed by a least cost corridor analysis. The maps of habitat potential and cost surface for the Crimson Sunbird (*Aethopyga siparaja*) are as shown (Fig. 12.4).

Least cost corridors from the Central Catchment Nature Reserve (CCNR), which was taken as the source key patch of the populations of various species, were modelled using the derived cost surface and the *corridor* algorithm in ArcMap. The corridors lead to ten vegetated patches as destinations around the perimeter of Singapore, and two at mid-points between the CCNR and the eastern and western tips, as shown for the Crimson Sunbird (*Aethopyga siparaja*) (Fig. 12.5) The

Table 12.1 Proposed indicators for ecological network monitoring

No.	Indicator	Explanation
1	Number of hexagons with network cohesion (NC) that are connected to at least one more	The interconnected hexagons form a key patch which may allow a minimum viable population of species, which is the population size with a probability of exactly 95% to survive 100 years under the assumption of 1 immigrant per population unit
2	Number of hexagons with network cohesion(NC) but not connected to any others	The hexagon contains a small patch or a group of patches, not large enough to allow a key population
3	Number of hexagons which are not part of any sustainable network	There are no cohesive habitat networks within these hexagons, with enough carrying capacity to render them sustainable
4	Biodiversity potential for 10 species in any part of the region	The resultant value taken at any point on the composite map of all the maps of habitat potential for 10 species. It is a function of the factors and weights applied by the various experts as to what constitutes a suitable habitat for their respective species under study or care
5	Number of species along least-cost corridors	The overlap of least-cost corridors of more species implies that these ones are more suitable for more species over others. Each least-cost corridor is a function of the factors and the weights applied by various experts

habitat potential and least cost corridor maps for each species represent its 'spatial cohesion' in relation to the whole island of Singapore, which not only includes forest fragments and reserves, but also parks and urban greenery found in urban infrastructure such as street trees and buildings.

Finally, a collation of the habitat potential and least cost corridors maps for the ten selected species into a composite map of 'landscape cohesion' enabled the determination of key ecological patches and corridors for Singapore (Opdam et al. 2003). The derived maps are a 'biodiversity potential map' which is a composite of all ten habitat potential maps, and a map of 'dispersal potential for biodiversity' which is a composite of all ten least cost corridor maps (Fig. 12.5). Two more indicators listed in Table 12.1 can be used to gauge the cohesion of the derived ecological network for the purposes of landscape planning, as well as benchmarking sustainable landscapes.

12.3.4 Communication and Decision Making Tools: Analytic Hierarchy Process and Delphi Approach

Taking from the 'decomposition' step in the Analytic Hierarchy Process (AHP), two hierarchies are proposed that explicitly links biodiversity conservation to taxonomic classes for Hierarchy 1, and to other potential land uses for Hierarchy 2 (Saaty 1980, 2001, 2005, 2008). Hierarchy 1 has four levels with the goal of biodiversity conservation at level 1 (Fig. 12.6). The actors themselves make up the second level. In Singapore, for land use management and biodiversity management, these actors are the government agencies and organizations. In this study, the following were included: Urban Redevelopment Authority (URA), which is the national land use planning agency, JTC Corporation (JTC), the national agency for commercial and industrial land management, Housing and Development Board (HDB), the national agency for public housing which houses 83% of the total population, National University of Singapore (NUS), representing views of a social science expert, and Nature Society Singapore (NSS), representing views of conservation experts. This outcome is influenced by perceptions of individual taxonomic classes, which constitutes the third hierarchical level. The assumption is that the perception to all organisms within each taxonomic class is the same, due to familiarity to taxonomic classes (such as mammals or reptiles) rather than individual species.

Hierarchy 2 is made up of three levels, and is related to Hierarchy 1 with a parallel set of objectives (Fig. 12.7). The goal at level 1 is to determine the land use preference for the sites. There may be more than one scenario for possible land use depending on the judgements assigned by the five experts, who are at the second level. The third level are thirteen land uses classified under four headings, namely *biodiversity conservation*, *social functions and amenities*, *residential*, and *development*. Figure 12.7 shows how Hierarchy 1 is linked to Hierarchy 2 via the biodiversity conservation category, which enables a conflict analysis to be carried out.

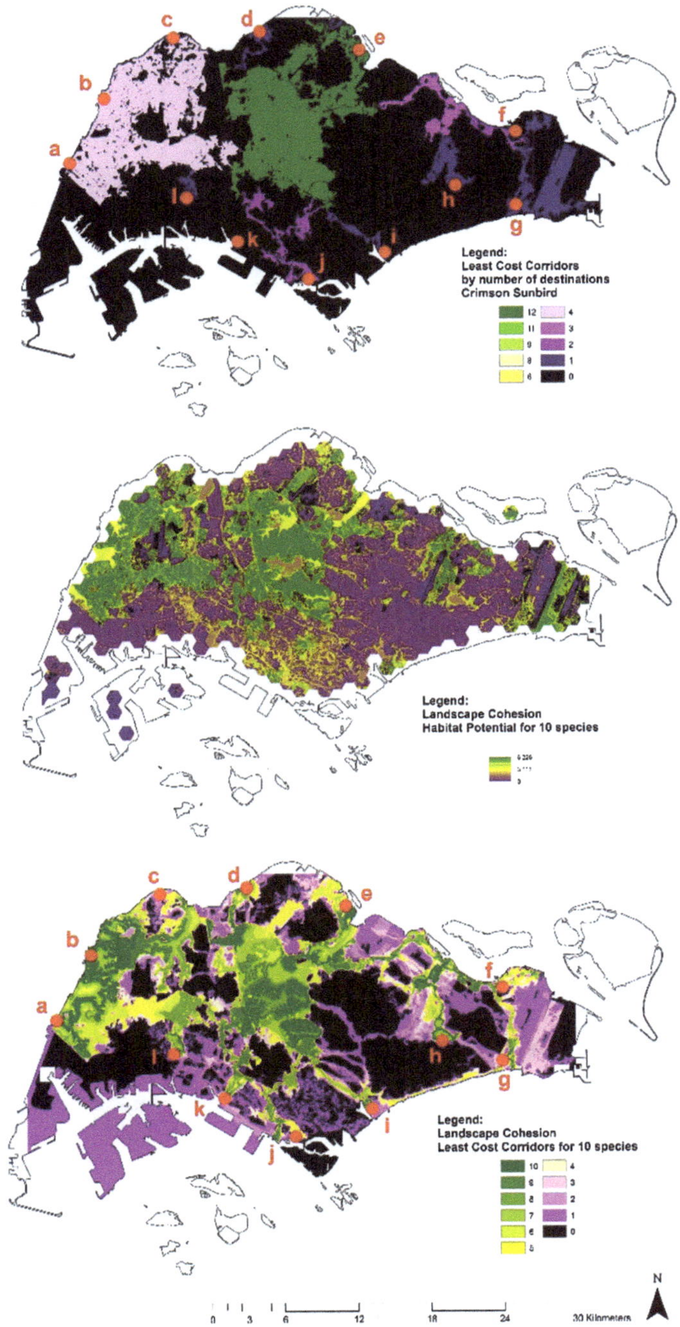

Fig. 12.5 *Top* Least-cost corridors from the Central Catchment Nature Reserve (CCNR) according to the number of various destination patches served around Singapore for the Crimson Sunbird (*Aethopyga siparaja*). Key for destination patches: *a* Western Catchment, *b* Sarimbun Reservoir, *c* Sungei Buloh, *d* Woodlands, *e* Yishun, *f* Pasir Ris, *g* Changi, *h* Bedok Reservoir, *i* Marina Bay, *j* Labrador Park, *k* Clementi Woods, *l* Jurong Lake. *Middle* Biodiversity potential map, based on the aggregation of the habitat potential maps for 10 ecological profiles. *Bottom* Map of dispersal potential for biodiversity employing the landscape cohesion concept by incorporating 10 ecological profiles, based on the aggregation of all least-cost corridors

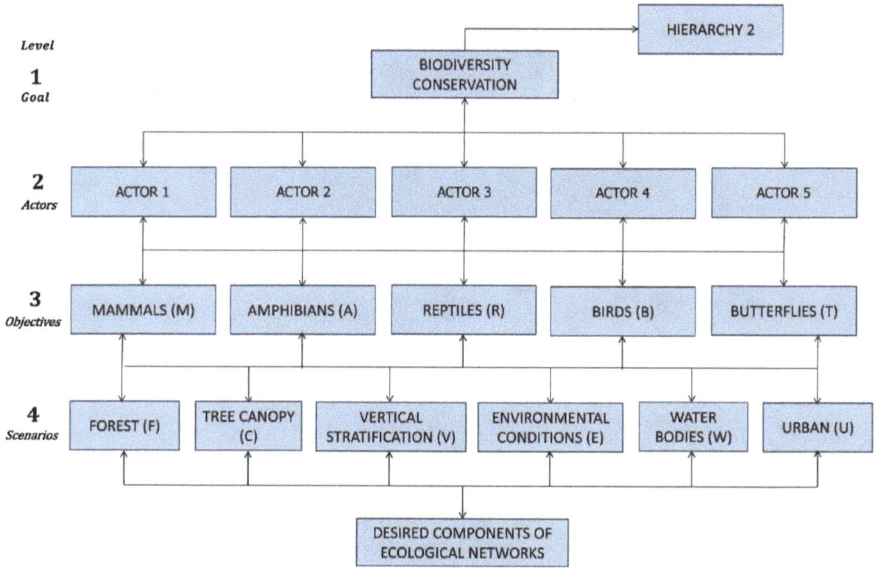

Fig. 12.6 Hierarchy 1 linking biodiversity conservation goal with land cover outcomes

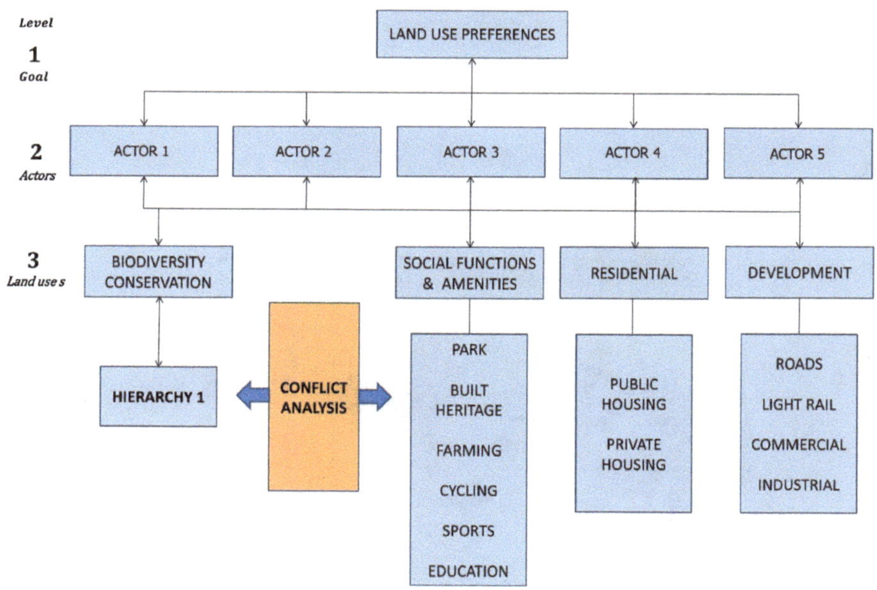

Fig. 12.7 Hierarchy 2 linking actors to land use preferences

The two hierarchies are meant to facilitate the communication between ecologists on the one hand which influences level 4 of Hierarchy 1, and planners which are *actors* in both Hierarchies 1 and 2. They clarify the link between scientists and planners, thus opening up more possibilities for multidisciplinary discourse and collaborations. Together with the judgement stage of the AHP and an application of linear algebra, the hierarchies form a tool kit which can be applied to any consultation process between multiple stakeholders regarding biodiversity conservation, and the implications on various land uses. This tool kit can facilitate the discussion of an ambition level for conservation, examine potential synergies between land uses and the interactions between social-ecological systems (SES) (Redman et al. 2004; Pickett et al. 2011), address potential conflicts, and choose from alternative scenarios.

To test this tool in a communication and decision-making process, experts from the government agencies and organizations mentioned previously were consulted. By referring to a particular site, they were requested to assign judgements to pairwise comparisons between biodiversity conservation versus other land uses, and then to comparisons between pairs of taxonomic classes in order to explore their preferences for certain classes over others. The weights for each category were computed for the site, and the results were presented on a spider diagram to each expert in a Delphi approach. Figure 12.8 is an example of the results for the Bukit Batok Nature Park.

12.4 Implications of Results

The learning outcomes from a study of ecological networks for cities can be grouped into the *5-multis*: *multispecies*, *multilevel*, *multiscalar*, *multifunctionality*, and *multidisciplinarity*. The ten chosen ecological profiles enable biodiversity conservation proposals to be developed for *multispecies*, beyond a single-taxon focus. This also illustrates the efficiency of the top-down approach to conservation. Although the ten are focal species, others with similar habitat requirements and dispersal capabilities will be taken into account in the network. The characterisation of the vegetation cover map into maps of network cohesion, spatial cohesion and landscape cohesion at various spatial scales, and the incorporation from the species to the landscape scales of biological organisation (Noss 1992), clearly illustrates how the *multiscalar* is embedded into the ecological network framework. This will enable cities to move up from a protected areas approach to biodiversity conservation, and incorporating ecological networks by identifying areas with high network cohesion, key habitat patches, and dispersal corridors for biodiversity. This is in turn part of an 'ecosystems management approach' to biodiversity conservation (Jongman 2004). The landscape cohesion maps, which are composites of the network and spatial cohesion maps of the ten ecological profiles, integrate the *multispecies* with the *multiscalar*. *Multilevel* refers to the incorporation of both top-down and bottom-up approaches for conservation, as well as the three levels of organization, namely 'structure', 'function', and 'composition' (Noss 1994; Walz and Syrbe 2013). The two related hierarchies which have been developed in Stage 3 of the research clarify the link between scientists and planners, thus opening up more

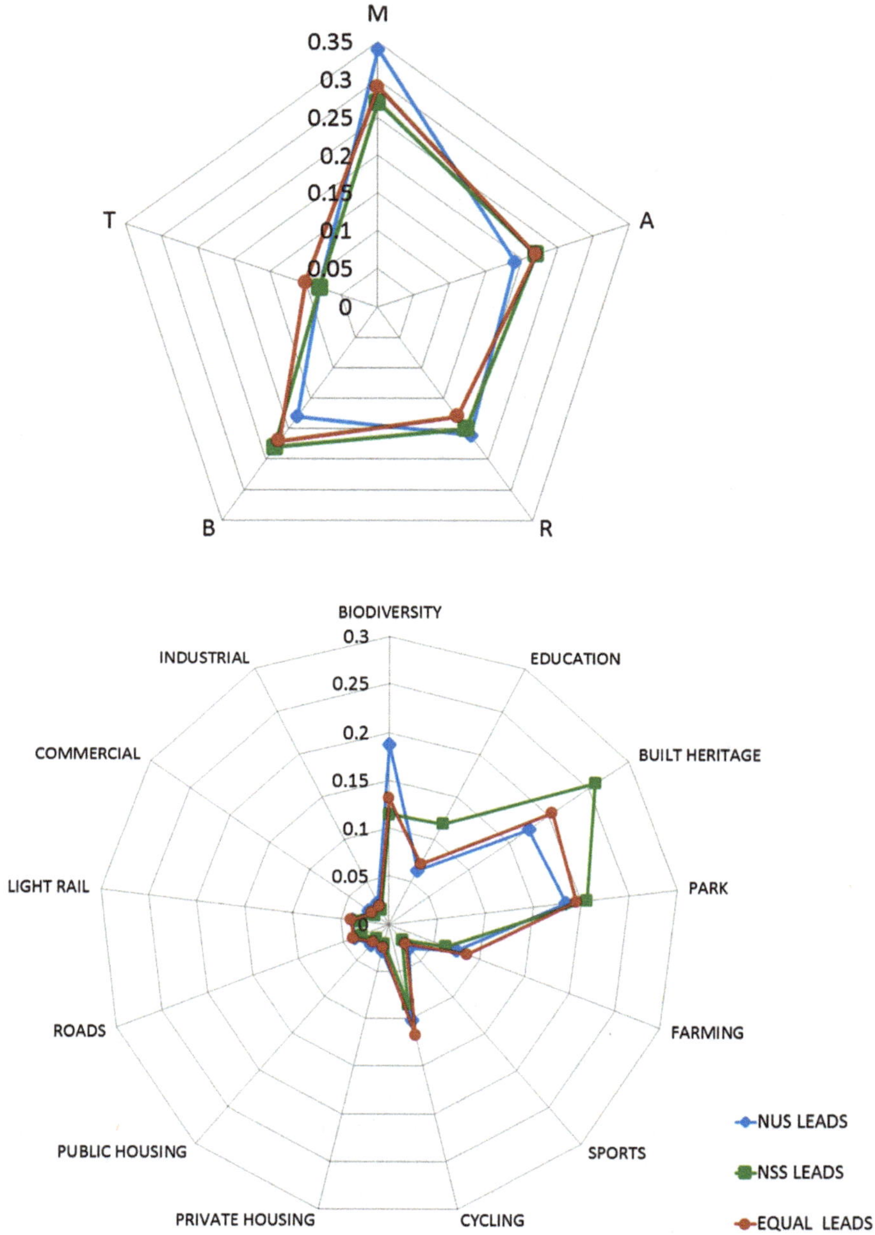

Fig. 12.8 *Top* Spider diagram showing the relative weights assigned to various taxonomic classes for the Bukit Batok Nature Park. The *letters* represent taxonomic classes: *M* mammals, *A* amphibians, *R* reptiles, *B* birds, *T* butterflies. Three scenarios are given: Expert 1 from the National University of Singapore (NUS) assigned higher power as a control, Expert 2 from the Nature Society (Singapore) (NSS) assigned higher power due to nature conservation priorities, and all actors or experts assigned equal power. *Bottom* Spider diagram showing the relative weights assigned to various land uses for the Bukit Batok Nature Park

possibilities for *multidisciplinary* discourse and collaborations. The greenery found in the city has been shown to play a part in biodiversity conservation by being incorporated into the ecological network, thus blurring the boundary between forest fragments and the urban matrix. This reinforces the idea that *multifunctional* landscapes in an urban context can be a strategy for biodiversity conservation (Opdam et al. 2006), and for reversing the negative impact of fragmentation (Tian et al. 2011).

With the aim of developing an interdisciplinary process for determining an ecological network for cities by integrating different methods, our research findings can become a launch pad to other areas of research. Firstly, although Opdam et al. (2003) stated that species distribution data is not necessary when using the ecological network framework for landscape planning, such data when available, should be used to validate the maps that have been derived from various methods, or be added onto the modelling process. They include detection data, relocation data, pathway data, and genetic data for habitat patches involving actual species observations, and these could be obtained via telemetry, or capture, mark and recapture methods. Conversely, the maps obtained using the ecological network approach can assist the scientists researching specific species locate bottlenecks in their dispersal. This engenders many opportunities for collaborations between different researchers from different fields or studying different species, rendering the ecological network approach truly multidisciplinary.

Secondly, it is not clear how satellite image resolution and the methods for deriving the vegetation cover map may affect the spatial features of the ecological network. Since image resolution is often a function of cost, a more affordable remote sensing method can also determine the willingness or ability of city governments to dedicate resources needed for such studies. Another factor to consider is the amount of fieldwork involved in ground truthing the vegetation cover maps, particularly in terms of measuring percentage canopy cover, as well as assessing the level of vertical stratification at the planning unit scale assessed visually and measured with the 'Domin' scale (Sutherland 2006). On-going studies attempt to determine if there is a correlation between NDVI derived from satellite imagery, and percentage canopy cover, leaf area index or vertical stratification in order to potentially reduce the amount of fieldwork that city governments may have to carry out if they were to adopt this approach to biodiversity conservation.

Thirdly, a further development for the ecological network framework will be to superimpose urban patterns and natural patterns as suggested by Vrijlandt and Kerkstra (1994). This can be done by overlaying the ecological network onto other types of networks found in cities, such as infrastructure networks incorporating roads and buildings, and social networks incorporating parks and park connectors (Fig. 12.9). Only then could the ecological network can be used to plan concurrently with other land use types. In addition, the location of the intersections between the three networks of greenery can inform a landscape architect as to how these nodes can be treated to enable the inclusion of biodiversity conservation objectives into developments or park design. Thus, various initiatives and projects, such as the 'ABC Waters' initiative (Public Utilities Board 2014) and the 'Park

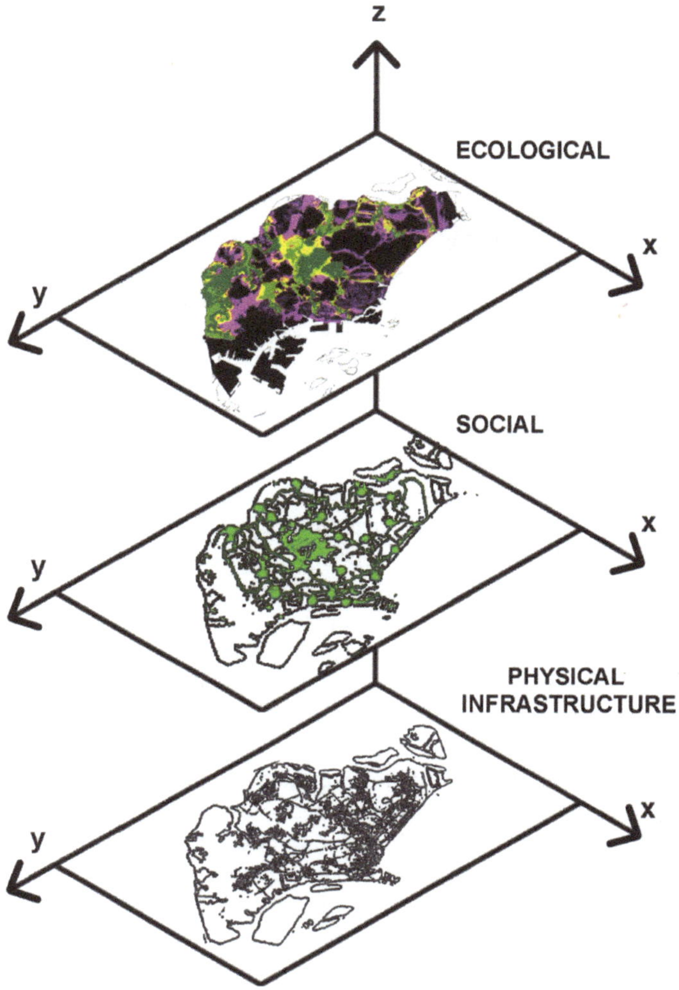

Fig. 12.9 The ecological, social and physical layers of urban greenery represented by an ecological network, parks and park connector network, and a network of roads and buildings (developments)

Connector Network' (Tan 2006), can then be integrated with the formulated ecological network, so as to synergise the objectives of all the different projects with the benefit of having a fully characterised ecological network for biodiversity, and to maximise the benefits of each project with the potential locations of high biodiversity. Further research can include the mapping of such nodes for the whole island of Singapore, as well as to explore planning and design options for these nodes leading to development control and design guidelines as recommended by Van Teeffelen et al. (2012), in order to bridge the gap between 'available knowledge' and 'required knowledge' in ecological network planning.

Finally, a long term challenge is to find a methodology that could enhance the cooperation between parallel ecological and administrative scales, whose boundaries need to considered (Opdam et al. 2006). If this is successful, ecological networks have a major role to play in linking across landscapes, regions, nations, and continents, thus a role in climate change mitigation.

12.5 Conclusion

An ecological network approach for urbanised areas can influence the outcome of cities' efforts in urban biodiversity conservation. The development of appropriate methodology coincides with an exciting era whereby cities have begun to recognize the need to conserve biodiversity, parallel to the growth of urban ecology and landscape ecology. This is illustrated in this chapter using Singapore as a case study.

This approach is novel in its application of theoretical concepts and methodologies of developing ecological networks for landscape planning in dense urban areas and cities. It looks at the intersection of three disciplines of conservation biology, landscape ecology and urban ecology for a new way of framing urban biodiversity conservation efforts for dense tropical cities such as Singapore. It not only synthesizes the methodologies of other research, but also pushes the envelope further by proposing three novel steps. The first is the Patchiness Index as a measure of 'network cohesion' which can be applied to the species level. The second is to combine the habitat potential, network cohesion, and least cost corridors maps for ten species into a composite map using the 'landscape cohesion' concept to incorporate multiple species. The third is the hierarchy that links biodiversity conservation with not only taxonomic classes, but also various land uses.

The resultant maps and diagrams form part of the toolkit for discussions between stakeholders early in the planning process, which may include the spatial distribution of the ecological network, the alignment of key corridors, the selection of key areas for conservation, the exploration of species and land use preferences, and finally the ambition level of biodiversity conservation that stakeholders are willing to undertake given finite economic resources, as well as different perceptions of biodiversity and conservation. It is hoped that further research and efforts could lead to the eventual implementation or inclusion of the ecological network framework into urban planning in the form of conservation policies, development control guidelines, and design strategies at specific sites.

Acknowledgements The authors thank A.T.K. Yee for providing the GIS shapefiles of forests in Singapore.

References

Abdullah SA, Nakagoshi N (2006) Changes in landscape spatial pattern in the highly developing state of Selangor, peninsular Malaysia. Landsc Urban Plann 77(3):263–275

Balvanera P, Pfisterer AB, Buchmann N, He J-S, Nakashizuka T, Raffaelli D, Schmid B (2006) Quantifying the evidence for biodiversity effects on ecosystem functioning and services. Ecol Lett 9(10):1146–1156

Benedict MA, McMahon ET (2006) Green infrastructure: linking landscapes and communities. Island Press, Washington, DC

Blaustein R (2013) Urban biodiversity gains new converts: cities around the world are conserving species and restoring habitat. Bioscience 63(2):72–77

Bolck M, de Togni G, van der Sluis T, Jongman RHG (2004) From models to reality: design and implementation process. In: Jongman R, Pungetti G (eds) Ecological networks and greenways: concept, design, implementation. Cambridge University Press, Cambridge

Bouwma I, Opdam P, Schrevel A (2003) Ecological networks: linking protected areas with sustainable development. Brochure Alterra, Wageningen

Brown C, Grant M (2005) Biodiversity and human health: what role for nature in healthy urban planning? Built Environ 31(4):326–338

Bruinderink GG, van der Sluis T, Lammertsma D, Opdam P, Pouwels R (2003) Designing a coherent ecological network for large mammals in Northwestern Europe. Conserv Biol 17 (2):549–557

Cardinale BJ, Duffy JE, Gonzalez A, Hooper DU, Perrings C, Venail P, Narwani A, Mace GM, Tilman D, Wardle DA, Kinzig AP, Daily GC, Loreau M, Grace JB, Larigauderie A, Srivastava DS, Naeem S (2012) Biodiversity loss and its impact on humanity. Nature 486 (7401):59–67

Chan L, Djoghlaf A (2009) Invitation to help compile an index of biodiversity in cities. Nature 460 (7251):33

Chou LM (2010) Biodiversity in sustainable cities. In: Ooi GL, Yuen B (eds) World cities: achieving liveability and vibrancy. World Scientific Publishing, Singapore, pp 135–153

Cook EA, van Lier HN (eds) (1994) Landscape planning and ecological networks. Elsevier Science B. V, Amsterdam

Coppolillo P, Gomez H, Maisels F, Wallace R (2004) Selection criteria for suites of landscape species as a basis for site-based conservation. Biol Conserv 115(3):419–430

Corlett RT (1992) The ecological transformation of Singapore, 1819–1990. J Biogeogr 19:411–420

Dearborn DC, Kark S (2010) Motivations for conserving urban biodiversity. Conserv Biol 24 (2):432–440

Dennis RLH, Dapporto L, Dover JW, Shreeve TG (2013) Corridors and barriers in biodiversity conservation: a novel resource-based habitat perspective for butterflies. Biodivers Conserv 22 (12):2709–2734

Dover J, Settele J (2008) The influences of landscape structure on butterfly distribution and movement: a review. J Insect Conserv 13(1):3–27

Farinha-Marques P, Lameiras JM, Fernandes C, Silva S, Guilherme F (2011) Urban biodiversity: a review of current concepts and contributions to multidisciplinary approaches. Innov Eur J Soc Sci Res 24(3):247–271

Fiedler PL, White PS, Leidy RA (1997) The paradigm shift in ecology and its implications for conservation. In: Pickett STA, Ostfeld RS, Shachak M, Likens GE (eds) The ecological basis of conservation: heterogeneity, ecosystems, and biodiversity. Chapman & Hall, New York

Forman RTT (1995) Land mosaics: the ecology of landscapes and regions. Cambridge University Press, Cambridge

Gutman G, Ignatov A (1998) The derivation of the green vegetation fraction from NOAA/AVHRR data for use in numerical weather prediction models. Int J Remote Sens 19(8):1533–1543

Haines-Young R (2000) Sustainable development and sustainable landscapes: defining a new paradigm for landscape ecology. Fennia 178:7–14

Haines-Young R, Green DR, Cousins S (1993) Landscape ecology and spatial information systems. In: Haines-Young R, Green DR, Cousins S (eds) Landscape ecology and geographic information systems. Taylor & Francis, London

Hilty JA, Lidicker WZ Jr, Merenlender AM (2006) Corridor ecology: the science and practice of linking landscapes for biodiversity conservation. Island Press, Washington, DC

Hoctor TS, Carr MH, Zwick PD, Maehr DS (2004) The Florida Statewide Greenways Project: its realization and political context. In: Jongman R, Pungetti G (eds) Ecological networks and greenways: concept, design, implementation. Cambridge University Press, Cambridge

Horwitz P, Kretsch C, Jenkins A, Abdul Hamid ARb, Burls A, Campbell K, Carter M, Henwood W, Lovell R, Malone-Lee LC, McCreanor T, Moewaka-Barnes H, Montenegro RA, Parkes M, Patz J, Roe JJ, Romanelli C, Sitthisuntikul K, Stephens C, Townsend M, Wright P (2015) Contribution of biodiversity and green spaces to mental and physical fitness, and cultural dimensions of health. UNEP, World Health Organization and Secretariat of the Convention on Biological Diversity

Hough RL (2014) Biodiversity and human health: evidence for causality? Biodivers Conserv 23 (2):267–288

Jaeger JAG (2000) Landscape division, splitting index, and effective mesh size: new measures of landscape fragmentation. Landscape Ecol 15:115–130

James P, Tzoulas K, Adams MD, Barber A, Box J, Breuste J, Elmqvist T, Frith M, Gordon C, Greening KL, Handley J, Haworth S, Kazmierczak AE, Johnston M, Korpela K, Moretti M, Niemelä J, Pauleit S, Roe MH, Sadler JP, Ward Thompson C (2009) Towards an integrated understanding of green space in the European built environment. Urban For Urban Greening 8 (2):65–75

Jongman RHG (2004) The context and concepts of ecological networks. In: Jongman RHG, Pungetti G (eds) Ecological networks and greenways: concept, design, implementation. Cambridge University Press, Cambridge

Koh LP, Sodhi NS (2004) Importance of reserves, fragments, and parks for butterfly conservation in a tropical urban landscape. Ecol Appl 14:1695–1708

Kostyack J, Lawler JJ, Goble DD, Olden JD, Scott JM (2011) Beyond reserves and corridors: policy solutions to facilitate the movement of plants and animals in a changing climate. Bioscience 61(9):713–719

Lambeck RJ (1997) Focal species: a multi-species umbrella for nature conservation. Conserv Biol 11:849–856

LaPoint S, Balkenhol N, Hale J, Sadler J, van der Ree R (2015) Ecological connectivity research in urban areas. Funct Ecol 29(7):868–878

Laurance WF et al (2012) Averting biodiversity collapse in tropical forest protected areas. Nature 489(7415):290–294

Linstone HA, Turoff M (1975) The Delphi method: techniques and applications. Addison-Wesley Pub. Co., Reading

McGarigal K, Marks BJ (1995) FRAGSTATS: spatial pattern analysis program for quantifying landscape structure. Forest Science Department, Oregon State University

Miller JR (2006) Restoration, reconciliation, and reconnecting with nature nearby. Biol Conserv 127(3):356–361

Moss MR (2000) Interdisciplinarity, landscape ecology and the "Transformation of Agricultural Landscapes". Landscape Ecol 15:303–311

Naeem S, Duffy JE, Zavaleta E (2012) The functions of biological diversity in an age of extinction. Science 336(6087):1401–1406

Noss RF (1992) Issues of scale in conservation biology. In: Fiedler PL, Jain SK (eds) Conservation biology: the theory and practice of nature conservation, preservation and management. Chapman & Hall, New York, pp 240–241

Noss RF (1994) Hierarchical indicators for monitoring changes in biodiversity. Essay 4A. In: Meffe GK, Carroll CR (eds) Principle of conservation biology. Sinauer Associates, Sunderland, pp 79–80

Oh K, Lee D, Park C (2011) Urban ecological network planning for sustainable landscape management. J Urban Technol 18(4):39–59

Opdam P (2002) Assessing the conservation potential of habitat networks. In: Gutzwiller KJ (ed) Applying landscape ecology in biological conservation. Springer, New York

Opdam P, Verboom J, Pouwels R (2003) Landscape cohesion: an index for the conservation potential of landscapes for biodiversity. Landscape Ecol 18:113–126

Opdam P, Steingröver E, Sv Rooij (2006) Ecological networks: a spatial concept for multi-actor planning of sustainable landscapes. Landsc Urban Plann 75(3–4):322–332

Phua M-H, Minowa M (2005) A GIS-based multi-criteria decision making approach to forest conservation planning at a landscape scale: a case study in the Kinabalu Area, Sabah, Malaysia. Landsc Urban Plann 71(2–4):207–222

Pickett STA, Zhou W (2015) Global urbanization as a shifting context for applying ecological science toward the sustainable city. Ecosyst Health Sustain 1(1):1–15

Pickett STA, Cadenasso ML, Grove JM, Boone CG, Groffman PM, Irwin E, Kaushal SS, Marshall V, McGrath BP, Nilon CH, Pouyat RV, Szlavecz K, Troy A, Warren P (2011) Urban ecological systems: scientific foundations and a decade of progress. J Environ Manage 92 (3):331–362

Pimm SL, Jenkins CN, Abell R, Brooks TM, Gittleman JL, Joppa LN, Raven PH, Roberts CM, Sexton JO (2014) The biodiversity of species and their rates of extinction, distribution, and protection. Science 344(6187):1246752

Public Utilities Board (2014) Active, Beautiful, Clean Waters design guidelines. Public Utilities Board (PUB)

Redman CL, Grove JM, Kuby LH (2004) Integrating social science into the long-term ecological research (LTER) network: social dimensions of ecological change and ecological dimensions of social change. Ecosystems 7(2):161–171

Reza MIH, Abdullah SA, Nor SBM, Ismail MH (2013) Integrating GIS and expert judgment in a multi-criteria analysis to map and develop a habitat suitability index: a case study of large mammals on the Malayan Peninsula. Ecol Ind 34:149–158

Rosenberg DK, Noon BR, Meslow EC (1997) Biological corridors: form, function, and efficacy. Linear conservation areas may function as biological corridors, but they may not mitigate against additional habitat loss. Bioscience 47(10):677–687

Saaty TL (1980) The analytic hierarchy process. McGraw-Hill Inc., New York

Saaty TL (2001) Decision making for leaders: the analytic hierarchy process for decisions in a complex world. RWS Publications, Pittsburgh

Saaty TL (2005) The analytic hierarchy and analytic network processes for the measurement of intangible criteria and for decision-making. Int Ser Oper Res Manag Sci 78:345–407

Saaty TL (2008) Decision making with the analytic hierarchy process. Int J Serv Sci 1(1):83–98

Saaty TL, Vargas LG (2001) Models, methods, concepts & applications of the analytic hierarchy process. Kluwer Academic Publishers, Boston

Savard J-PL, Clergeau P, Mennechez G (2000) Biodiversity concepts and urban ecosystems. Landsc Urban Plann 48:131–142

Schooley RL, Branch LC (2011) Habitat quality of source patches and connectivity in fragmented landscapes. Biodivers Conserv 20(8):1611–1623

Secretariat of the Convention on Biological Diversity (2006) Global biodiversity outlook 2. Secretariat of the Convention on Biological Diversity

Secretariat of the Convention on Biological Diversity (2012) Cities and biodiversity outlook. Action and policy. A global assessment of the links between urbanization, biodiversity, and ecosystem services. Secretariat of the Convention on Biological Diversity

Seto KC, Güneralp B, Hutyra LR (2012) Global forecasts of urban expansion to 2030 and direct impacts on biodiversity and carbon pools. Proc Natl Acad Sci USA 109(40):16083–16088

Sodhi NS, Brook BW, Bradshaw CJA (2007) Tropical conservation biology. Blackwell, Malden

Steudel B, Hector A, Friedl T, Löfke C, Lorenz M, Wesche M, Kessler M (2012) Biodiversity effects on ecosystem functioning change along environmental stress gradients. Ecol Lett 15 (12):1397–1405
Sutherland WJ (ed) (2006) Ecological census techniques: a handbook. Cambridge University Press, Cambridge
Tan KW (2006) A greenway network for singapore. Landscape and urban planning 76(1–4):45–66
Tian Y, Jim CY, Tao Y, Shi T (2011) Landscape ecological assessment of green space fragmentation in Hong Kong. Urban For Urban Greening 10(2):79–86
Van Teeffelen AJA, Vos CC, Opdam P (2012) Species in a dynamic world: consequences of habitat network dynamics on conservation planning. Biol Conserv 153:239–253
Verboom J, Pouwels R (2004) Ecological functioning of ecological networks: a species perspective. In: Jongman R, Pungetti G (eds) Ecological networks and greenways: concept, design, implementation. Cambridge University Press, Cambridge
Verboom J, Froppen R, Chardon P, Opdam P, Luttikhuizen P (2001) Introducing the key patch approach for habitat networks with persistent populations: an example for marshland birds. Biol Conserv 100:89–101
Vos CC, Verboom J, Opdam PFM, ter Braak CJF (2001) Towards ecologically scaled landscape indices. Am Nat 183(1):24–41
Vrijlandt P, Kerkstra K (1994) A strategy for ecological and urban development. In: Cook EA, van Lier HN (eds) Landscape planning and ecological networks. Elsevier Science, Amsterdam
Walz U, Syrbe R-U (2013) Linking landscape structure and biodiversity. Ecol Ind 31:1–5
Wiens JA (2002) Central concepts and issues of landscape ecology. In: Gutzwiller KJ (ed) Applying landscape ecology in biological conservation. Springer, New York
Wu J, Loucks OL (1995) From balance of nature to hierarchical patch dynamics: a paradigm shift in ecology. Q Rev Biol 70(4):439–466
Zonneveld I (1994) Landscape ecology and ecological networks. In: Cook EA, van Lier HN (eds) Landscape planning and ecological networks. Elsevier Science, Amsterdam

Chapter 13
Urban Heritage Trees: Natural-Cultural Significance Informing Management and Conservation

Chi Yung Jim

Abstract Trees with outstanding traits have always attracted human attention, echoed by 60 epithets harvested from the literature. They have been formally designated as heritage trees using diverse criteria, such as size, tree form, historical-cultural associations, and sacred-mythical connotations. Ancient trees with veteran features offer varied micro-habitats to support a surprising assemblage of companion organisms. Other large trees furnish keystone structures with far-reaching ecological impacts. Inventory and scientific data can reinforce community awareness and improve management. Engaging citizens and the business sector could cultivate ownership and muster support. Assessing their economic value could explain multiple benefits, strengthen public-funding justifications, and raise prestige and value of property development. Preserving initial genial site conditions is critical for tree survival in the urban setting. The hitherto neglected soil domain deserves meticulous protection and improvement. Harmful grade change of tree sites should be avoided. Badly degraded sites could be rehabilitated using tailor-made site-specific techniques. The ageing tree-population structure demands proactive nurturing of younger successors to sustain the lineage. The statutory approach is advocated for assured protection and conservation. Overzealous and aggressive tree care should be replaced by a sympathetic and dedicated approach. The frequent omission of lightning protection should be promptly rectified. Conflicts with developments should be settled by in situ preservation, and transplanting should be the last resort. Sentimental and emotional responses towards tree loss could be carefully massaged employing public-relation skills. Heritage-tree conservation could be enhanced by transgenerational urban forestry, precision arboricultural practices, and joint venture of government and citizens.

Keywords Heritage tree · Veteran tree · Ancient tree · Tree rehabilitation · Transgenerational urban forestry · Urban tree ecology

---

C.Y. Jim (✉)
University of Hong Kong, Pokfulam, Hong Kong
e-mail: hragjcy@hku.hk

P.Y. Tan and C.Y. Jim (eds.), *Greening Cities: Forms and Functions*,
Advances in 21st Century Human Settlements,
DOI 10.1007/978-981-10-4113-6_13

13.1 Introduction

The conditions for tree growth in urban areas can vary greatly from natural-genial to artificial-stressful. The wide range of variations has been denoted as a spectrum of naturalness on a biotope scale (Kunick 1990; Sukopp 2008). Natural sites can be inherited from pre-urbanization ecosystems and subsequently embedded in built-up areas. They could remain relatively intact despite the surrounding urban development. Such locales are more likely to inherit and preserve remnant meritorious trees, or permit their development from the existing tree stock. However, site quality may degrade due to intrusion and perturbation activities, which reduce the capacity to accommodate fine trees. Occasionally, high-quality natural sites could be created within urbanized areas to emulate nature and enhance the nature-in-city repertoire.

Urban-tree habitats are often beset by diverse restrictions above and below the ground. The stress level could increase in response to urban redevelopment which brings incursion into occluded green fields, infilling of relatively low density areas and overall densification. The plethora of physical and physiological constraints would compromise tree performance. In extreme cases, existing trees could be partly or wholly eliminated. The roadside strip and dense residential and commercial land uses, especially in compact cities, suffer more from such damages. Many urban trees could not express their genotypic biological potentials in terms of final dimensions, tree structure and form, longevity, health and vigour. Outstanding trees bequeathed from pre-urbanization vegetation could be degraded, and trees planted in the post-urbanization period could be deprived of the necessary conditions to excel and join the elite tree group. Thus it is uncommon to find exceptionally superior trees in cities.

The quantity and quality of the most outstanding specimen trees echo a city's natural history and human attitudes and actions towards the notable living companions. This valuable tree cohort deserves special attention and care. This chapter explains: (1) designation criteria and the process for evaluating the ecological, cultural and historical significance of outstanding trees in cities; and (2) tree growth problems, management challenges and conservation measures. These have been developed through local and overseas research and field studies supplemented by literature review.

13.2 Designation Criteria and Quality Assessment

13.2.1 Dimensional Emphasis

The key tree dimensions, namely trunk diameter at breast height of 1.4 m from the ground (dbh), tree height and crown spread, are widely used as the fundamental size criteria to judge whether a tree should be officially designated. For instance, since the 1940s, the American Forestry Association (undated, recently renamed American

Forests) in the United States has used the three tree measures to calculate respective points, and summed them as an aggregate score to identify nominees for the National Register of Big Trees. The scoring method has been converted to the SI unit with the Imperial unit given in brackets:

- One point for every 2.54 cm (1 in.) trunk circumference at 1.37 m (4.5 feet) above the ground.
- One point for every 30.5 cm (1 foot) of tree height measured to the nearest foot.
- 1/4 point for every 30.5 cm (1 foot) of average crown spread measured to the nearest foot.

The quantitative system aims squarely at identifying the largest specimen tree of each species. Thus the contest is amongst members of the same species (intraspecific) rather than between (interspecific). It includes both native and exotic species. Some forestry commissions would solicit nomination at the state level (May 1990; Van Pelt 1996; Mississippi Forestry Commission 2005), from which the biggest state trees could join the national competition.

The tree with the highest aggregate score denoting the largest tree for a given species in the state will be designated as the state's champion tree of the species. Where two or more trees have very close aggregate scores within 5 points, co-champions can be declared. Where a tree has accumulated points in excess of those of the same species in the National Register of Big Trees, the scores will be submitted to American Forests to compete for the national champion.

The city of Chicago has conducted a similar exercise labelled the 'Treemendous Tree Program' (Barro et al. 1997). The project also studied the roles and values offered by citizens in making the nominations. Of the four physical characteristics, history of the tree ranked first, followed by notable physical features, condition and health, and lastly size. Regarding functional and aesthetic values, positive assets were predominant with few mentioning the negative side. Concerning the symbolic or emotional meanings, ties to the general history of the site or region ranked first, followed closely by ties to individuals and family, and then emotions or feelings about the tree. Overall, the historical and emotional associations were given notably greater weights than other attributes.

Other countries adopted similar size criteria to select champion trees. Scotland has developed a similar system based on superlative bulk in addition to finding heritage trees (Rodgers et al. 2003). Ireland also differentiates between champion trees based on record height and girth, and heritage trees due to other considerations (Boylan 2010).

13.2.2 Other Pertinent Selection Criteria

Outstanding trees have been warmly cherished by different peoples in different times and places. Human responses to trees have engendered a wide spectrum of feelings since ancient times, including notably respect, admiration, awe, wonderment, strength, endurance, longevity, companionship, reverence, and utility. They echo that people are intimately connected to nature, and trees are intricately interwoven with human culture (Hu et al. 2011). Unfortunately, the association has become more detached if not divorced in modern times.

In the course of identifying the prominent trees, it has been found that some targets may not be particularly large, yet they are unusually attractive due to other features worthy of enjoyment and appreciation by present and future generations. Broadly labelled as champion trees in Hong Kong (Jim 1994a, b) and in Bangkok (Thaiutsa et al. 2008), the targets were judged based on size as well as other heritage yardsticks.

Judging mainly by inherent features plus some site factors, the splendid trees have been accorded commonly the accolade of heritage trees (Figs. 13.1 and 13.2). Nominated by citizens and government professionals, a panel of experts would usually evaluate the submissions and make recommendations to the authority. The candidates are often treated as living legacies or natural-cum-cultural heritage (Jim 1994a), which excel by virtue of the pedigree traits (Jim 1994a; Browne 2001; Rodgers et al. 2003; Boylan 2010; Asciuto et al. 2015; City of New York, undated; London Tree Forum, undated; Pennsylvania Urban and Community Forestry Council, undated). A healthy tree which can satisfy one or more of the following criteria could warrant heritage-tree status:

- Sheer size (a common rule of thumb is >1 m trunk diameter)
- Advanced age (often adopting the anthropocentric centenarian threshold)
- Impressive or exemplary tree form for the species
- Exceptional aesthetic interest
- Landscape contribution to the locality
- Landmark or sense-of-place icon
- Documented association with notable historical event, personality, building or monument
- Documented important cultural, traditional or folklore connotations
- Considered to have sacred, mythical, idolatrous or spiritual significance
- Functional or aesthetic association with natural features (such as a riverbank or a ridgeline)
- Unusual or rare habitat
- Important ecological contributions
- Uncommon or rare species
- Notable botanical interest
- Sentimental and emotional attachment of the community
- Social and economic values

The heritage-tree recognition can extend from individual entities to a healthy and mature grove of trees (Goldberg et al. 2007) or an avenue with trees. The compound unit can be considered for collective designation following the criteria for individual trees (City of Vancouver undated).

Fig. 13.1 This ancient heritage tree near the Notre Dame in Paris has an inclined trunk, truncated branches, wood decay, cavities and other veteran features. Supported by props that emulate natural trunks, the *Robinia pseudoacacia* planted in 1601 has been protected as an 'arbre remarquable' by the municipal authority (*Photo credit* C.Y. Jim)

The monumental trees recognized in some countries are comparable to heritage trees in terms of the principal characters. The primary qualities are sacred and historical-cultural significance transmitted through generations as folklores and stories. Some may include the utilitarian functions as selection yardsticks, such as protection

Fig. 13.2 Some banyan trees (*Ficus* spp.) can grow spontaneously on stone retaining walls in tropical cities such as this fine example in Hong Kong. Seeds brought by droppings of birds and bats can lodge in the gaps between masonry blocks. A tiny proportion of seeds could germinate successfully, literally defying gravity to reach large dimensions on the artificial cliffs. Such heritage trees clinging on the unique ruderal habitat against the odds denote natural-cum-cultural heritage features that deserve heritage-tree designation (*Photo credit* C.Y. Jim)

of wildlife habitat, regulation of stream flow, and mitigation of erosion and avalanche (Asan 2001). Based on special or unique traits and after passing the basic requirements of age, size and health, four types of monumental trees have been recognized, namely historical, folklore, mystical, and dimensional (Gnec and Guner 2001). They may include trees dwelling in remote natural areas, cultivated lands, villages or urban areas (Golabek and Tukiendorf 2002; Arnan et al. 2012). For all intents and purposes, they could be considered as the equivalent of heritage trees.

13.2.3 Multiple Ecological Contributions

The elderly trees could offer multiple ecological services that deserve special attention (Fig. 13.1) Tree age is a relative concept to be gauged against the normal life expectancy of the species. Old trees tend to develop veteran features with trunk and main branches being gnarled, twisted, deformed, fissured, jagged, broken, beset by common presence of branch loss, dieback, dead wood and fungal fruiting bodies, unnaturally short and squat, and in general a vivid display of the ageing look (Davis et al. 1996; Read 2000; Woodland Trust 2008).

The structural and morphological complexity embodies a rich collection of microhabitats exploited by an equally rich assemblage of flora, fauna and micro-organisms (Woodland Trust 2009a). The dead or decayed trunks, branches and framework roots can be riddled with cavities, cracks, seams and various nooks and crannies to accommodate surprisingly diverse companion creatures. They may contain the full complement of wood conditions to suit different preferences, ranging from sound to partly decayed and thoroughly rotted, and from relatively dry to moist and wet.

Some hollows or indentations may allow water to pond to create aquatic microhabitats of small pools at different elevations and with varied exposure to sunshine and wind. The relatively drier chambers could offer nesting abodes to small mammals, birds and other wildlife. The loose and partly detached bark at different moisture states and oozing sap engender various surface conditions and hiding places for small animals, insects, fungal mycelia, lichens and epiphytic plants which include climbers and semi-parasites.

The numerous small focal sites collectively expand the ecological amplitude and capacity of the constituent ecosystem. They allow veteran trees especially the sizeable ones to play the role of keystone structures, offering habitats and resources to a sizeable community of organisms (Stagoll et al. 2012). Whereas some arboreal dwellers are generalists, many are specialists demanding rather unique and precise habitat conditions to survive and reproduce. The cardinal management principle is to sustain the microhabitat variability and quality. Such roles could be effectively fulfilled even if the old trees have a scattered distribution pattern (Manning et al. 2006). Many saproxylic organisms can secure sustenance from and flourish on and inside the multivariate niches of the gentle, prodigious, generous, altruistic and elderly doyens (Alexander et al. 1998).

Trees of one to several centuries of age can literally form and foster its own complex community. Some younger trees, however, may prematurely develop veteran symptoms, especially due to abuses by humans. Tree managers can take proactive measures to preserve the intricate habitats and their companion creatures which may include some rare, protected or endangered species. The practice of arboricultural care, especially pruning, should be sympathetic to such critical functions. Instead of routine or standard methods, they could be appropriately adjusted to preserve and enhance the ecological roles. Arborists could learn about tree qualities that can enhance biodiversity and biological processes which are integral components of the urban forest. Active and skilful management could deliberately create veteran features to enhance ecological services in a process known as 'veteranization' (Read 2000).

The annual rings of the old trees keep a faithful record of the past climate and environmental changes. They have frozen the changes in the form of ring thickness, wood microstructure, and wood chemistry. Wood samples from elderly trees can be analyzed to re-construct the past climate and environment records, as well as historical trends of human treatment and mistreatment of trees and their site conditions.

Large maiden trees with little veteran symptoms can offer a sizeable biomass and surface area to accommodate many plant and animal companions. Their large crown, high live crown ratio, dense branching and foliage density, and high leaf-area index provide a rich source of sustenance for dependent organisms, which may include commensalism and mutualism interactions with the host tree (Fig. 13.2). The complex tree structure can generate its own microclimate with diverse conditions at different parts to accommodate wildlife with different habitat requirements. Other forms of old trees such as coppice stools, pollards and lapsed pollards can enrich the micro-habitat diversity to welcome wildlife.

Trees that grow up spontaneously to attain heritage calibre on post-urbanization ruderal sites present a different kind of meritorious nature in the urban fabric. Often established on unusual and rare habitats, they denote the biological receptivity of urban installations to nature. Trees growing on old stone retaining walls could be subsumed under this special category (Jim 1998a). The seeds brought by birds or bats could lodge incidentally in the crevices between masonry blocks. Despite the odds, a tiny number of seeds that managed to germinate may grow into huge wall-clinging trees that defy gravity (Jim 2010, 2013). Some members of the genus *Ficus* (Banyans) are pre-adapted to grow successfully on such artificial cliffs (Jim 2014). They denote an unusual urban ecological treasure and natural-cum-cultural heritage deserving conservation and appreciation (Fig. 13.3). Sometimes, they can establish on old buildings to enhance urban ecology and urban landscape (Jim and Chen 2011).

13.2.4 Importance to Tree Conservation

Heritage-tree designation can trigger consequential improvements in their management. It may raise awareness amongst the public and managers to better appreciate their fine qualities and reinforce their determination to safeguard them. In jurisdictions without a dedicated or effective tree ordinance to protect the notable doyens, legislative initiatives could be triggered. More resources could be secured for their assessment, monitoring and protection. The iconic trees can attract donations and sponsorships to augment management (Torbay Council 2011). A tree adoption programme can kindle enthusiasm and support in the community for the respectable cause.

The knowledge and data gleaned by tree managers and researchers could be fed into the management stream to inform decisions. The detailed tree inventory, compiled in the form of reports, books and websites, could inform nature conservation and property development (English Nature 2000a), who can adopt appropriate and timely measures to shield affected trees from harm. Tree appreciation could be promoted as an educational resource to strengthen teaching and learning endeavours about nature. The same resource base can develop ecotourism activities, to be merged with interests on local history and culture (Cannizzaro and Corinto 2014).

Relict trees may represent key members of a former ecosystem that has since been destroyed. Often existing as rare or endangered species, they have high

Fig. 13.3 Elaborate protective measures including wooden fences and many wooden props have been installed to sustain an ancient heritage tree in the grounds of the Nishi Hongwan Temple in Kyoto. The *Ginkgo biloba* doyen has dwelt blissfully at the site for about 400 years (*Photo credit* C.Y. Jim)

conservation value. Such trees may be planted ex situ in botanical gardens or humanized landscapes, or exist as in situ remnants of natural vegetation (Sierra Club Sequoia Task Force 2000; Schlawin and Zahawi 2008; Kozlowski et al. 2012). Such large old trees, often serving as keystone structures, are vulnerable to damage and loss. Tailor-made protective policies and practices are needed for proactive management intervention to sustain critical ecological functions (Lindenmayer et al. 2014).

The social and cultural benefits of heritage trees could be more earnestly promoted to arouse community interest in their living monuments. By emphasizing the human-related values, the conventional conservation cause could be complemented by a new dimension with the potential to secure wider support (Blicharska and Mikusinski 2014). The universality of the sacredness and symbolism of some core-cultural tree species (Dafni 2006), such as Linden (*Tilia* spp.) in Europe and Fig (*Ficus* spp.) in the pan-tropical realm, could well exemplify the dominant status of heritage trees to many people (Wilson and Wilson 2013; Tenche-Constantinescu et al. 2015). Such deep human feelings can be mobilized to muster strong support for conservation.

13.2.5 Economic Value and Valuation

As non-market goods, the value of trees and greenery in cities has been assessed using several well-tested innovative methods. They include mainly the contingent valuation method (CVM) and hedonic pricing method (HPM). The wide range of ecosystem services offered by urban forests can be evaluated in monetary terms (Center for Urban Forest Research 2004; Jim and Chen 2006, 2009, 2010). The dollar values assigned to trees by citizens echo positive perception of their benefits to society and of their existence value (Asciuto et al. 2011). These methods assess the value of trees from the opinions or consumption behaviours of citizens, which may differ from the judgment of tree managers and specialists.

Many studies investigate the value of urban green spaces rather than trees specifically, and few focus on heritage trees. In a study in Guangzhou (China), the decision whether to pay or not to conserve heritage trees is influenced by the overall tree value. However, the decision on the payment amount is contingent upon income level. Citizens are willing to pay more to protect rare species in comparison with common ones (Chen 2015). Unwillingness to pay (protest responses) is due to distrust of the government or limited understanding of heritage trees (Chen and Hua 2015). To win the trust and hence support of the people, the government can provide more opportunities to access heritage trees, high-quality information, effective management, and public engagement.

A Formulaic Expert Method (FEM) was developed specifically to assess the monetary value of heritage trees (Jim 2006). It converts key tree attributes into numerical scores based on a basket of objective criteria, and uses the current average residential property value as the baseline to calculate tree monetary value. In other words, it combines detailed professional tree evaluation with valuation, thus giving due weight to the expert judgment of tree specialists. The basic tenet of FEM is that property development poses the major threat to heritage trees in cities. Linking tree value to property price can serve as an effective deterrent against indiscriminate damage. The results can offer hints to the court to determine a fair and realistic level of compensations and penalties in case of tree damages.

On the other side of the coin, heritage trees can contribute to property development. A fine tree embedded in a development site should be regarded as a blessing rather than a burden or hindrance. It could be sympathetically merged with the artificial structure to add value to the property, and bestow a unique character and a sense of place (Woodland Trust 2011). The prestige of the project could be enhanced by sympathetic blending with the respectable antiquity of distinction. For the benevolent and noble deed, the corporate social responsibility of the developer could be recognized and respected by the community.

Patches of greenery or nature as well as individual trees have been studied, and the monetary values have been estimated (Becker and Freeman 2009). The results offer a more tangible way to understand and promote the importance of heritage trees in the urban setting. The dollar worth accorded to the precious trees can provide a convincing basis to justify public funding on the key community assets, and compete with other claims (Asciuto et al. 2015). Appreciation of tree values in time can be monitored to reflect the progressive investment returns and to inform the community of the continually growing environmental and ecological functions. The same information can reinforce the conservation cause and strengthen the case for conservation efforts.

13.2.6 Vivid and Diverse Terminology

As many as 60 epithets have been proposed and used in the literature to describe trees that have drawn human attention. The target trees can dwell in natural, rural and urban areas. Some would focus on a specified trait or a related group of traits, and others may be more eclectic and encompassing. The plethora of terminology echoes the diversities of arboreal features that are appreciated by humans, and the importance attached to them that traverse geographical, temporal and cultural divides. They have been classified into eight groups and matched with their relevance to the criteria (Table 13.1):

- Paramountcy of tree dimensions
- Advanced age traversing human generations
- Superlative performance amongst peers of the species
- Special ecological functions
- Emphasis on visual or scenic delights
- Association with prominent personalities or historical events
- Natural or cultural bequests
- Spiritual or mythical connotations.

13.3 Management and Conservation Challenges

13.3.1 Preserving Genial Initial Site Conditions

Site quality is a clinching factor in nurturing and sustaining high-calibre heritage trees (Jim 2005a; Table 13.2). An excellent site that has nurtured an excellent tree should not be allowed to degrade. In the course of urban development, outstanding trees could be designated for preservation. Their subsequent performance to a large extent is contingent upon primarily the initial site treatment and secondarily the quality of tree care. Improper site preparation could destine the tree to a chronically-stressed state, thus contradicting if not defeating the preservation decision. It could lead to long-term and premature tree decline and a heavy tree-care burden with a low benefit-cost ratio.

For trees inherited from the pre-urbanization landscape, the initial site treatment is a key determinant of its continued survival. Wide difference between original and modified site conditions may stress the tree and jeopardize its survival. The original

Table 13.1 Sixty epithets for trees that have attracted special human attention and their association with eight classification criteria

	Epithet	A Outstanding dimension	B Advanced age	C Superlative performance	D Special ecological function	E Scenic-visual dominance	F Personality-event association	G Natural-cultural bequest	H Spiritual-mythical connotation
1	Big tree	Primary	Secondary	Tertiary	Secondary	Tertiary	Tertiary	Tertiary	Tertiary
2	Great tree	Primary	Secondary	Tertiary	Secondary	Tertiary	Tertiary	Tertiary	Tertiary
3	Large tree	Primary	Secondary	Tertiary	Secondary	Tertiary	Tertiary	Tertiary	Tertiary
4	Giant tree	Primary	Secondary	Tertiary	Secondary	Tertiary	Tertiary	Tertiary	Tertiary
5	Old tree	Secondary	Primary	Tertiary	Secondary	Tertiary	Tertiary	Tertiary	Tertiary
6	Ancestral tree	Secondary	Primary	Tertiary	Secondary	Tertiary	Tertiary	Tertiary	Tertiary
7	Ancient tree	Secondary	Primary	Tertiary	Secondary	Tertiary	Tertiary	Tertiary	Tertiary
8	Long-living tree	Secondary	Primary	Tertiary	Secondary	Tertiary	Tertiary	Tertiary	Tertiary
9	Pre-urbanization tree	Secondary	Primary	Tertiary	Secondary	Tertiary	Tertiary	Tertiary	Tertiary
10	Champion tree	Secondary	Tertiary	Primary	Tertiary	Secondary	Tertiary	Tertiary	Tertiary
11	Elite tree	Secondary	Tertiary	Primary	Tertiary	Secondary	Tertiary	Tertiary	Tertiary
12	Magnificent tree	Secondary	Tertiary	Primary	Tertiary	Secondary	Tertiary	Tertiary	Tertiary
13	Outstanding tree	Secondary	Tertiary	Primary	Tertiary	Secondary	Tertiary	Tertiary	Tertiary
14	Remarkable tree	Secondary	Tertiary	Primary	Tertiary	Secondary	Tertiary	Tertiary	Tertiary
15	Exceptional tree	Secondary	Tertiary	Primary	Tertiary	Secondary	Tertiary	Tertiary	Tertiary
16	Treemendous tree	Secondary	Tertiary	Primary	Tertiary	Secondary	Tertiary	Tertiary	Tertiary
17	Veteran tree	Tertiary	Secondary	Tertiary	Primary	Tertiary	Tertiary	Secondary	Tertiary
18	Keystone tree	Tertiary	Secondary	Tertiary	Primary	Tertiary	Tertiary	Secondary	Tertiary
19	Old-growth tree	Tertiary	Secondary	Tertiary	Primary	Tertiary	Tertiary	Secondary	Tertiary
20	Signature tree	Secondary	Tertiary	Secondary	Tertiary	Primary	Tertiary	Tertiary	Tertiary
21	Accent tree	Secondary	Tertiary	Secondary	Tertiary	Primary	Tertiary	Tertiary	Tertiary
22	Scenic tree	Secondary	Tertiary	Secondary	Tertiary	Primary	Tertiary	Tertiary	Tertiary
23	Aesthetic tree	Secondary	Tertiary	Secondary	Tertiary	Primary	Tertiary	Tertiary	Tertiary
24	Focal tree	Secondary	Tertiary	Secondary	Tertiary	Primary	Tertiary	Tertiary	Tertiary
25	Natural monument tree	Secondary	Tertiary	Secondary	Tertiary	Primary	Tertiary	Tertiary	Tertiary
26	Famous tree	Tertiary	Secondary	Tertiary	Tertiary	Tertiary	Primary	Secondary	Tertiary
27	Renowned tree	Tertiary	Secondary	Tertiary	Tertiary	Tertiary	Primary	Secondary	Tertiary
28	Named tree	Tertiary	Secondary	Tertiary	Tertiary	Tertiary	Primary	Secondary	Tertiary
29	Commemorative tree	Tertiary	Secondary	Tertiary	Tertiary	Tertiary	Primary	Secondary	Tertiary
30	Memorable tree	Tertiary	Secondary	Tertiary	Tertiary	Tertiary	Primary	Secondary	Tertiary
31	Identity tree	Tertiary	Secondary	Tertiary	Tertiary	Tertiary	Primary	Secondary	Tertiary
32	Valuable tree	Tertiary	Secondary	Tertiary	Tertiary	Tertiary	Primary	Secondary	Tertiary
33	Special tree	Tertiary	Secondary	Tertiary	Tertiary	Tertiary	Primary	Secondary	Tertiary
34	Notable tree	Tertiary	Secondary	Tertiary	Tertiary	Tertiary	Primary	Secondary	Tertiary
35	Tree-membrance tree	Tertiary	Secondary	Tertiary	Tertiary	Tertiary	Primary	Secondary	Tertiary
36	Historical tree	Tertiary	Tertiary	Tertiary	Tertiary	Tertiary	Secondary	Primary	Secondary
37	Historic tree	Tertiary	Tertiary	Tertiary	Tertiary	Tertiary	Secondary	Primary	Secondary
38	Tree of history	Tertiary	Tertiary	Tertiary	Tertiary	Tertiary	Secondary	Primary	Secondary
39	Heritage tree	Tertiary	Tertiary	Tertiary	Tertiary	Tertiary	Secondary	Primary	Secondary
40	Legacy tree	Tertiary	Tertiary	Tertiary	Tertiary	Tertiary	Secondary	Primary	Secondary
41	Monumental tree	Tertiary	Tertiary	Tertiary	Tertiary	Tertiary	Secondary	Primary	Secondary
42	Remnant tree	Tertiary	Tertiary	Tertiary	Tertiary	Tertiary	Secondary	Primary	Secondary
43	Relict tree	Tertiary	Tertiary	Tertiary	Tertiary	Tertiary	Secondary	Primary	Secondary
44	Legendary tree	Tertiary	Tertiary	Tertiary	Tertiary	Tertiary	Secondary	Primary	Secondary
45	Cultural tree	Tertiary	Tertiary	Tertiary	Tertiary	Tertiary	Secondary	Primary	Secondary
46	Culturally-important tree	Tertiary	Tertiary	Tertiary	Tertiary	Tertiary	Secondary	Primary	Secondary
47	Tree of special interest	Tertiary	Tertiary	Tertiary	Tertiary	Tertiary	Secondary	Primary	Secondary
48	Tree of knowledge	Tertiary	Tertiary	Tertiary	Tertiary	Tertiary	Secondary	Primary	Secondary
49	Tree of wisdom	Tertiary	Tertiary	Tertiary	Tertiary	Tertiary	Secondary	Primary	Secondary
50	Tree of life	Tertiary	Tertiary	Tertiary	Tertiary	Tertiary	Secondary	Primary	Secondary
51	Sacred tree	Tertiary	Secondary	Tertiary	Tertiary	Tertiary	Tertiary	Secondary	Primary
52	Religious tree	Tertiary	Secondary	Tertiary	Tertiary	Tertiary	Tertiary	Secondary	Primary
53	Venerable tree	Tertiary	Secondary	Tertiary	Tertiary	Tertiary	Tertiary	Secondary	Primary
54	Worshipped tree	Tertiary	Secondary	Tertiary	Tertiary	Tertiary	Tertiary	Secondary	Primary
55	Revered tree	Tertiary	Secondary	Tertiary	Tertiary	Tertiary	Tertiary	Secondary	Primary
56	Blessing tree	Tertiary	Secondary	Tertiary	Tertiary	Tertiary	Tertiary	Secondary	Primary
57	Wishing tree	Tertiary	Secondary	Tertiary	Tertiary	Tertiary	Tertiary	Secondary	Primary
58	Symbolic tree	Tertiary	Secondary	Tertiary	Tertiary	Tertiary	Tertiary	Secondary	Primary
59	Mystic tree	Tertiary	Secondary	Tertiary	Tertiary	Tertiary	Tertiary	Secondary	Primary
60	Mythological tree	Tertiary	Secondary	Tertiary	Tertiary	Tertiary	Tertiary	Secondary	Primary

Relevance: Primary; Secondary; Tertiary

site qualities should be preserved as far as possible, or be emulated in the altered site to bring a good match with the original. The change from a natural vegetation ambience such as woodland or forest to occlusion in a highly-urbanized and small plot could incur drastic degradation of site conditions. Conversion from a village or farmland landscape or a suburban domestic garden to built-up land use would bring less drastic changes, as the tree would have pre-adapted to the partly humanized circumstances.

Site matching could extend from the target tree per se to its living companions. Preserving the tree in association with its original neighbour trees, undergrowth and other plants could keep the soil, water and microclimatic conditions to enhance survival. The best option is to protect a sufficiently large pocket of the original ecosystem that includes the target tree. The worst choice is to trap it solitarily in an alien and confined niche. For particularly valuable heritage trees, preservation in an urban park rather than built-up land use may enhance survival. The planting site could be carefully designed to emulate the original natural or semi-natural state to improve the chance of adapting to the otherwise novel environment.

13.3.2 Extending Protection to Soil Domain

A main cause of tree decline is acute site and environmental degradation. Soil disturbance and root damage associated with roadwork and building construction constitute the principal predisposing factors of heritage-tree decline and demise (Jim 2005b; Table 13.2). Tree conservation should protect the tree per se as well as the growth conditions in the subaerial and subterranean domains (Jim and Zhang 2013). The failure of tree preservation endeavours in development projects could often be attributed to inadequate attention to the critical but widely neglected soil component. The original soil volume and the constituent roots should be kept as much as possible. The soil conditions amenable to various soil processes could be maintained to avoid stressing the critical edaphic-health attributes. The organic O horizon and the organo-mineral A horizon, jointly constituting the topsoil, should particularly be protected in its original state. The activities of the soil organisms and their key roles as decomposers in nutrient cycling could be retained.

Based on recent tree-root research findings, the minimum diameter of the soil protection zone (SPZ) should be defined by the tree's drip line, and the minimum depth should be 1 m. If site conditions allow, an additional safety belt of 5 m beyond the drip line should be added to the SPZ. For particularly large, precious or vulnerable trees, the SPZ should be larger than the minimum provision. The entire soil volume within the SPZ in the form of a soil-root disc has to be guarded against excavation, ploughing, filling and trenching activities. To prevent soil compaction and attendant serious soil degradation, human and animal trampling and vehicle movement and parking should be excluded from the SPZ. Sites with degraded soil could be rehabilitated (Fig. 13.4).

Compact cities with proximity of trees to the built-up fabric could induce frequent and serious conflicts. If protected trees are situated within or near a construction site, extra precautions are needed to prevent other forms of soil disturbances. They include paving, stockpiling of earth, construction materials or machinery, spillage of harmful fluids, and intrusion by construction and other wastewater. The use of pesticides, herbicides, chemical fertilizers, road salts and other types of agrochemicals should be prevented within and in the vicinity of the SPZ.

13.3.3 Avoiding Destructive Grade Change

Urban sprawl into green fields commonly involves extensive vegetation removal and site formation to create developable land at a certain elevation. The affected lands are subject to earth filling or cutting to create a uniform level surface to receive buildings, roads and other facilities. Changing the grade of an existing mature-tree site could gravely stress the soil and roots (Table 13.2). The tree will respond by dieback in the short term, and retrenched growth and poor health in the long run. Large trees demanding an extensive SPZ are more prone to the impacts of grade change. Developers are less inclined to maintain the grade within the SPZ of large trees due to cost considerations.

Table 13.2 Critical factors, precautions and solutions related to the management and conservation of urban heritage trees

Critical factor	Precaution or solution
(1) *Preserving genial initial site conditions*	
(a) Keeping original site quality	Retain key site attributes that nurture meritorious trees
(b) Modifying site conditions	Prevent materials and processes that may degrade site quality
(c) Improving site conditions	Introduce measures to emulate enabling natural conditions
(d) Maintaining woodland habitat	Forestall loss of valuable woodland ecosystem traits
(e) Replanting companion vegetation	Compensate for loss of companion vegetation
(2) *Extending protection to soil domain*	
(a) Protecting soil and roots	Protect holistic subaerial and subterranean growth conditions
(b) Guarding soil volume and root integrity	Aim at preserving the bulk of the root system
(c) Designating soil protection zone (SPZ)	Demarcate SPZ within dripline plus buffer margin
(d) Conserving topsoil	Preserve integrity of critical O and A soil horizons
(e) Safeguarding soil quality	Maintain soil structure, moisture and nutrient capability

(continued)

Table 13.2 (continued)

Critical factor	Precaution or solution
(f) Precluding soil compaction	Adopt effective measures to prevent soil compaction
(g) Averting soil contamination	Stop ingress of polluted effluents and materials into SPZ
(h) Blocking intrusion into SPZ	Forbid stockpiling of construction materials and machines in SPZ
(3) *Avoiding destructive grade changes*	
(a) Avoiding grade change	Prohibit grade change within SPZ
(b) Understanding grade change impacts	Recognize that even small grade changes could seriously harm trees
(c) Adopting sympathetic site design	Innovate to accommodate needs of trees despite development
(4) *Rehabilitating degraded sites*	
(a) Recognizing rehabilitation feasibility	Realize that poor site conditions are amenable to rehabilitation
(b) Rebuilding truncated topsoil	Add soil mix enriched with organic matter to rebuild topsoil
(c) Mulching soil surface	Cover bare soil surface with living vegetation or organic mulch layer
(d) Regenerating degraded soil structure	Amend with organic aggregating agents to improve soil structure
(e) Ameliorating heavily compacted soil	Apply air-spade method to replace compacted soil with good soil mix
(5) *Nurturing potential heritage trees*	
(a) Evaluating generation gap	Study age distribution of heritage-tree population structure
(b) Defending old trees	Take measures to keep trees at elderly end of the spectrum
(c) Nurturing potential heritage trees	Identify potential heritage-tree candidates for special care
(d) Creating conditions for new heritage trees	Provide sustained suitable site conditions to rear new heritage trees
(e) Sustaining genetic superiority	Ensure that superior genes of heritage trees are inheritable
(6) *Statutory designation and governance*	
(a) Developing effective tree ordinance	Enact or amend tree ordinance to augment heritage tree protection
(b) Adopting citizen nomination system	Engage citizens to nominate heritage trees
(c) Involving citizens in designation	Include citizens and experts in designating heritage trees
(d) Cultivating tree ownership	Bestow recognition and pride to heritage-tree owners
(e) Helping tree owners in tree care	Offer subsidies and technical support in tree maintenance
(f) Building partnership with community	Involve corporate sector and tree-care companies in joint endeavour
(g) Boosting publicity and public education	Strengthen efforts to enhance awareness and knowledge base

(continued)

Table 13.2 (continued)

Critical factor	Precaution or solution
(7) *Overzealous tidying of old trees and risk assessment*	
(a) Comprehending veteran tree functions	Learn key ecological contributions of ancient veteran trees
(b) Retaining veteran features	Refrain from tidying up old trees and eliminate veteran features
(c) Learning veteran tree maintenance	Acquire knowledge and practical skills in caring for veteran trees
(d) Stabilizing site conditions for old trees	Avoid disturbing site and soil conditions of old trees
(e) Initiating tree risk assessment	Refine tree risk assessment method dedicated to heritage trees
(f) Introducing tree risk management	Strike an appropriate balance between conservation and tree risk
(8) *Lightning protection*	
(a) Appreciating lightning hazard	Evaluate trees regarding need for lightning protection
(b) Using professional protection gears	Employ professionals to install high-quality protection gears
(c) Ensuring long-term protection	Inspect and adjust regularly the protection system
(9) *Transplanting as the last resort*	
(a) Understanding tree transplanting	Appreciate the procedures, costs and harms of transplanting
(b) Discouraging transplanting large trees	Recognize firmly that transplanting is the last resort
(c) Assembling multidisciplinary crew	Establish tree and engineering professional team to move large trees
(d) Reducing transplant shock	Optimize all steps to minimize harmful impacts of transplant shock
(e) Making decision on root ball	Determine root ball size and strong containment method
(f) Embracing phased root pruning	Implement phased root pruning with intervening recuperation periods
(g) Fostering replacement roots	Facilitate development of new replacement roots in root ball
(h) Providing suitable reception site	Prepare reception site to match donor site conditions
(10) *Community sentiment towards tree loss*	
(a) Gauging public reaction to tree felling	Anticipate reaction of the community towards heritage-tree felling
(b) Following public relation skills	Design appropriate public relation strategy to ameliorate impacts
(c) Supplying quality information	Disseminate high-quality, professional-scientific timely information
(d) Notifying the public effectively	Convey messages o citizens through multiple means

Grade lowering around a tree could bring deleterious loss of rootable soil volume and a notable proportion of roots. The worst scenario is trapping the preserved tree in an undersized container standing above the surrounding areas. As most tree roots are concentrated in the top 50 cm or so of the soil, lowering the ground surface even by an apparently small amount could incur massive root destruction. Besides losing the water and nutrient absorption capability, the ill-treatment could seriously weaken tree anchorage to induce instability. Grade raising could bury the original soil surface and the root system to induce root decline and suppress new root growth. Grade change could be prevented by sympathetic site design aiming at accommodating the preserved trees. If it cannot be avoided, comprehensive precautionary measures should be adopted to minimize the short- and long-term impacts.

13.3.4 Rehabilitating Degraded Site

Trees trapped in built-up areas often face site degradation due to urban-fabric densification (Jim 2005c). Good initial site conditions could gradually be dampened. Rehabilitation schemes could be applied to mitigate the constraints (Fig. 13.4). Some problems are amenable to amelioration whereas others may not. The treatments and methods should be tailor-made and site-specific (Table 13.2).

If the soil surface is bare and the organic O horizon is missing or too thin, compost could be added as mulch to rebuild it. The protective veneer can offer a surprising range of benefits: reduce topsoil temperature, suppress evaporation, conserve soil moisture, shield the soil from rain splash impact to forestall soil crust formation and erosion, provide a genial environment for decomposer organisms to release available nutrients from organic substances, release humic substances to enhance soil structure, and promote healthy and strong root growth.

Some trees suffer from poor soil structure often caused by organic-matter deficiency. Organic materials could be added to increase the supply of aggregating agents to improve soil organization and induce better aeration, infiltration, drainage, storage of plant-available moisture, and soil environment for microbial and decomposer activities. Mature compost or other organic substances could bring relief and supply nutrients. For routine soil management, organic fertilizers can be applied instead of chemical types to promote a healthy soil structure.

If soil compaction is not serious and confined to the topsoil layer, organic-matter amendments could re-build soil aggregates, increase porosity and lower bulk density (Jim 1993, 1998b). The top 3 cm of the soil could be scarified so that the organic matter could be mixed with the loosened mineral soil. For more serious compaction that has lifted the bulk density to over 1.8 Mg/m^3, and is affecting a deeper soil layer, more drastic measures are necessary. The cultivation or deep ploughing method cannot be used because it will damage too many roots and jeopardize tree survival and stability.

The air spade method could be adopted to blow away the compacted soil whilst keeping most of the roots (Ames and Dewald 2003). The removed soil could be replaced with a prepared soil mix enriched with organic matter to maintain soil structure and prevent future compaction. This soil replacement method should be

Fig. 13.4 The site condition of this old but robust *Cinnamomum camphora* is being rehabilitated by expanding the soil and root protection zone well beyond the drip line and improving the soil quality within the entire area. It can be appreciated in the Tokyo University campus (*Photo credit* C.Y. Jim)

applied in phases to a tree. Treating the entire SPZ in one exercise will impose too much stress on the tree. The SPZ could be divided into eight sectors. Each treatment could involve a pair of two opposite sectors. After applying one treatment, one year should elapse to allow the tree to recover from the shock before the next phase is implemented.

13.3.5 Nurturing Potential Heritage Trees

Outstanding trees tend to be inherited in old neighbourhoods of old cities (Jim 1994a, 2004a). Their existence could be attributed to fortuitous combination of circumstances before and after designation. The vicissitudes of the sites over the years would determine tree wellness and survival. Unfortunately, many trees face premature decline due to progressive site degradation which is a particularly acute in compact cities. As existing trees gradually die, the urban milieu meanwhile is unable to provide the necessary and sufficient conditions to foster outstanding replacement trees (Table 13.2).

The existing heritage tree stock thus tends to suffer from an ageing population structure. Most trees are old and relatively younger successors are difficult to find. The age discontinuity has failed to sustain the lineage. In effect, a generation gap has emerged which could foretell gradual demise of older trees accompanied by dwindling of the heritage-tree cohort. Something proactive could be done to arrest this trend by ensuring a steady pace of replenishment. Proactive steps are needed to rear new recruits.

Two approaches could be adopted to assure continuity. The short-term strategy can identify trees with qualities marginal to the heritage-tree calibre. They have the potential to be designated in due course, such as falling somewhat short of the age or dimension thresholds. Such trees could be given special care to ensure that they will soon join the elitist register. Site conditions could be maintained and preferably improved to prevent tree decline.

The long-term strategy is to create the necessary site conditions to rear a new crop of heritage trees, complemented by conscious selection of species to match site qualities and sourcing robust planting materials. The planting effort should be spaced over time to establish trees of different ages (Woodland Trust 2009b). The procedures are akin to the nurturing of sports champions. It demands sustained efforts of high-grade and long-range cross-generational arboriculture to ascertain delivery of meritorious products in the extended endeavour (Jim 2005a). The critical path is to provide a town plan that is conducive to nurturing of future champions (Jim 2004b).

The individual heritage trees denote a reservoir of superior genes which are irreplaceable. They furnish nature's treasure trove expressed in the form of disease resistance, exceptional vigour, unique growth form, and unusual longevity. The germ-plasm pool should be sustained and inherited by future generations. For new or replacement trees, seeds could be collected from outstanding mother trees to nurture a meritorious crop of descendants with genetic continuity with their progenitors. If the old trees do not produce viable seeds any more, the self-seeded saplings and young trees situated nearby could furnish members of the next generation.

13.3.6 Statutory Designation and Governance

For places with an urban tree or urban forestry ordinance, the policies, legal mechanism, selection, designation and protection of heritage trees can be subsumed under the statutory umbrella (e.g. South Australia Government 2000; Table 13.2). Some cities such as Portland, OR, has a dedicated Heritage Tree Ordinance (Portland Parks and Recreation 2016). The City of Vancouver (undated), WA, offers a fine exemplary. It includes trees lying in public and private lands. The nomination requires the agreement of the private property owner before moving to the assessment and determination stages. Once the consent is given, it is binding on future property owners, successors or heirs. Before making decisions, a heritage-tree board with public-service and citizen representatives will assess the nomination and hold at least one public hearing to glean the community's views.

The cardinal objective of the designation is to protect and community's heritage-calibre assets, and to ascertain their long-term welfare. The listed trees are protected by law from unnecessary felling, damage and aggressive maintenance. Designation has to be accompanied by a tree preservation plan to ensure proper and timely professional care as well as healthy and robust state. Tree removal with good justifications such as dead, dying or dangerous, has to be approved by the urban forest authority. Pruning affecting over 20% of the existing crown also requires permission. Pruning works must adhere to high arboricultural standard by a qualified professional. Some cities require a permit before any tree work. In some jurisdictions without the tree-law instrument, administrative procedures and other tangentially related laws would shoulder this onus (Jim and Chan 2016).

Most cities with official designation of heritage trees have established an online tree register to record pertinent facts such as location, species, dimensions and performance. Some city authorities would include maps, tree photographs, as well as historical and cultural information. To connect with the community, various measures have been adopted. Nomination procedures and forms are provided for citizens to propose trees (e.g. Mississippi Forestry commission 2005; City of Vancouver undated). Partnerships or sponsorship of green groups, corporations and tree care companies are sought to ensure proper tree maintenance. Property owners may get financial or professional help in tree care.

To strengthen public support, a holistic package of publicity and public education could include awareness, recognition, designation, protection, management, education and training (Woodland Trust 2007). Protection could be augmented by cultivating community ownership. Formal education about heritage trees could be included in the school curriculum so that kids can learn and appreciate their importance to nature and culture. Public education could remind citizens of the multiple benefits and values. An online official heritage-tree register easily accessed by the public would propagate relevant knowledge and raise awareness. A well-designed and attractive plaque placed at or near the heritage tree would provide vivid and direct information.

13.3.7 Overzealous Tidying of Old Trees and Risk Assessment

The oldest heritage trees, sometimes labelled ancient trees, demand higher-level and more meticulous arboricultural attention (Table 13.2). They tend to display veteran features requiring non-conventional care and high-order expertise and experience. They would have lost some absorption and framework roots to depress its anchorage, stability and vigour. The trunk and main branches generally suffer from different structural defects (see Sect. 13.2.3). The height:dbh ratio, live crown ratio, and leaf area index are usually low due to widespread dieback, loss of large branches and general height reduction. Growth may become sluggish, but some can continue to add annual wood increments at the trunk and main branches. However, there is a tendency for the growth to decrease with age.

The veteran traits are the raison d'etre of heritage-tree status. Their performance and welfare could be judged by dedicated standards rather than the routine arboricultural yardstick. The temptation to tidy them up to conform to the norm should be resisted. Rectifying such 'defects' by 'corrective pruning' would deprive them of their inherent characters and precious ecological services (see Sect. 13.2.3). For instance, overzealous crown cleaning and hollow draining and filling should be avoided. No attempt should be made to rejuvenate or sanitize such elderly trees. Instead, the tree managers could identify and preserve the veteran features (Davis et al. 1996).

Measures could be adopted to prevent site degradation that may harm old trees (Read 2000). The water supply to the soil should not be unduly increased or decreased. The existing water table and its seasonal changes should be maintained. Competition for space and resources from adjacent non-heritage trees could be appropriately regulated. Removal of surrounding trees or structures may adversely modify microclimate conditions for tree growth. Possible damages due to spreading of fire and pollutants from nearby areas should be prevented. Bark injuries, trampling and soil compaction incurred by people, animals and vehicles could be avoided.

The site access and use frequency may impose constraints on retention of potentially hazardous tree parts (Fay 2001). The tree owner has to shoulder the legal responsibility of maintaining the tree in a safe state to avoid harming people and properties. The duty of care includes foreseeing dangers and providing reasonable care to abate them as far as reasonably possible. Taking no action or taking inappropriate action could incur legal liability (English Nature 2000b). Tree risk assessment should be conducted regularly by a specialist. The assessment method should be tailor-made for different species or species groups which demonstrate different expressions of veteran and risk features (Jim and Zhang 2013).

The desire to conserve old trees should be balanced by the need to abate the danger to people and properties. This aspect of tree management is especially pertinent in compact urban areas with high probability of tree failure hurting people. Skilful treatment and pruning can be applied to abate the risks and avoid felling. A cordon could be demarcated under the tree canopy to limit access to a possible branch dropping zone. An arborist with special training in veteran tree care and tree surgery should be enlisted to provide delicate care. The overarching principles are regular inspection, preventive maintenance, and enlisting surgery only if it is necessary.

13.3.8 Lightning Protection

Lightning protection is an essential preventive-care tool that has been commonly omitted (Table 13.2). This is particularly necessary where the subject tree is tall, solitary, or the tallest amongst a tree group. A lightning strike can send a huge electrical charge through the tree body to the ground. As a result, a tremendous amount of heat is generated inside the stems and roots to hurt tree tissues. The passage of the powerful electrical energy may be accompanied by steam explosion inside the branches and the trunk to shatter the wood, detach the bark and damage the roots. Some trees may die instantly, whereas others may decline gradually due to secondary infection by pests and diseases.

The standard lightning protection gears begins as air terminal installed at the top of the tree. It is connected to a strip of copper conductor to transmit rapidly the electrical pulse to the ground via an earthing or grounding device. They have to be designed and installed by professionals, and be inspected on a regular basis.

13.3.9 Transplanting as Last Resort

Some urban development projects conflict with existing heritage trees. The normal expectation is modifying development plan to allow tree preservation, such as shifting or moulding the building footprint or road alignment. Innovative design backed by determination could allow in situ preservation (Jim 1988). If the conflict could not be avoided and the tree is considered too precious to fell, the undesirable alternative of transplanting has to be contemplated (Table 13.2). It should be stressed that transplanting a large tree is an elaborate and expensive exercise, and that it could incur irreversible damages. Old and feeble trees may not survive the move, or may be driven into a decline spiral. It should therefore be considered as the last resort.

Moving a large tree requires meticulous planning and implementation to ensure success (Jim 1995). The complex exercise demands teamwork of the arboricultural and engineering professions. The most critical step is the preparation of a root ball held in a strong box with a size and hence weight that is manageable. If it is too small, too many roots are lost and the tree may not survive the shock. If it is it too big, it is difficult to lift and transport. The capacity of these two crucial engineering manoeuvres could be constrained by site and terrain factors at both the source and destination sites. The optimal root ball size adopted by most large-tree transplants usually incurs 80–90% root loss.

The cardinal consideration is whether the tree can tolerate the tremendous stresses of the 'massive controlled injuries' as a result of losing so many roots. To reduce the transplant shock, root pruning has to be conducted in phases separated by long enough recuperation periods. Besides cutting roots that will not be included in the root ball, root pruning serves the important function of stimulating the growth of new absorption roots at the cut faces and within the root ball. These new roots held in the root ball could compensate partly for the lost roots and reduce the shock impact.

Another key consideration is the estimated duration of useful tree life span after the move, which is expected to be not less than several decades. To facilitate adaptation to the new home, the recipient site should as far as possible match the donor site conditions, or appropriately modified to match them. It is essential to provide sufficient soil of good quality around the deposited root ball at the recipient site. The post-transplant new root growth into the new site soil will determine largely whether the tree can recover from the shift.

13.3.10 Community Sentiment Towards Tree Loss

The loss of heritage trees could induce strong and sometimes unexpected community responses, often considered to be equivalent to the loss of loved ones. The diverse feelings and emotions could cover sadness, grief, distress, melancholy, anguish, anger, and agony. The tree management authority has to sooth the upset minds and disturbed psyche in a timely manner (Table 13.2). Measures developed by the public relations profession (Munson 1993) could be adapted to prevent unduly passionate reactions and contain misunderstanding and discontent.

For the removal of particularly important heritage trees to which the local community has developed deep sentimental attachment, it is necessary to provide high-quality, science-based, professional-led, succinct information in sufficient details and in plain language. The publicity work should be delivered before, during and after the takedown event. The communication topics could anticipate the spectrum of critical issues: *why* (justifications), *when* (timetable), *how* (safe and efficient procedures), *who* (skillful and experienced professional arborists), and *what* (recompense as post-felling landscape improvement plan).

The messages could be conveyed at critical junctures through announcements, notices, press releases and meetings. Forthright and honest explanations could amend an apparently unpopular decision and unfavourable publicity. If it is deemed necessary, a memorial service could be conducted by local religious leaders to conclude the succession of actions. The depth and extent of such public-relation measures could be adjusted in proportion to the significance of the subject tree to citizens. Above all, the important trees must not be removed in surprise without establishing a prior, frank and timely dialogue with the local community.

13.4 Conclusion

Heritage trees are the fortuitous and cherished bequests of a bygone era. As valuable natural-cum-cultural and living public assets, the inherited treasures should be given a long service life and rendered inheritable. The present generation has the obligation to ensure that the meritorious trees can continue to thrive and be appreciated and enjoyed. To fulfil this goal, it is worthwhile to nurture the mentality of transgenerational urban forestry, and translate it into precision arboricultural practices to care for the prized doyens. Especially for compact cities with increasing if not relentless pressure on the heritage-tree stock, it is quite necessary to transform tree care from the probabilistic to the deterministic mode. Generation of relevant scientific knowledge and their translation into general knowledge, professional practices and enabling policies could augur well for the welfare of our beloved living companions.

Trees are living creatures with a finite service life span which tends to be more constrained in the stressful high-density urban milieu. Whatever we do to help them, they will eventually succumb to the combined vagaries of natural and human impacts. The pertinence of ensuring tree safety would demand systematic tree removal near the terminal stage of their life or in response to serious and incorrigible tree defects. It is necessary to continue to furnish the right conditions in the town plan to nurture the next crop of heritage trees to sustain their welcomed amenities in our neighbourhoods. After all, it is the community's collective attitude, behaviour, expectation, and value attached to heritage trees that will determine their welfare, continued existence of the present stock, and emergence of the relay cohort. The success or otherwise of the endeavour is contingent upon a fruitful joint venture between the government and citizens.

References

Alexander K, Green T, Key R (1998) Managing our ancient trees. Tree News (Spring 1998):10–13

American Forestry Association (undated) The national register of big trees. Washington, DC

Ames B, Dewald S (2003) Working proactively with developers to preserve urban trees. Cities 20:95–100

Arnan X, López BC, Martínez-Vilalta J, Estorach M, Poyatos R (2012) The age of monumental olive trees (*Olea europaea*) in northeastern Spain. Dendrochronologia 30:11–14

Asan Ü (2001) Monumental trees and forests of Turkey. Third Balkan scientific conference on study, conservation and utilisation of forest resources, Sofia, Bulgaria, 2–6 Oct 2001, vol II, pp 389–399

Asciuto A, D'Acquisto M, Passarello S (2011) Monetary valuations of monumental trees and other natural resources between demand for conservation and recent requirements for outdoor activities: some case studies in the Madonie and Nebrodi Regional Parks of Sicily. In: Conference on present and future role of forest resources in the socio-economic development of rural areas, Rome, 23–24 June 2011, pp 6–22

Asciuto A, Borsellino V, D'Acquisto M, Di Franco CP, Di Gesaro M Schimmenti, E (2015) Monumental trees and their existence value: case study of an Italian natural park. J Forest Sci 61:56–61

Barro S, Gobster PH, Schroeder HW, Bartram SM (1997) What makes a big tree special? Insights from the Chicagoland Treemendous trees program. J Arboric 23(6):239–249

Becker N, Freeman S (2009) The economic value of old growth trees in Israel. Forest Policy Econ 11:608–615

Boylan C (2010) Champion and heritage trees of Ireland. 22nd IFPRA World Congress, Hong Kong, 15–18 Nov 2010, pp 1–18

Blicharska M, Mikusinski G (2014) Incorporating social and cultural significance of large old trees in conservation policy. Conserv Biol 28:1558–1567

Browne D (2001) Our remarkable trees: a selection of Northern Ireland's special trees. Conservation Volunteers Northern Ireland, Belfast, UK

Cannizzaro S, Corinto GL (2014) The role of monumental trees in defining local identity and in tourism: a case study in the Marches Region. Geoprogress J (s Humanities 1) 1(1):29–48

Center for Urban Forest Research (2004) The large tree argument: the case for large-stature trees vs small-stature trees. US Department of Agriculture Forest Service, Southern Center for Urban Forestry Research and Information, Athens, GA

Chen WY (2015) Public willingness-to-pay for conserving urban heritage trees in Guangzhou, South China. Urban Forest Urban Greening 14:796–805

Chen WY, Hua J (2015) Citizens' distrust of government and their protest responses in a contingent valuation study of urban heritage trees in Guangzhou, China. J Environ Manage 155:40–48

City of New York (undated) The great trees of New York City. Parks and Recreation, New York

City of Vancouver (undated) Heritage trees. Vancouver, WA. http://www.cityofvancouver.us/publicworks/page/heritage-trees. Accessed 21 Apr 2016

Dafni A (2006) On the typology and the worship status of sacred trees with a special reference to the Middle East. J Ethnobiol Ethnomed 2:26–40

Davis C, Fay N, Mynors C (1996) Veteran trees initiative: specialist survey method. English Nature, Peterborough, UK

English Nature (2000a) The future for veteran trees. Veteran Trees Initiative, Peterborough, UK

English Nature (2000b) Veteran trees: a guide to risk and responsibility. Veteran Trees Initiative, Peterborough, UK

Fay N (2001) Old trees: liability for habitat. Tools for preserving woodland biodiversity. Treework Environmental Practice, Bristol, UK

Hu L, Li Z, Liao WB, Fan Q (2011) Values of village fengshui forest patches in biodiversity conservation in the Pearl River Delta, China. Biol Conserv 144:1553–1559

Gnec M, Guner ST (2001) A new method to select monumental tree among the forest tree species of Turkey: an application. In: International conference on forest research: a challenge for integrated European approach, Thessaloniki, Greece, 27 August–1 September 2001, pp 1–6

Golabek E, Tukiendorf A (2002) Growth in thickness of monumental English Oaks *Quercus robur*, and their age, health status and dust fall in Bayesian approach. Polish Journal of Environmental Studies 11:331–337

Goldberg E, Kirby K, Hall J, Latham J (2007) The ancient woodland concept as a practical conservation tool in Great Britain. J Nat Conserv 15:109–119

Jim CY (1988) Preservation of a large Chinese Banyan on a construction site. J Arboric 14:176–180

Jim CY (1993) Soil compaction as a constraint to tree growth in tropical and subtropical urban habitats. Environ Conserv 20:35–49

Jim CY (1994a) Champion trees in urban Hong Kong. Hong Kong Flora and Fauna Series, Urban Council, Hong Kong

Jim CY (1994b) Evaluation and preservation of champion trees in urban Hong Kong. Arboricultural J 18:25–51

Jim CY (1995) Transplanting two champion specimens of mature Chinese Banyans. J Arboric 21:289–295

Jim CY (1998a) Old stone walls as an ecological habitat for urban trees in Hong Kong. Landscape Urban Plan 42:29–43

Jim CY (1998b) Soil compaction at tree planting sites in urban Hong Kong. In: Watson GW, Neely D (eds) The Landscape below ground *II*. International Society of Arboriculture, Champaign, IL, pp 166–178

Jim CY (2004a) Spatial differentiation and landscape-ecological assessment of heritage trees in urban Guangzhou (China). Landscape Urban Plan 69:51–68

Jim CY (2004b) Evaluation of heritage trees for conservation and management in Guangzhou city (China). Environ Manage 33:74–86

Jim CY (2005a) Outstanding remnants of nature in compact cities: patterns and preservation of heritage trees in Guangzhou city (China). Geoforum 36:371–385

Jim CY (2005b) Monitoring the performance and decline of heritage trees in urban Hong Kong. J Environ Manage 74:161–172
Jim CY (2005c) Floristics, performance and prognosis of historical trees in the urban forest of Guangzhou city (China). Environ Monit Assess 102:285–308
Jim CY (2006) Formulaic expert method to integrate evaluation and valuation of heritage trees in compact city. Environ Monit Assess 116:53–80
Jim CY (2010) Old masonry walls as ruderal habitats for biodiversity conservation and enhancement in urban Hong Kong. In: Muller N, Werner P, Kelcey JG (eds) Urban biodiversity and design. Blackwell, Oxford, pp 323–347
Jim CY (2013) Drivers for colonization and sustainable management of tree-dominated stonewall ecosystem. Ecol Eng 57:324–335
Jim CY (2014) Ecology and conservation of strangler figs in urban wall habitats. Urban Ecosyst 17:405–426
Jim CY, Chan MWH (2016) Urban greenspace delivery in Hong Kong: spatial-institutional limitations and solutions. Urban For Urban Greening 18:65–85
Jim CY, Chen WY (2006) Impacts of urban environmental elements on residential housing prices in Guangzhou (China). Landscape Urban Plan 78:422–434
Jim CY, Chen WY (2009) Value of scenic views: hedonic assessment of private housing in Hong Kong. Landscape Urban Plan 91:226–234
Jim CY, Chen WY (2010) External effects of neighbourhood parks and landscape elements on high-rise residential value. Land Use Policy 27:662–670
Jim CY, Chen WY (2011) Bioreceptivity of buildings for spontaneous arboreal flora in compact city environment. Urban For Urban Greening 10:19–28
Jim CY, Zhang A (2013) Defect-disorder and risk assessment of heritage trees in urban Hong Kong. Urban For Urban Greening 12:585–596
Kozlowski G, Gibbs D, Huan F, Frey D, Gratzfeld J (2012) Conservation of threatened relict trees through living ex situ collections: lessons from the global survey of the genus *Zelkova* (Ulmaceae). Biodiv Conserv 21:671–685
Kunick W (1990) Spontaneous woody vegetation in cities. In: Sukopp H, Hejný S, Kowarik I (eds) Urban ecology: plants and plant communities in urban environments. SPB Academic, The Hague, pp 167–174
Lindenmayer DB, Laurance WF, Franklin JF, Likens GE, Banks SC, Blanchard W, Gibbons P, Ikin K, Blair D, McBurney L, Manning AD, Stein JAR (2014) New policies for old trees: averting a global crisis in a keystone ecological structure. Conserv Lett 7:61–69
London Tree Forum (undated) Celebrating the great trees of London. London Tree Forum, London
Manning AD, Fischer J, Lindenmayer DB (2006) Scattered trees are keystone structure: implications for conservation. Biol Conserv 132:311–321
May DM (1990) Big trees of the midsouth forest survey. US Department of Agriculture Forest Service. Southern Forest Experiment Station, New Orleans, LA
Mississippi Forestry Commission (2005) Champion trees of Mississippi. Mississippi Forestry Commission, Jackson, MS
Munson RH (1993) Handling the demise of historic trees: a problem of public relations. J Arboric 19(1):48–50
Pennsylvania Urban and Community Forestry Council (undated) Trees of historic Philadelphia: lessons from the past for our future. Mechanicsburg, PA
Portland Parks and Recreation (2016) Heritage trees of Portland. https://www.portlandoregon.gov/parks/40280. Accessed 21 Apr 2016
Read H (2000) Veteran trees: a guide to good management. Veteran Trees Initiative, English Nature, Peterborough, UK
Rodgers D, Stokes J, Ogilvie J (2003) Heritage trees of Scotland. Tree Council, London
Schlawin J, Zahawi RA (2008) 'Nucleating' succession in recovering neotropical wet forests: the legacy of remnant trees. J Veg Sci 19:485–492
Sierra Club Sequoia Task Force (2000) Giant Sequoia National Monument: a citizen's guide. http://vault.sierraclub.org/ca/sequoia/monument/. Accessed 20 Apr 2016

South Australia Government (2000) Development Act 1993: significant urban trees. Plan Bull, Planning SA, Adelaide, Australia

Stagoll K, Lindenmayer DB, Knight E, Fischer J, Manning AD (2012) Large trees are keystone structures in urban parks. Conserv Lett 5:115–122

Sukopp H (2008) The city as a subject for ecological research. In: Marzluff JM, Endlicher W, Alberti M, Bradley G, Ryan C, ZumBrunnen, Simon U (eds) Urban ecology: an international perspective on the interaction C between humans and nature. Springer, New York, pp 281–298

Tenche-Constantinescu A-M, Varan C, Fl Borlea, Madoşa E, Szekely G (2015) The symbolism of the linden tree. J Horticult For Biotechnol 19:237–242

Thaiutsa B, Puangchit L, Kjelgren R, Arunpraparut W (2008) Urban green space, street tree and heritage large tree assessment in Bangkok, Thailand. Urban For Urban Greening 7:219–229

Torbay Council (2011) Tree-membrance commemorative trees programme. Devon, UK

Van Pelt R (1996) Champion trees of Washington state. University of Washington Press, Seattle, WA

Wilson D, Wilson A (2013) Figs as a global spiritual and material resource for humans. Human Ecol 41:459–464

Woodland Trust (2007) Ancient trees in the UK: securing their future. Grantham, Lincolnshire, UK

Woodland Trust (2008) What are ancient, veteran and other trees of special interest? Ancient tree guide 4. Grantham, Lincolnshire, UK

Woodland Trust (2009a) The special wildlife of trees. Ancient tree guide 6, Grantham, Lincolnshire, UK

Woodland Trust (2009b) Ancient trees for the future. Ancient tree guide 7, Grantham, Lincolnshire, UK

Woodland Trust (2011) Trees and development. Ancient tree guide 3, Grantham, Lincolnshire, UK

Chapter 14
Conservation and Creation of Urban Woodlands

Chi Yung Jim

Abstract Many cities especially compact ones are deprived of natural elements. High-quality pre-urbanization natural ecosystems such as forests are often obliterated in the course of city growth. Surveys of the key ecological and environmental benefits of urban woodlands provide the basis to advocate conservation and creation. Urban woodlands tend to be isolated or fragmented remnant pockets enveloped by built-up areas. They are threatened by urban sprawl, and degraded by pollutant penetration, recreational impacts, inappropriate management, detachment from propagule sources, declining regeneration capacity, exotic invasion, and native-species pauperization. Sustainable management should be based firmly on ecological principles, to restore natural factors and processes, introduce minimum inputs, guard against intrusions, and foster spontaneous rehabilitation of degraded sites. Conservation strategy can aim at preserving large woodland patches, enlarging existing patches, fusing or connecting small woodlots with habitat corridors, and merging with adjacent natural areas. New woodlands can be proactively created at suitable green, brown and grey (rooftop) fields. Spontaneous colonization could trigger and sustain woodland succession to deliver urban woodlands on green and brown fields without human help. Afforestation with ameliorative treatments could be applied to harsh sites especially with poor substrate properties and scanty seed arrivals. On intractable sites, innovative techniques such as assisted relay floristics using an initial exotic nurse crop and direct plantation of grey fields could pump-prime woodland establishment. As a hybrid urban green space amalgamating nature and human influences in the novel urban setting, urban woodland conservation and management demand innovative and fusion solutions.

Keywords Urban woodland conservation · Urban woodland creation · Woodland ecology · Natural colonization · Urban afforestation · Re-naturalization

C.Y. Jim (✉)
University of Hong Kong, Pokfulam, Hong Kong
e-mail: hragjcy@hku.hk

P.Y. Tan and C.Y. Jim (eds.), *Greening Cities: Forms and Functions*,
Advances in 21st Century Human Settlements,
DOI 10.1007/978-981-10-4113-6_14

14.1 Introduction

Urbanization can proceed at different intensities, bringing differential ratios of natural to impervious artificial coverage. Some cities have managed to keep extensive stretches of natural areas, whereas others may have eliminated them. The compact development mode can bring massive obliteration or drastic modification of nature. To compensate for the loss of nature, cities often create new urban green spaces (UGS) as surrogates of nature. Some UGS adopt the naturalistic or ecological design to create vegetated areas with a high degree of naturalness that can emulate the organization and functions of natural ecosystems. More often than not, most UGS follow the routine and non-descript parkland design reflecting domination by the urban mentality. They are beset with limited green cover, scattered and sparse tree distribution, simplistic tree structure, excessive hard paving or impervious surfaces, and meagre provision of ecosystem services.

In the course of urbanization, some natural pockets could be spared from development by default or by design to leave a legacy of interstitial wilderness (Jorgensen and Tylecote 2007). Areas with rugged terrain and steep slopes which are less amenable to building and road construction could be bypassed or set aside for the time being. In relative terms, the easily developable lands would often be preferentially mobilized. Lands covered by dense woodlands or other vegetation with high nature content and ecological value could be intentionally preserved. Mature forests, especially old growth ones, would often be designated as protected areas. Sites with cultural significance such as sacred woods and *feng shui* (oriental geomancy belief) woodlands, and others with historical association, would normally be respected and preserved.

These valuable natural enclaves may be situated originally at the urban fringe wrapped by rather natural or rustic environs. Relentless urban sprawl could literally bring the city to their doorsteps, and sometimes intruding into the woodland periphery. The converted site would often be used for high-end residential or leisure purposes. Further urban growth may engulf them amidst the built-up matrix. With proximity to residents, they would increasingly be used as outdoor recreational venues. The pull of nature in conjunction with the push of biophilia instinct would engender the visitor stream. Especially where UGS provision is inadequate in dense and poorly-planned cities, the urban woodlands take up the role of surrogate urban parks of a different genre.

It is important to respect the innate ecological endowment of urban woodlands. High-quality nature is in itself a valuable and rare community asset that should not be diluted or squandered by mistaken management inputs. The overzealous and ill-conceived idea to transform them to the manicured urban-park mode is a waste of their inherent and precious qualities (Jim 2003a). The temptation to introduce silvicultural overkill and to tame the wild should be eschewed. The proximal and accessible venues, whilst serving important salubrious functions to enhance the health of human users, could have their own ecological health damaged by excessive and improper treatments.

The most important natural enclaves in cities have to be the urban woodlands. As the most complex ecosystem in the sprawling sea of urbanization, they offer critical ecosystem services to residents and enhance the quality of the urban environment at large (Tregay 1979). Due to inherent complexity, they are vulnerable to degradation by human impacts and general stresses brought by urbanization such as microclimate and air quality decline, and various forms of recreational impacts. Moreover, the threat of conversion to built-up uses would always loom large. The continued urban growth and shortage of developable land would generate relentless pressure to usurp the natural lands that are located at hand. Under enlightened administrative and planning regimes, however, urban woodlands would be warmly valued and assiduously protected. In addition, new urban woodlands would be created from scratch to enrich the stock.

This study surveys the state of knowledge on urban woodlands with respect to their ecological importance in the urban context, the approaches that can be taken to preserve them whilst serving conservation and recreational functions, and the methods to facilitate their creation. The key research questions are: (1) Why is it imperative to have woodlands in cities? (2) What can be done to protect existing woodlands from urbanization impacts? (3) How can new urban woodlands be created within cities?

14.2 Why Do Cities Need Woodlands?

Urban woodland refers to the most complex and natural UGS which is dominated by assemblages of trees forming the elevated main canopy. The trees are accompanied by subcanopy strata of undergrowth and ground vegetation, as well as companion fauna. In most parts, the trees grow closely together to form a rather continuous canopy with touching or interlocking crowns. The soil usually remains rather natural, albeit with some signs of human disturbance. The woodland floor is commonly covered by natural organic litter at different stages of decomposition to form the continuous O horizon. The underlying A horizon with minerals enriched by organic matter, mainly humic substances originated from the above layer, tends to remain rather intact. The ecosystem is characterized by slight or no management input, allowing natural processes to operate with little interference. The complex biomass structure, species heterogeneity, spatial variations, moderated microclimate, attractive odour and sound, collectively connote an idyllic perception of naturalness (Jim 2011a).

City dwellers harbour an innate desire to escape from the excessive artificiality and aggressiveness of urbanity. The formally designed stereotype urban parks, being manicured, neat, regular, predictable, deprived of species and biomass diversities, and symptomatic of detachment from if not suppression of nature, could only partly satisfy people's basic want to be close to nature. They need diversion from the poor emulation of nature by seeking solace in genuine and high-order natural ambience. With UGS dominated by urban parks composed of over-designed and poor emulation of nature, the alternative of preserving or creating natural areas could be promoted (Grosse-Bächle 2005). The informal outdoor recreational use and associated indirect environmental benefits of urban woodlands have long been recognized (Payne 1983).

To meet this demand, domination of UGS by formal urban parks could be relaxed. Instead, a biotope spectrum with different kinds of nature-in-city renditions could be instituted. They could range from simple lawn to parkland and woodland. However, people often have to travel long distance outside city boundaries in order to reach woodlands with high-calibre nature. Woodlands have often been mistakenly regarded as incompatible with cities. There is general misconception that woodlands demand special habitat requirements and care that can hardly be satisfied in urban areas. Countryside and city do not need to be mutually exclusive. Woodlands should not be accommodated only in the countryside; they are equally at home in urban areas where they are very much wanted.

The city as an urban ecosystem could include a diverse range of UGS to support different assemblages of flora and fauna and to achieve a mosaic of habitat heterogeneity. Woodlands as the most complex terrestrial ecosystem in cities could be an integral and pivotal member of this nature spectrum. Diversity in habitats can enhance the biodiversity of the city, and woodlands can make notable contributions to this role. As a quintessential wildscape, it has a relatively high carrying capacity for wildlife. The complex food webs nurtured by equally complex flux of energy and cycling of nutrients would provide a natural self-sustaining system that demands little human inputs for its continued operation. In the face of tight municipal budget, this relatively low-cost UGS in terms of both establishment and maintenance could be more actively promoted in lieu of the expensive urban parks. Urban greening programmes do not need to be stalled or trimmed due to inadequate funding. They could be steered towards the more natural woodlands.

Some cities harbour a higher plant species diversity than their surrounding countryside (Stewart et al. 2004; Sukopp 2004). This phenomenon could be attributed to the complex mosaic of habitats composed of different types, sizes, stages of ecological succession and disturbance regimes. They offer heterogeneous conditions for floristic life with a mixture of native and exotic components. The relatively high alpha (intra-patch) and beta (inter-patch) species diversities of woodlands can contribute notably to this biotic richness. The presence of rare species calls for special attention. Rarity could be intrinsic of the species, or due to new arrivals that have not had the chance to spread and multiply. Those that are relicts of the former larger and more natural population in the parent forest could indicate the deteriorating conditions for sensitive members. Critically tiny species populations could be threatened by local extinction.

Besides ecological functions, the urban woodland can provide important environmental benefits. Woodlands with dense tree cover and open soil are effectively cooled by evapotranspiration. With the least surface soil sealing of urban land cover types, woodlands in cities can fulfil this function well (Pauleit and Duhme 2000). It can create its own microclimate characterized by dampening of extremes to bring smaller diurnal and seasonal temperature amplitudes. The resulting lower maximum and higher minimum temperatures incur a pleasant moderating impact. The cooled air coming from woodlands is also fresh, aromatic and clean, thus serving as an airshed with health-enhancing capability for citizens.

The cooling can help to suppress the urban heat island (UHI) effect which is generated by large stretches of built-up areas absorbing and storing solar energy as heat in the urban fabric. With looming climate-change impacts, the presence of urban woodland can make cities more resilient and ready to meet the challenges. The cooling effect can spill over to adjacent urban areas to provide a climatic amelioration footprint larger than the woodland area. The export of cool air is more evident on the downwind side of the vegetated patch. As most cities are suffering from UHI which is compounded by the superimposition of climate-change consequences, installation of urban woodlands can offer a cost-effective and durable solution that is accompanied by a plethora of other collateral benefits.

Trees in an urban woodland can trap suspended particulate matters and reduce their amount especially PM_{10} and $PM_{2.5}$ that remain floating in the air. The particles deposited on leaves by gravity and wind impaction can be washed down to the soil by rainfall. In the process of air exchange associated with photosynthesis and respiration, gaseous pollutants could be absorbed. Overall, the urban woodland can continually cleanse the air and mitigate air pollution in cities. These works are performed without human input of energy or material. They are driven and sustained by free and clean solar energy. The woodland setting allows trees to congregate at one location to bring synergistic environmental functions.

14.3 How to Protect Existing Woodlands?

14.3.1 Understanding Vulnerability of Woodland Ecology

Urban woodlands have been continually degraded due to various stresses and forces of destruction (Hedblom and Söderström 2008). Decline can be induced by urban growth into green fields in the urban fringe. Urban densification in existing built-up areas can occur in different modes. Intra-urban remnant green fields could be converted to urban use by infilling. Small and scattered green sites situated in developed sites could be built upon by in situ intensification. The small woodland pockets could be literally squeezed out of the urban matrix.

The urban core has already experienced extensive elimination of nature, resulting in the least woodland cover and native species (Clarkson et al. 2007). The urban fringe often has more woodland cover, and the patches are usually larger. They are prone to the pressure of new urban growth that may degrade or obliterate existing green fields including woods (Pauleit et al. 2005; Kim and Pauleit 2009). Attempts could be made to preserve some woodland plots within the future urban areas (Fig. 14.1). The countryside envelope around cities could experience woodland loss and fragmentation due to farmland expansion. With continued urban

Fig. 14.1 The Hassamu nature reserve in the west part of Sapporo is a fine example of protecting a remnant primary forest patch in the course of urban sprawl into a pristine forest area (*Photo credit* C.Y. Jim)

sprawl, such rural lands could be incorporated into the city and subject to urban growth. Less woodlands could be inherited by the future urban areas.

Some woodlands can decline insidiously by in situ degradation due to direct disturbance or general degradation in ecological conditions (Lehvävirta and Hannu 2002). Abiotic factors can include deleterious changes in microclimate, air quality and water-table. The biotic constraints can involve harmful edge effect, habitat fragmentation and isolation, reduction in pollination, dispersal agents, and propagule sources. Anthropogenic factors include recreational impacts such as trampling, in situ pollution, intrusion of pollutants into the site, vegetation damage, and uncleared dog excrement.

The lack of natural regeneration is a common problem facing some urban woodlands, with implications on their restoration and long-term sustainability. The seed rain may fail to arrive due to long distance to seed sources or physical or physiological obstructions to the dispersal process. A small daughter woodland detached from the large parent forest could be starved of external propagule supply. The seed banks in the soil may be deficient or defunct. The existing seeds in or new arrivals to the seed bank may lose viability due to unfavourable site conditions. The topsoil may be degraded such that it can no longer offer a suitable seedbed for germination. Successfully germinated seedlings may not survive due to human disturbance, aggressive herbivory and abiotic stresses.

Accessibility to visitors could curtail woodland growth and performance. Human impacts due to excessive or inappropriate recreational use can degrade woodland habitat conditions, making them less able to support ecological processes. The management inputs to cater to the need of visitors could inadvertently impose harms. Removal of undergrowth, commonly practised as a crime prevention measure, can bring collateral impacts on normal woodland ecosystem functions. Frequent removal of dead wood from living trees or woodland floor would bring similar negative effects. Intensively used sites tend to demand more management incursion to impose more undesirable impacts. More attractive sites with good tree cover and large old trees tend to lure heavier patronage. The best sites could run the risk of being loved to sickness and even death.

Road and building construction adjacent to woodlands can impose off-site impacts such as water and air pollution and lowering of the water table. The construction and post-construction noise can drive away sensitive fauna which may include essential pollinators and dispersers. The light pollution during and after construction would impose a similar negative effect. Whereas improper management inputs can introduce damages to urban woodlands, the lack of effective management measures to prevent intrusion of impacts could be equally harmful. Without legal protection and enforcement, important woodlands may not be guarded against perturbations.

The factors conducive to high species richness and native species growth could be studied to find ways to sustain, improve or introduce them (Stewart et al. 2009). They include large patch size, proximity to large natural forest area, connection via habitat corridors or stepping stone sites to diaspore sources, relatively undisturbed natural organic litter on the woodland soil surface, higher vegetation cover, and taller canopy height. As the government tends to plant more exotics in urban parks, public gardens and other public lands than citizens' private yards, it is necessary for officers to wean their preference or entrenched habit for introduced species in managing woodlands.

14.3.2 Respecting Nature by Sustainable Management

A comprehensive survey of urban woodlands in a city can establish a database and to understand the resource base, inform management, and provide a baseline for continual monitoring of their condition and performance (Ode and Fry 2006). Sustainable management in the spontaneous or ecological mode has been advocated as the preferred tactic (Gustavsson et al. 2005). Urban woodlands can be better off with less human inroads by adopting the non-intervention approach (Stewart et al. 2004). The tendency to sanitize or manicure woodlands using unnecessary, excessive or disruptive treatments can be weaned. The common and ingrained urban-park mentality and practice should not be applied to natural sites.

Natural woodlands with high ecological value deserve more attention. A typology based on development stage and species diversity can assist the evaluation endeavour (Fig. 14.2). Natural woodland could be declining due to stresses to become degraded. With the abatement of disturbance, the degraded woodlands

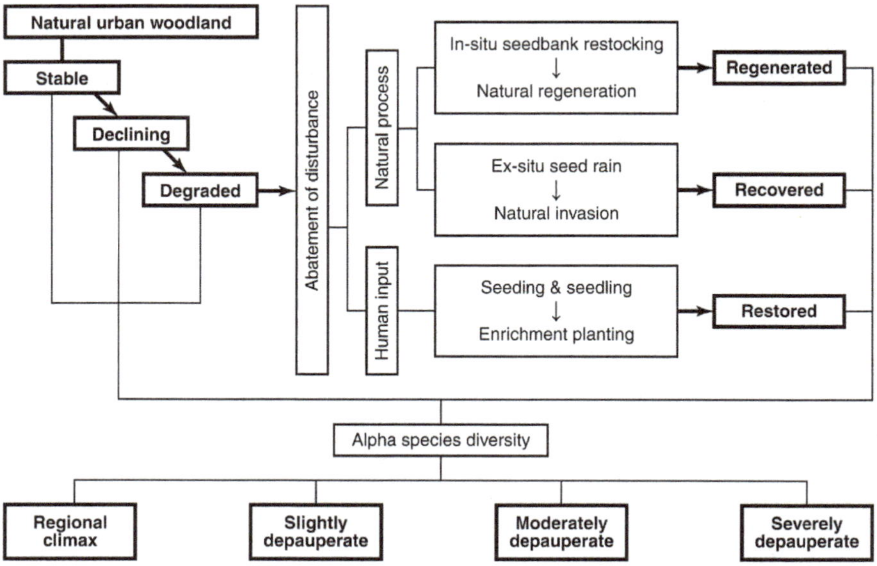

Fig. 14.2 Typology of natural urban woodlands based on developmental stage and species diversity (*Figure credit* C.Y. Jim)

can improve by natural processes of in situ seedbank restocking to bring regeneration, or ex situ seed rain and natural invasion to bring recovery. If natural processes cannot help, enrichment planting with seeds and seedlings could contribute to restoration.

To realize sustainable management of previous economic forests, the production management and harvest practices could be superseded by the naturalistic ones. Measures can be adopted to gradually replace the narrow species composition by thinning and enrichment planting using native species. For non-production woodlands with high biodiversity and naturalness, the temptation to apply human input and care can be suppressed. It is particularly important to minimize interference in pristine woodlands. The guiding principle is to allow spontaneous ecological processes and conditions to operate without interference, and to foster domination by native species (Fig. 14.3). Such an approach can reduce management cost whilst conditioning the woodland towards the self-maintaining and self-sustaining mode.

The size of woodland patches has an important bearing on species richness (Godefroid and Koedam 2003a). The semi-log species-area relationship aligns to a certain extent with the island biogeography theory. More species have a higher frequency in larger sites. Species groups with high conservation value, such as ancient forest species and rare species, are more frequently encountered in bigger sites. Bigger sites can accommodate large equilibrium species populations and reduce species extinction rate. Some small sites can hold more species per area unit than large sites, and accommodate more species that can survive in a small woodlot. The possibility of enlarging existing woodlands could be explored to maximize

Fig. 14.3 The urban forest at the northern edge of Celje in Solvenia, denoting a remnant natural forest that has escaped urban disturbance or annihilation, is an important part of the city's urban green infrastructure (*Photo credit* C.Y. Jim)

ecological benefits. In identifying sites for woodland conservation, the in situ site qualities as well as the suitability of surrounding areas for woodland restoration could be attempted (Lee et al. 2002). The chance to integrate urban woodlands with surrounding natural habitats can be explored (Toni and Duinker 2015).

Conservation strategy could target preserving or creating large woodland patches to sustain biodiversity and serve as a gene pool to supply other sites (Hill 1985). The role played by small woodlots should not be neglected, as some of them may harbour relict communities to form biodiversity hotspots (Croci et al. 2008). The opportunity to merge small proximal patches into a large one could be considered. If contiguous coalescence cannot be realized, habitat corridors suitable for dispersal and movement of constituent species could provide other links between woodlots. Alternatively, small and closely-spaced patches could be preserved or established between others to provide stepping stones. Urban sprawl into forested areas should aim at preserving large and connected patches in the future urban matrix (Fig. 14.3).

For woodlands with degraded species composition, skewed age distribution dominated by old trees, deprivation of spontaneous regeneration, and unnatural stand structure, tailor-made measures could be applied to assist restoration. The choice of species in afforestation programmes can facilitate biodiversity enhancement, aiming at emulating the biotic richness of the regional climax forest. The initial assistance can focus on establishing long-term self-regeneration capability. In situ and ex situ sources of seeds can be explored to sustain natural regeneration and to maintain an equitable spread of tree age. Intra-site spatial variations can be created to generate a range of habitat conditions, development stages and corresponding species combinations. Besides differences due to the core-edge dichotomy, intra-patch variations at the lateral and vertical dimensions and at the macro- and micro-habitat scales can be provided. A mixture of parcels with small openings, medium coverage, dense coverage and water bodies can generate wide permutations of main and ecotone communities to raise niche and species diversities.

The woodland edge interfacing with built-up areas can be conceived as a unique biotope to be labelled the city-woodland ecotone. Its species assemblage differs from that in the woodland interior. With a higher level of disturbance and different abiotic conditions, the edge tends to favour a collection of disturbance indicators, reflecting invasion of opportunistic and invasive pioneers. The exotic members tend to be generalists with reference to environmental tolerance range (Godefroid and Koedam 2003b). By minimizing the edge length and area, the woodland interior can preserve more of the woodland specialists. However, some species that are rare or have high conservation value can dwell at the edge. Thus woodland management should take care of both interior and edge habitats.

The long-term management plan can be based firmly on sound ecological principles (Millward and Sabir 2010). Whereas unwanted and disruptive human stresses should be discontinued, some natural stresses that are integral and essential to ecosystem operation should not be modified, suppressed or removed. A hierarchy of urban woodlands can be established based on ecological importance and vulnerability to disturbance. The management strategy can aim at differential treatments to match the spectrum of site characteristics, and protect cherished woodlands from undue recreational and other impacts. Within a large and heterogeneous site, a zonal management regime can regulate visitor movements and distributions to steer them away from the most sensitive and valuable parts. Subtle methods and materials can be adopted to influence visitor behaviour with minimum impact on woodland ecology and landscape quality.

14.3.3 Improving Degraded Woodlands

After studying the causes of decline, specific measures can be applied to restore degraded urban woodlands (Millward and Sabir 2010). They include raising species richness, increasing native species relativity, reducing exotic presence, adjusting the dominance of old trees, fostering natural regeneration by nurturing seedlings and young trees, mitigating human impacts and disturbance, and aiming at the long-term goal of a self-sustaining woodland ecosystem.

Planting native trees in small scattered colonization foci in the degraded patch could attract dispersers from nearby natural woodlands to facilitate recruitment and natural regeneration (Robinson and Handel 2000). This nucleation approach can offer a cost-effective technique to accelerate the rate of woodland succession. The cardinal principle is to provide the minimum amount of assistance to help nature to recover from past degeneration. Once this is achieved, further inputs should be pared down to the minimum.

In some countries, production forests are progressively shifted to non-productive uses such as conservation, recreation and landscape enhancement to become people's forests (Gundersen et al. 2005). The management approach has been changed in tandem with the passive silvicultural mode with drastic reduction in inputs and reverting to domination by natural processes. The patches situated in the urban fringe could be incorporated in the UGS system catering to the increasing recreational demands of urban residents. The intrinsic ecological value of such ex-economic woodlands may be relatively low, but native invasion and enrichment planting may gradually raise their naturalness.

In woodland management, the soil component is often the most neglected. The importance of maintaining soil integrity and quality in nature conservation could be more emphatically brought home. Whereas woodlands provide sanctuaries for flora and fauna, they also serve as a refuge for soil. A healthy soil can support vigorous plant growth as well as serving as the seed-bank repository and germination seedbed. Large woodlots can offer better protection to soil, especially in sustaining the organic matter and nutrient stock (Jim 2003b). The health of the soil is often a key determinant of woodland succession and establishment. Many urban woodlands carry the tell-tale signs of soil disturbance. Up to a certain threshold, soil could be quite resistant to human perturbation (Jim and Chan 2004). Beyond the critical tipping point, soil may degrade considerably to a stage that is difficult to restore.

Badly degraded woodlands are usually accompanied by serious degradation or loss of the pertinent topsoil O and A horizons, accompanied by decline in fine materials, nutrients, soil structure, soil porosity, and water and nutrient holding capacities. Woodland restoration could allow to a certain extent soil recovery mainly in fertility-related attributes (Jim 2003a). In woodland restoration, if the site soil is largely intact and has reasonably good quality, overzealous soil disturbance or improvement inputs should be avoided. For sites with degraded soil, a package of ecologically-oriented soil improvement methods should be tailor-made to meet the specific needs and nature of the replanting project (Jim 1993).

14.4 How to Create New Urban Woodlands?

14.4.1 Starting with Green, Brown and Grey Fields

Where urban woodlands are in short supply, new ones can be proactively created on suitable sites. Three kinds of initial site conditions could provide a pool of surfaces for woodland creation (Fig. 14.4). Scattered *green fields* composed mainly of natural vegetation at different stages of disturbance could be left by default within

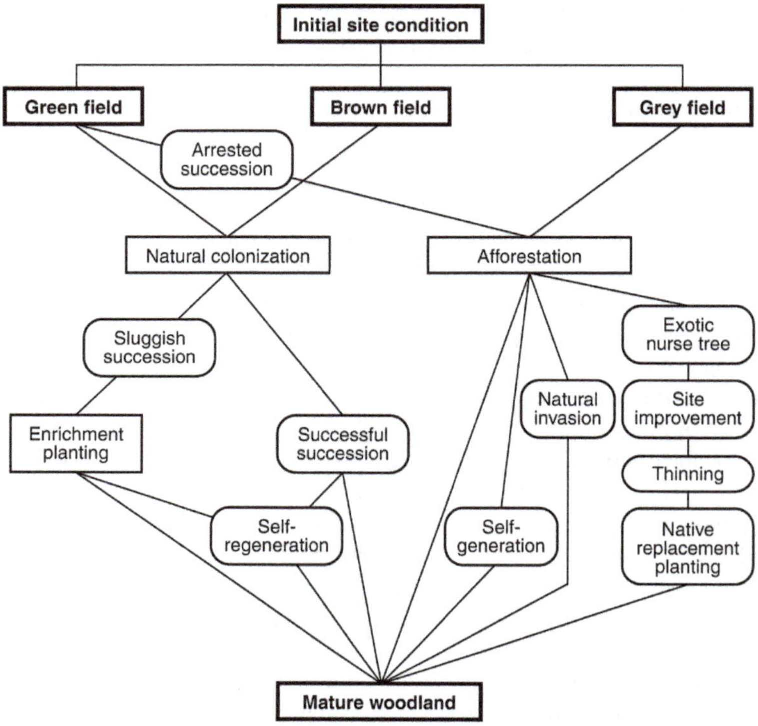

Fig. 14.4 Development of new urban woodlands on three kinds of sites, namely green, brown and gray (rooftop) fields (*Figure credit* C.Y. Jim)

the urban matrix. The abandoned urban-industrial lands, known as *brown fields* or industrial fallow lands, are increasingly common in post-industrial cities. Numerous rooftops of buildings offer *grey fields* that denote a wasted resource ripe for green roof installation. These sites could be filled by trees to develop into mature woodlands in due course through two ways. *Natural colonization*, depending on nature's seed rain to trigger and sustain ecological succession, is suitable for green and brown fields. Human assistance in the form of *afforestation* can accelerate the revegetation process in all three kinds of fields.

14.4.2 Letting Nature Run Its Course

Instead of treating brown fields as an idle resource, they offer opportunities for nature to return to the city (Kowarik 2005). What is forsaken by humans, nature will take over and work towards a comeback by naturalization (Toni and Duinker 2015). The spontaneous ruderal growth can spring up in the fallow nature sites in the form of 'overgrowth' (Rink 2005). With gradual successional changes from pioneers to shrubs and trees, post-industrial urban woodlands can be created serendipitously by nature literally at our doorsteps (Jorgensen et al. 2005). The diversity of anthropogenic substrate types with novel properties offers a wide range of habitat conditions that can be colonized by a broad spectrum of plant assortments. Thus habitat heterogeneity can breed species diversity and ecological varieties. The new urban wilderness could be added to the traditional UGS repertoire to enrich urban ecology and visitor experience (Kowarik and Langer 2005).

Humans provide the unintentional or accidental niches, whereas nature's sieve would screen and selectively admit candidates to fill the niches by natural colonization. The resulting floristic assortments would be contingent upon the general macro-climate and micro-climate of the site, soil and water conditions of the substrate, and solar access. As ecological succession proceeds, dynamic changes in the biotic components would modify abiotic habitat factors to change continually the nature of the sieve. The microclimatic, soil and water conditions can be progressively enhanced in the course of woodland succession (Oldfield et al. 2014). The resultant species combinations could be distinctive if not unique and seldom found in nature, and they have been likened to recombinant community (Angold et al. 2006). Restrictive site conditions, especially poor soil quality of the anthropogenic substrate, could straitjacket species admission and result in low species diversity (Hodge and Harmer 1996).

Woodland establishment on brown fields may happen spontaneously. The changes by default require no planning, management, input or interference. The benign negligence could generate pleasing products to contribute to smart ecological planning. At hardly any cost to the community, cities can just let nature run its own course to create high-calibre wilderness. The findings suggest that brown fields can play key roles in enriching urban ecology and biodiversity. The general belief that brown fields are derelict wastelands with no useful functions could be rectified. Green fields have less restrictions and more potential for spontaneous woodland succession. Suitable sites could be zoned as UGS to allow woodland establishment. Planning could adopt an innovative ecological dimension to proactively keep such sites to enrich the city's nature contents.

14.4.3 Creating Woodlands by Afforestation

Some cities especially in developing countries have grown rapidly in recent decades to bring compact developments with inadequate UGS provision and poor environmental quality and quality of life. Despite increase in UGS planning standards, the goals often cannot be met because inserting green areas into the existing tightly-packed urban matrix is usually impossible. Abandoned farmlands, degraded forest lands, deserted large industrial sites, and disused railway yards usually situated at the city edge can provide solution space to create new urban woodlands. In some shrinking cities, obsolete residential, commercial and institutional sites within the city could offer opportunities for UGS installation.

For harsh sites not amenable to natural succession to establish woodlands, or where the succession process is arrested or predicted to be too sluggish, human assistance in the form of afforestation can be attempted. They offer a novel kind of UGS with more activity space, denser vegetation cover and higher level of naturalness than urban parks. Located not too far from densely-populated areas, they can be reached easily by private or public transport. Nature near people's homes, at about 300 m or 15 min of walking distance, is particularly valuable and more likely to be patronized (Natural England 2010). For large and attractive sites, people are willing to travel longer distance to reach them.

The Un Bosco in Città (Wood in the City) in Milan provides an exemplar of an afforested urban woodland (Anon 1985). The city has many densely-packed neighbourhoods with meagre supply of UGS. Despite the lifting of UGS standard to 8 m^2/person in 1968, little relief came due to the lack of suitable sites in built-up areas. In 1974, the NGO Italia Nostra (Our Italy) decided to initiate its own greenspace programme. Successfully leasing 34 ha of abandoned farmland near the city from the government, a woodland was planted with the help of NGO, school and citizen volunteers, and with seeds donated by the forestry authority. It serves multiple functions of re-naturalization, reforestation, conservation, recreation and education. It represents an excellent adaptive use of otherwise unused agricultural land, and large-scale and sustained implementation of the participatory approach. The site has since been progressively increased to 110 ha. The project signifies a fruitful private-public partnership to tackle a nature-deficit situation, and to tap the ideas, skills, resourcefulness, energy and enthusiasm of citizens (Van Herzele et al. 2005a).

Other similar urban woodland creation projects have been attempted in different cities. For instance, the Parco Nord (North Park) with a notable woodland component was developed in Milan on a large brown field site of 640 ha in the urban fringe (Fig. 14.5). Back in the 1850s, the sprawling Bois Bologne (Bologne Woodland) covering 845 ha was established on a green field, a remnant ancient oak forest site, in the western edge of Paris (Fig. 14.6). In the 1920s, a forest covering 70 ha with 100,000 trees was nurtured at a green field site in Tokyo to honour Emperor Meiji and Empress Shoken (Meiji Jingu undated). It forms the Yoyogi Park around the Meiji Jingu in the heart of Tokyo, as the most centrally-located mature and sizeable urban woodland in the world (Fig. 14.7). More recently, the Grünen Bogen (Green Bow) including urban woodlands was developed on 120-ha of green fields in Paunsdorf in Leipzig as a large-scale urban green infrastructure for the city (Fig. 14.8).

Fig. 14.5 The Parco Nord in northern fringe of Milan has been established since the late 1960s on a 640-ha brown-field site, including woodlands established by afforestation (*Photo credit* C.Y. Jim)

Fig. 14.6 The Bois Boulogne in the western edge of Paris, a huge urban woodland occupying 845 ha, was established in the 1850s on a site occupied by remnant ancient oak forest (*Photo credit* C.Y. Jim)

Fig. 14.7 The 54-ha Yoyogi Park in Tokyo, filled with tall and dense mature trees, is one of the world's largest urban forest situated in the heart of a metropolis; originally a green-field site, it occupies the world's most expensive land for an urban green space (*Photo credit* C.Y. Jim)

The choice between natural colonization and afforestation can be a trade-off between cost and efficacy (Oldfield et al. 2013). For urban afforestation, species choice presents a cardinal concern. In recent years, the use of native species has been increasingly advocated and practised in tree planting in both cities and the countryside. Good sites may be receptive to a wider range of both native and exotic species, but poor ones may only accommodate the hardy ones which contain many exotics (Willoughby et al. 2007). The compromise is to use ecosystem service as the yardstick, and permit co-existence of both exotics and natives on poor sites, aiming at offering an acceptable level of service in comparison with exclusive native plantations (Oldfield et al. 2013).

Landscape ecological principles and associated landscape metrics can provide guidance for a spatially explicit woodland creation plan (Lee and Thompson 2005). It can aim at creating sites with a large patch size and an equant shape which can benefit from a low perimeter-area ratio and hence less disturbance at the edge. To counteract the common problem of fragmentation, sites should be located and demarcated to avoid isolation and maximize proximity and connectivity. The ultimate aim is to ensure high biodiversity, ecosystem services and sustainability. The systematic spatial guidelines can avoid the pitfall of the routine random site selection and design approaches.

Fig. 14.8 The Grünen Bogen Paunsdorf in northeast Leipzig is a large urban green infrastructure project initiated in 1999; it includes an afforested area (Paunsdorfer Wäldchen) catering to the nature-in-city ecological and recreational needs of the city (*Photo credit* C.Y. Jim)

14.5 Alternative Approaches to Woodland Establishment

14.5.1 Innovative Woodland Creation

On drastically disturbed sites where natural regeneration is sluggish or even impossible, a modified afforestation approach can pump-prime the ecological rehabilitation. The dual-stage *assisted relay floristic* technique is designed to short-circuit the woodland succession process, especially to circumvent the most difficult and critical initial establishment constraints (Jim 2012). Planting can begin with seedlings of exotic pioneers that can tackle the harsh initial site conditions, such as barren surface, exposure to strong sunshine and wind, skeletal soil with little organic matter and nutrients, soil surface unsuitable as seed bed, and absence of a viable seed bank. Where conditions allow, direct seed sowing can be attempted to reduce the cost and increase efficiency. The exotic pioneers are carefully chosen to establish at a relatively fast pace despite the odds. They serve as a *nurse crop* to accelerate the ecosystem recovery process, most importantly to expedite nutrient accumulation, improvement in soil properties and water-holding capacity, and microclimatic amelioration.

Once the site conditions have been improved to a suitable state, the exotic transient residents can be systematically thinned to give way to enrichment planting of natives as permanent residents. Spreading the enrichment-planting sites in scattered loci can expedite the subsequent process of natural regeneration. If parent sites are available in the vicinity, the seed rain can accelerate natural enrichment by natives in a *natural infilling process*. Gradually, the exotics role will become obsolete and redundant as they are progressively replaced by natives. The eventual mature woodland will be dominated by equilibrium indigenous species with little traces of the pioneers.

In compact cities, extensive areas can be completely filled with buildings and roads with no nature in sight. The lack of ground-level space to insert UGS is rampant and irresolvable. The rooftops of numerous buildings meanwhile remain largely vacant and barren as wasted resource. This pool of elevated *grey fields* can be enlisted to install *sky woodlands* as an innovate strand of urban woodlands. The successful recent establishment of rooftop urban woodlands in dense urban areas of Hong Kong, using native tree species, can serve as role models (Jim 2008). The cases suggest that the varied ingredients necessary to set up a woodland on building tops can be planned and realized without much technical difficulties. The main limitations are the lack of planting materials due to domination of exotic supply in the region's tree nursery industry. The more fundamental and intangible obstacle is the unwillingness to accept the innovative solution, and the reluctance to relax the plethora of bureaucratic and regulatory restrictions to installation.

A refined urban afforestation scheme can move away from the preoccupation with maximizing the tree coverage and total species richness. It could extend to the vertical dimension, especially aiming at emulating the multiple-layered stand structure (Richnau et al. 2012). Suitable combinations of species have to be chosen to fit each of the layers. For tropical woodlands, the epiphytes and lianas which traverse the vertical strata, and the emergent trees which shoot above the main tree canopy, can be included in the species palette for different layers.

Urban developments may remove some woodlands which under some planning regimes require *habitat compensation* or *offset* (Morris et al. 2006). It stipulates creation at an off-site location a comparable habitat in terms of area, biotic composition and ecological functions. At present, the relevant knowledge and techniques are inadequate regarding evaluating the site destined for elimination, establishing the substitute habitat, and monitoring the performance. For complex woodlands, it is pertinent to take measures to ensure delivery of the expected outcomes. If solution space is available, the replacement should be located near the impacted site (Quétier and Lavorel 2011), and preferably near residential areas to encourage patronage and increase beneficiaries for environmental benefits (Van Herzele et al. 2005b).

Catering to different natural and human needs, the created woodlands can adopt *zonation design and management* with different levels of naturalness (Jorgensen et al. 2005). The wilderness zone has minimal human inputs and dominated by natural woodland. The remaining zones could meet human demands for facilities and feeling of safety. They could include the moderately humanized zone and intensively humanized zone, with gradational changes in vegetation density and cover and provision of recreational facilities. Such design principles aim partly at meeting the safety concern of users, who tend to harbour somewhat ambivalent attitude towards nature. On the one hand, nature is warmly embraced as salubrious, whereas on the other it is considered as unsafe and unwelcomed. In this sense, some parts of afforested urban woodlands can be enlisted as a surrogate urban park.

14.5.2 Tackling Intractable Substrate Constraints

Whether the creation of urban woodland is due to spontaneous colonization or afforestation, similar constraints could be encountered. Understanding the multiple limitations can throw lights on their improvements and solutions. The history of previous land use is a key determinant of the ground conditions for vegetation establishment. Some heavy industrial, dockyard, mining, landfill, incinerator, and railway sites could be contaminated with rather intractable pollutants to restrict spontaneous plant growth. The presence of an impermeable and durable surface cover made of concrete or asphalt, and the heavily compacted subbase below the paving, would hinder the colonization process. Breaking up and removing the hard paving would be necessary to facilitate nature's work.

The poor quality substrates could restrict plant growth due to undesirable and rather persistent physical and chemical properties (Weiss et al. 2005). They are often beset by high contents of coarse materials such as gravels, stones, sand and construction rubble, and paucity of fine particles in the silt and clay particle-size fractions. The porosity of the coarse-textured soil is dominated by macro-pores (>60 μm diameter) with insufficient meso-pores (0.2–60 μm diameter) to hold plant-available moisture (Jim and Peng 2012). Rain or irrigation water infiltrates quickly into the soil but also drains swiftly away from the rooting zone, leaving little to support plant growth. The presence of too much macro-pores also induces excessive aeration, resulting in fast decomposition of organic matter and fast drying up of the soil by evaporation.

In chemical terms, the coarse substrate has little nutrient reserve and meagre available nutrients for plant absorption. The lack of organic matter would limit the supply of key macro-nutrients such as nitrogen and phosphorus which are mainly derived from organic sources. The shortage of fine particles and humic substances would curtail the nutrient-holding capacity which can be measured by the cation exchange capacity. Deficiency in organic constituents would restrict the development of a strong soil structure which is critical to the capacity for vegetation development. In extreme cases with multiple limitations that are not amenable to improvement by natural or artificial means, plant growth could be retarded and stunted and ecological succession could be arrested at the herb or shrub stage, and woodland may fail to emerge.

The soil pH value could impose an obstacle to plant growth. Either too acidic or too alkaline reaction would be inimical. The alkaline problem in cities is often attributed to the liberal occurrence of lime-rich construction rubbles (Jim 1998). The acid problem could be related to industrial or mining activities. Nutrient availability can be reduced under such conditions due to low solubility in the soil solution or fixation by other soil constituents. Under alkaline reaction, some metal ions such as iron and manganese could suffer from deficiency to induce the chlorosis symptom. Some heavy metals may become too soluble to induce toxicity problems (Jim 2001). At extreme levels, amendments could be applied to adjust the pH to an acceptable level. Species choice could be geared towards the pH condition and site quality.

For sites with particularly low-quality substrate that cannot be helped by treatments and amendments, woodland succession could not proceed. Excessively infertile, contaminated and droughty sites, and continual disturbance, would impede successional changes. The plant community would be dominated by persistent pioneers. A site assessment method could be developed to classify brown fields and identify those with the potential to nurture woodlands. For sites with enabling substrate but which fail to receive the right assortment of diaspores in the seed rain, assistance by afforestation could be contemplated. The transfer of natural woodland soil from nearby sites may provide an innovative way to re-stock the seed bank and facilitate germination and species recruitment (Nakamura et al. 2005). Where the substrate constraints are intractable and incorrigible, the more drastic soil replacement process may have to be contemplated.

14.5.3 Resolving Biotic and Abiotic Limitations

The lack of propagules can suppress the normal course of woodland succession on unmanaged sites. The absence of parents presents a fundamental constraint to species recruitment (Hodge and Harmer 1996). The presence of physical obstructions or distance to source areas could retard woodland succession. Abiotic dispersal agents may fail to reach the subject site, and biotic agents could be absent or present in tiny numbers to reduce the success of dispersal. Small and isolated urban woodlands surrounded by the built-up matrix and remote from seed sources are more vulnerable to scanty propagule arrival (Smale and Gardner 1999). The original indigenous vascular members in fragmented woodland pockets, deprived of regeneration chances, are prone to pauperization. For native species depressed to a small population, the prospect of local extinction could loom large.

Unwanted propagules of exotic and invasive species could be brought into the woodland by visitors who carry unintentional seed passengers. Seeds picked up from other places, attached to shoes, socks, clothing, bags and belongings, could be deposited in the woodlands. Some species with invasive or aggressive traits, especially exotics, could become highly successful in the adopted new homes. Trampling and other disturbances introduced by recreationists could alter soil properties to become conducive to the growth of pioneers. Exotic plant invasion has become a common urban influence on woodlands (Aronson et al. 2004). Intrusion by aggressive aliens, adventives and garden escapees could dilute the woodland's native species composition.

The use of native species has been advocated for afforestation projects to establish new urban woodlands, as well as enrichment planting of existing ones. The choice of species can be refined to match the need of different strata in the stand structure, and different locations with reference to the core-edge dichotomy. It can be further fine-tuned to match the intra-site micro-habitat variations such as soil quality, water table level, exposure to wind, solar access, and proximity to human disturbances. Comprehensive and excellent scientific information about indigenous trees is widely available in developed countries to guide such sophisticated species selection. However, in the developing world, such knowledge base is lacking and piecemeal, imposing an obstacle to native tree planting.

Urban development adjacent to urban woodlands could generate harmful impacts. Roads that run along or through woodlands can spread gaseous and particulate pollutants on vegetation, some of which may be washed into the soil by rainfall (Fowler et al. 2004). Thus woodlands play the important role of air-pollutant scavenger and sink to improve air quality within and around the site, on condition that the capacity is not overburdened. The surface runoff water from roads and other urban land uses, containing pollutants and sediments, could intrude into the woodlands. Roads can spread seeds which can be carried and pushed by the air currents generated by moving vehicles.

14.6 Conclusion

God made the country and humans made the town. Smart humans could take the hint and put the country into the town. Urban woodlands as the quintessential surrogate and ambassador of nature embedded in cities can amply fulfil this role. Urban woodland enclaves existing in the built-up areas can be proactively studied and protected. Urban sprawl into natural areas can identify valuable wooded sites for designated as a special protected area to enrich the urban green space repertoire. Where nature is in deficit in cities, solution space could be sought in degraded green fields, brown fields and grey fields to allow new woodlands to be generated by natural colonization or human-assisted afforestation. The mindset of decision-makers would need to be enlightened regarding the pertinent benefits that urban woodlands can bring to both people and nature to contribute to the liveability and sustainability quests.

Urban woodlands are valuable living assets situated at our doorsteps, yet they tend to be belittled if not neglected and ignored. The resources poured into formal urban parks in capital and recurrent terms are disproportionally large in comparison with meagre sums directed to urban woodlands. Yet urban woodland protection, creation and maintenance are relatively less costly than formal urban green spaces. Such natural enclaves are particularly precious due to two critical attributes, namely the intrinsic ecological endowment, and the extrinsic urban location. The juxtaposition of nature and city, which suffers chronically from nature deficit, should have generated a warm welcome to the proximal woodlands. Perhaps their inadequate attention echoes the common human folly of familiarity breeds contempt. It is high time that these embedded ecosystems of a high order could be properly studied, understood, appreciated, respected and protected.

The task of urban woodland conservation, creation and management demands knowledge and skills from a basket of disciplines. As the most complex natural systems in the urban realm, urban woodlands embody and inherit the consequences of both natural and cultural influences. They are neither truly natural nor truly synthetic in their biotic and abiotic traits. Instead, they present a unique if not fusion type of ecosystem engulfed by the urban matrix and intruded by urban impacts. As the synergistic products of nature-culture interactions, and as venues for continual nature-culture interplays, they present distinct characteristics that demand different treatments. Much remain to be learnt about their hybrid intricacies. Identifying the knowledge gaps can steer in-depth multidisciplinary research to refine the task (Jim 2011b). The gem of nature and the pride of the community straddling the urban domain deserves the better to fulfil the rising expectations.

References

Anon (1985) The wood in the city: a first example of urban afforestation. Bosco in Citta, 3 p

Angold PG, Sadler JP, Hill MO, Pullin A, Rushton S, Austin K, Small E, Wood B, Wadsworth R, Sanderson R, Thompson K (2006) Biodiversity in urban habitat patches. Sci Total Environ 360:196–204

Aronson MFJ, Hatfield CA, Hartman JM (2004) Plant community patterns of low-gradient forested floodplains in a New Jersey urban landscape. J Torrey Bot Soc 131:232–242

Clarkson BD, Wehi PM, Brabyn LK (2007) A spatial analysis of indigenous cover patterns and implications for ecological restoration in urban centres, New Zealand. Urban Ecosyst 10:441–457

Croci S, Butet A, Georges A, Aguejdad R, Clergeau P (2008) Small urban woodlands as biodiversity conservation hot-spot: a multi-taxon approach. Landscape Ecol 23:111–1186

Fowler D, Skiba U, Nemitz E, Choubedar F, Branford D, Donovan R, Rowland P (2004) Measuring aerosol and heavy metal deposition on urban woodland and grass using inventories of ^{210}PB and metal concentrations in soil. Water Air Soil Pollut 4:483–499

Godefroid S, Koedam N (2003a) How important are large vs. small forest remnants for the conservation of the woodland flora in an urban context? Glob Ecol Biogeogr 12:287–298

Godefroid S, Koedam N (2003b) Distribution pattern of the flora in a peri-urban forest: An effect of the city-forest ecotone. Landscape Urban Plan 65:169–185

Grosse-Bächle L (2005) Strategies between intervening and leaving room. In: Kowarik I, Körner S (eds) Wild urban woodlands. Springer, Berlin, pp 231–246

Gundersen V, Frivold LH, Löfström I, Jørgensen BB, Falck J, Øyen BH (2005) Urban woodland management—the case of 13 major Nordic cities. Urban For Urban Greening 3:189–202

Gustavsson R, Hermy M, Konijnendijk C, Steidle-Schwahn A (2005) Management of urban woodland and parks—searching for creative and sustainable concepts. In: Konijnendijk C, Nilsson K, Randrup T, Schipperijn J (eds) Urban forests and trees. Springer, Berlin, pp 369–397

Hedblom M, Söderström B (2008) Woodlands across Swedish urban gradients: Status, structure and management implications. Landscape Urban Plan 84:62–73

Hill DB (1985) Forest fragment and its implications in Central New York. For Ecol Manage 12:113–128

Hodge SJ, Harmer R (1996) Woody colonization on unmanaged urban and ex-industrial sites. Forestry 69:245–261

Jim CY (1993) Ecological rehabilitation of disturbed lands in the humid tropics. In: Parham WE, Durana PJ, Hess AL (eds) Improving degraded lands: promising experiences from South China. Bishop Museum Bulletin in Botany 3, Bishop Museum Press, Honolulu, pp 217–299

Jim CY (1998) Urban soil characteristics and limitations for landscape planting in Hong Kong. Landscape Urban Plan 40:235–249

Jim CY (2001) Managing urban trees and their soil envelopes in a contiguously developed city environment. Environ Manage 28:819–832

Jim CY (2003a) Conservation of soils in culturally protected woodlands in rural Hong Kong. For Ecol Manage 175:339–353

Jim CY (2003b) Soil recovery from human disturbance in tropical woodlands in Hong Kong. Catena 52:85–103

Jim CY (2008) Ecological design of sky woodland in compact urban Hong Kong. In: Proceedings sixth annual greening rooftops for sustainable communities conference, Baltimore, MD, April 30–May 2, 2008, 15 p

Jim CY (2011a) Urban woodlands as distinctive and threatened nature-in-city patches. In: Douglas I, Goode D, Houck MC, Wang R (eds) The Routledge handbook of urban ecology. Routledge, New York, pp 323–337

Jim CY (2011b) Holistic research agenda for sustainable management and conservation of urban woodlands. Landscape Urban Plan 100:375–379

Jim CY (2012) Restoration of forests associated with new town development in Hong Kong. In: Stanturf J, Madsen P, Lamb D (eds) A goal-oriented approach to forest landscape restoration. Springer, Berlin, pp 129–148

Jim CY, Chan MWH (2004) Assessing natural and cultural influence on soil in remnant tropical woodlands. Area 36:6–18

Jim CY, Peng LLH (2012) Substrate moisture effect on water balance and thermal regime of a tropical extensive green roof. Ecol Eng 47:9–23

Jorgensen A, Hitchmough J, Dunnett N (2005) Living in the urban wildwoods: a case study of Birchwood, Warrington New Town, UK. In: Kowarik I, Körner S (eds) Wild urban woodlands. Springer, Berlin, pp 95–116

Jorgensen A, Tylecote M (2007) Ambivalent landscapes: wilderness in the urban interstices. Landscape Res 32:443–462

Kim KH, Pauleit S (2009) Woodland changes and their impacts on the landscape structure in South Korea, Kwangju City Region. Landscape Res 34:257–277

Kowarik I (2005) Wild urban woodlands: towards a conceptual framework. In: Kowarik I, Körner S (eds) Wild urban woodlands. Springer, Berlin, pp 1–32

Kowarik I, Langer A (2005) Natur-park Südgelände: linking conservation and creative in an abandoned railyard in Berlin. In: Kowarik I, Körner S (eds) Wild urban woodlands. Springer, Berlin, pp 287–299

Lee JT, Bailey N, Thompson S (2002) Using geographical information systems to identify and target sites for creation and restoration of native woodlands: a case study of the Chiltern Hills, UK. J Environ Manage 64:25–34

Lee JT, Thomspon S (2005) Targeting sites for habitat creation: an investigation into alternative scenarios. Landscape Urban Plan 71:17–28

Lehvävirta S, Hannu R (2002) Natural regeneration of trees in urban woodlands. J Veg Sci 13:57–66

Meiji J (undated) Meiji Jingu official website: nature at Meiji Jingu. Tokyo. http://www.meijijingu.or.jp/english/. Accessed 10 Aug 2016

Millward AA, Sabir S (2010) Structure of a forested urban park: implications for strategic management. J Environ Manage 91:2215–2224

Morris RKA, Alonso I, Jefferson RG, Kirby KJ (2006) The creation of compensatory habitat—can it secure sustainable development? J Nat Conserv 14:106–116

Nakamura A, Morimoto Y, Mizutani Y (2005) Adaptive management approach to increasing the diversity of a 30-year-old planted forest in an urban area of Japan. Landscape Urban Plan 70:291–300

Natural England (2010) Nature nearby: accessible natural greenspace guidance, London

Ode Å, Fry G (2006) A model for quantifying and predicting urban pressure on woodland. Landscape Urban Plan 77:17–27

Oldfield EE, Felson AJ, Wood SA, Hallett RA, Strickland MS, Bradford MA (2014) Positive effects of afforestation efforts on the health of urban soils. For Ecol Manage 313:266–273

Oldfield EE, Warren RJ, Felson AJ, Bradford MA (2013) Challenges and future directions in urban afforestation. J Appl Ecol 50:1169–1177

Pauleit S, Duhme F (2000) Assessing the environmental performance of land cover types for urban planning. Landscape Urban Plan 52:1–20

Pauleit S, Jones N, Nyhuus S, Pirnat J, Salbitano F (2005) Urban forest resources in European cities. In: Konijnendijk CC, Nilsson K, Randrup TB, Schipperijn J (eds) Urban forests and trees: a reference book. Springer, Berlin, pp 49–78

Payne BR (1983) Urban woodlands and their recreational uses. Trees in the 21st century. Arboricultural Association, AB Academic, Berkhamsted, UK, pp 77–85

Quétier F, Lavorel S (2011) Assessing ecological equivalence in biodiversity offset schemes: key issues and solutions. Biol Conserv 144:2991–2999

Richnau G, Wiström B, Nielsen AB, Löf M (2012) Creation of multi-layered canopy structures in young oak-dominated urban woodlands: the ecological approach revisited. Urban For Urban Greening 11:147–158

Rink D (2005) Surrogate nature or wilderness? Social perceptions and notions of nature in an urban context. In: Kowarik I, Körner S (eds) Wild urban woodlands. Springer, Berlin, pp 67–80

Robinson GR, Handel SN (2000) Directing spatial patterns of recruitment during an experimental urban woodland reclamation. Ecol Appl 10:174–188

Smale MC, Gardner RO (1999) Survival of Mount Eden Bush, an urban forest remnant in Auckland, New Zealand. Pacific Conser Biol 5:83–93

Stewart GH, Ignatieva ME, Meurk CD, Earl RD (2004) The re-emergence of indigenous forest in an urban environment, Christchurch, New Zealand. Urban For Urban Greening 2:149–158

Stewart GH, Meurk CD, Ignatieva ME, Buckley HL, Magueru A, Casee BS, Hudson M, Parker M (2009) Urban Biotopes of Aotearoa New Zealand (URBANZ) II: floristics, biodiversity and conservation values of urban residential and public woodlands, Christchurch. Urban For Urban Greening 8:149–162

Sukopp H (2004) Human-caused impact on preserved vegetation. Landscape Urban Plan 68:347–355

Toni SA, Duinker PN (2015) A framework for urban-woodland naturalization in Canada. Environ Rev 23:321–336

Tregay R (1979) Urban woodlands. In: Laurie IC (ed) Nature in cities. Wiley, New York, pp 267–295

Van Herzele A, Collins K, Tyrväinen L (2005a) Involving people in urban forestry—a discussion of participatory practices throughout Europe. In: Konijnendijk CC, Nilsson K, Randrup TB, Schipperijn J (eds) Urban forests and trees: a reference book. Spinger, Berlin, pp 207–228

Van Herzele A, De Clercq EM, Wiedemann T (2005b) Strategic planning for new woodlands in the urban periphery: through the lens of social inclusiveness. Urban For Urban Greening 3:177–188

Weiss J, Burghardt W, Gausmann P, Haag R, Haeupler H, Hamann M, Leder B, Schulte A, Stempelmann I (2005) Nature returns to abandoned industrial land: monitoring succession in urban-industrial woodlands in the German Ruhr. In: Kowarik I, Körner S (eds) Wild urban woodlands. Springer, Berlin, pp 143–162

Willoughby I, Stokes V, Poole J, White JEJ, Hodge SJ (2007) the potential of 44 native and non-native tree species for woodland creation on a range of contrasting sites in lowland Britain. Forestry 80:531–553

Chapter 15
Urban Waterfront Revivals of the Future

Swinal Samant and Robert Brears

Abstract Urban waterfronts form part of cities' critical intersection between the natural and man-made environment, linking the city and its inhabitants with water. In the context of high density urban environments, they are integral to the network of green and public spaces and have the potential to encompass a range of uses including residential, commercial, leisure, recreational, heritage and art offering a multitude of economic, social, environmental benefits. The cases of HafenCity Hamburg and Waterfront Toronto discussed in this paper demonstrate successful approaches to achieving social and environmental sustainability at the waterfronts, highlighting the importance of ensuring mixed uses, public access, sustainable design and construction of buildings and infrastructure including climate change adaptations. Integrated and incremental planning of waterfronts in conjunction with citywide planning alongside careful consideration for greening, urban ecology, biodiversity, and aquatic ecosystems is also critical. In an era of rapidly urbanizing and homogenized waterfront developments, distinctiveness and authenticity derived via meaningful engagement with the local context and via engaging in participatory design processes is of increasing relevance.

Keywords Urban waterfront · Environmental sustainability · Social sustainability · HafenCity Hamburg · Waterfront Toronto · High density

---

S. Samant (✉)
National University of Singapore, Singapore, Singapore
e-mail: akisama@nus.edu.sg

R. Brears
Mitidaption, Christchurch, New Zealand
e-mail: rcb.chc@gmail.com

P.Y. Tan and C.Y. Jim (eds.), *Greening Cities: Forms and Functions*,
Advances in 21st Century Human Settlements,
DOI 10.1007/978-981-10-4113-6_15

15.1 Introduction

Waterfront developments have seen a 'sea' of change over the past few centuries with a move from traditional port and industrial activities with factories, warehouses and related infrastructure dominating much of cities' waterfronts, to an increasing emphasis on leisure, relaxation, recreation, waterside living, heritage, commerce, art, entertainment and tourism related uses.

Growing interest in abandoned harbour fronts, ports, warehouses and industrial infrastructure led to the conservation and adaptive reuse of the historical structures encompassing entertainment, art, leisure and dining. These developments have together created distinctive environments in Europe and North America, and eventually led to the incorporation of residential and office buildings in places like Toronto and Vancouver. Asian cities such as Hong Kong, Dubai, Singapore and Seoul followed suit with many Chinese cities in the inland regions also harnessing the potential of river bodies by creating riverfront developments as a means to augment the area and the city's image.

Indeed, the developments are characterized by varying scales and ambitions, from daily (exercising/promenading) to communal (festivals, markets places) activities and other notable large-scale developments (expositions such as that in Lisbon and tourist attractions in Singapore) that gave a new purpose and identity to particular parts of these cities. With cities competing on an international scale in being livable and sustainable, regeneration has also been driven by a growing focus on greening, and in doing so, enhancing city aesthetics, environment, ecology, landscape, infrastructure, economy, social and communal fabric, health and well-being and more. Consequently, it is widely recognized that water has been instrumental to the culture, form, and function of urban settlements the world over (Sairinen and Kumpulainen 2006).

15.1.1 Problematization

Urban waterfronts have come under increased pressure due to urban migration and expanding population, with resultant densification as well as sprawl in urban environments with shrinking urban spaces. Whilst cities have largely moved away from privatization of waterfronts, this 'exclusive' edge is still largely characterized by high income housing with related social impacts of gentrification and segregation as highlighted by MacLeod and Ward (2002), Dovey (2005), Sairinen and Kumpulainen (2006), and Stevens (2009). Furthermore, waterfronts are contested territories whereby social and recreational needs of the local community are often overridden by real estate- driven and larger tourism-focused developments focusing primarily on economic gains. This is particularly evident for example, at Victoria Harbour in Hong Kong where existing local social and recreational needs were moved to the outer harbour areas whilst the inner harbour area is reserved for tourist needs (Cheung and Tang 2015). To overcome this, waterfronts across the world are striving to encompass a range of local community-oriented social, leisure (health) and recreational uses whilst also incorporating city-wide tourist-related uses. These efforts have achieved varying degrees of success due to the inherent difficulties of the competing and contrasting demands placed on them. Discriminatory development and prioritizing of real estate market demands have also led to continuous expansion and landfilling that pose immense stress to this vital yet limited resource of cities, potentially detrimentally affecting all aspects of public life.

15.1.2 Governance

Despite the challenges highlighted above, there are also positive examples of waterfront developments that have resulted from sensitive governance. For instance, given the land constraints in the city-state, Singapore's sustainable approach to the waterfronts focuses on land reclamation and optimization, sustainable water solutions for urban needs, encouraging mixed uses, integration of developments with the local community, and harnessing of tourism opportunities in specific waterfront areas. In addition, waterfronts, usually not considered in city planning, offer valuable opportunities for community led/focused developments to challenge those that are economically driven. One such notable project is the modest yet powerful, Bandra waterfront regeneration project in Mumbai's affluent suburb where a 3 km stretch of disused wasteland was transformed into a community-orientated public space. The success of this prompted further waterfront revitalization projects in Mumbai. Indeed, Oslo, Rotterdam and Gothenburg are more mature examples of public-private partnerships and civic/stakeholder participatory processes whereby public space and public realm developments, deemed as valuable for the community, form integral part of large scale visioning (Giblett and Samant 2011). The Waterfront Communities Project focuses on early 21st century waterfront regeneration initiatives in nine port cities around the North Sea of comparable socio-cultural and economic contexts and institutional frameworks. This research also highlights the value of citizen and stakeholder participation, and multidisciplinary and entrepreneurial public-private sector collaboration in improving city governance, policy making, provision of infrastructure, land control and public funds and in the implementation of a more balanced, socially, economically and physically inclusive regeneration (Samant 2014).

15.1.3 Globalisation and Identity

'Economic globalisation is the consolidation of capital, human resources and information and the increasing dependency between economic regions… Global markets have transposed our fundamental image of the world from boundaries defined by the state to the network of cities and the flows of people, goods, money, information and images' (Giblett and Samant 2011). Economic globalisation with concurrent prioritization of fast paced, economically driven developments in cities such as Hong Kong and Shanghai, in the pursuit of a marketable 'world class city' image and status, has resulted in homogenous waterfront environments. The reproduced landscapes created through standardized residential and commercial use worldwide where the previous industrial heritage is more or less wiped out is commonplace, resulting in a loss of cultural identity and distinctiveness.

Stevens (2009) studied 'how designers control geographic, climatological, hydrological and urbanistic dimensions of the waterscape to create idealized urban settings that optimize consumptive leisure and place promotion'. He examined international case studies including artificial lagoons in Brisbane and Cairns and beaches and floating swimming pools in Berlin, Vienna, Paris and London, urban riverfront leisure precincts in Melbourne, Brisbane and London and indoor water theme park near Berlin, two public fountains in London and Melbourne, and the reconstructed urban stream in Seoul, all of which seek to attract tourists and residents through place marketing. The analysis concentrated around four aspects of the illegitimacy of such urban waterfronts: 'taming the landscape to provide comfort and safety; augmenting the landscape to provide varied sensory stimulation; carefully positioning the waterfront within a wider climatic, thematic and functional context; and managing the temporal dimension of visitor experience', some or all of which are typically observed to varying degrees in waterfront revitalization projects. For example, Chang and Huang (2005) argue that the Singapore Tourism Board (STB) has attempted to remodel the national image such that it has physically and symbolically transformed the waterfront demonstrating a process described as 'creative destruction'.

Stevens (2009) points to the world-wide phenomenon of homogenized and artificial environments which he suggests amounts to 'channel surfing' for the pedestrians. These environments typically include the 'gathering together of various water-related activity spaces, often with quite clearly artificial and contrasting themes, serves consumers' appetite for novelty, and for the conceptual shallowness of escapist spectacle'. Whilst these contemporary urban waterfronts use natural elements and their associated atmospheres, the excessive enthusiasm for consumption and uniqueness, means many of these projects lead to subvergence of the meaning and importance of the water bodies. Steven's study warns of the negative impacts of 'radically unnatural ecologies' of many of the waterfront projects as they seek to serve consumptive behaviours and leisure activities with a focus on maximizing environmental comfort. Rightly, he draws attention to the environmental impacts of such practices and the possible creation of new 'degraded brownfield sites' of the future.

15.1.4 Planning and Land Reclamation

Furthermore, to meet the demands placed by rapid growth and higher densities, many Asian cities (Hong Kong, Shanghai, Dubai and Singapore) as mentioned earlier, have resorted to landfilling, reclamation and rebuilding of city coastlines. The expansion of Singapore's surface area for example, has increased 22% since the 1960s with sand being sourced from Vietnam, Malaysia and Cambodia and having devastating effects on coral reefs and sea-grass beds in the South China Sea (Giblett and Samant 2011). Such landfilling has led to new peri-urban waterfront areas that are now redefining cities with new infrastructure and accessible open spaces and natural environments, often with damaging effects on oceanic environments and ecology. For example, Dubai embarked on aggressive and extensive reclamation to support real estate and infrastructure development, and marketed itself as a major tourist and business hub. Such ambitious waterfront developments must give due consideration to ecological aspects and natural processes such that the manmade waterfront environments are in harmony with their natural marine counterparts.

15.1.5 Environment and Ecology

This delicate interface between land and water is integral to a city's wider network of open and green spaces and has the ability to impact upon environmental conditions (temperatures including urban heat island effects, flooding, humidity, noise, pollution, water and soil quality), biodiversity, and migration of fauna. The health and vitality of water bodies, waterfronts and their associated ecosystems are therefore fundamental resources of cities as they affect their economic vitality, social equity and public health (Chandran and Gowda 2014). Consequently, waterfront development plans are increasingly integrated with cities' water management plans with consideration for storm water management, the urban water cycle, watershed issues, regional environmental issues, and demands on infrastructure and city facilities (Chandran and Gowda 2014). Arabianranta in Helsinki, for example, adopts a unique planning approach and takes full advantage of the surrounding natural landscape and migratory birds in the area with the execution of a peaceful urban park that separates the built up area from the water, whilst ensuring views to the bay and the opportunity for people to get close to the water.

Changing climatic conditions and highly urbanized waterfronts have left many cities now facing higher flood risk (IPCC, 2007). Although a range of interventions are available, large-scale ones requiring significant funding and centralized management are considered to be less appropriate due to concerns over negative ecological impacts. Instead, small scale, local interventions and adaptions 'promoting flood-resilient architecture and local flood preventive measures, in combination with improved evacuation and disaster management' are recommended

(Veelen et al. 2015). Furthermore, resilient strategies should be based on a detailed understanding of local conditions including vulnerabilities, stakeholder engagement and the adoption of site-specific adaptation measures (Wardekker et al. 2010). For example, HafenCity, Hamburg discussed later, successfully implemented extensive flood-secure infrastructure in the form of raised plinths for buildings to overcome restrictions presented in the event of flooding and storm surges.

The adaptive pathway method, i.e. 'taking small-step interventions along shorter time lines in order to avoid future lock-ins, reduce potential regrets or to seize the advantage of possible adaptation opportunities' (Dessai and van der Sluijs 2007; Gersonius 2012; Haasnoot 2013) allows stakeholders to consider a wide range of adaptation actions as a means to ensuring resilience in urbanized deltas. Vollmer (2009) explores how urban waterfront rehabilitation has been used as a sustainable development strategy and a means to improved environmental quality and management and consequently improved quality of life in Chinese cities. The Qinhuai River Environmental Improvement Project in Nanjing, the Suzhou Creek Rehabilitation in Shanghai, and the Wuli Lake Rehabilitation in Wuxi demonstrate the value of incremental improvements as 'a necessary catalyst for sustainable urban development'. Whilst these developments do not restore the ecosystems to those of the pre-development conditions, the 'modified ecosystems still offer a great deal of value to the urban area' (Vollmer 2009).

Dyson and Yocom (2015) recognise the detrimental impacts of conventional infrastructures on aquatic ecosystems and highlight the opportunities ecologically designed urban waterfronts present for mitigating environmental impacts of urbanization. These include protection from coastal hazards and severe storms, recovering ecosystem functions, and habitat provision and supporting biodiversity, making them important assets for urban conservation and ecological rehabilitation. Novel approaches and ecological design and design modifications that respect processes and functions of (natural and urban shoreline) ecosystem are evident in urban design practices and in and near-water infrastructure.

To reduce the impacts of conventional waterfront infrastructure, the ecological modifications suggested include 'infrastructure that uses natural materials or mimics the physically properties of natural habitats to support the habitat requirements of native species and reduce the influence of non-native species (Lukens and Selberg 2004)' and that 'ecological infrastructural design should reference locally specific and ecologically intact shorelines, paying particular attention to microhabitat, surface orientation, and nearshore habitat area' (Dyson and Yocom 2015). They also suggest new approaches to the 'design of docks and seawalls' and 'adding microhabitat, creating more shallow water habitat, and reconstructing missing or altered rocky benthic habitats'.

Designing ecological infrastructure to address local effects of global climate change such as expected sea level rise, warming of surface water and acidification over the short and long term, for example, encompassing a range of climate change impact scenarios in the intervening period and consequent response from aquatic species, is a challenging task. However, as Dyson and Yocom (2015) suggest, upgrading existing infrastructure is a cost effective way to replace conventional infrastructure with ecological infrastructure.

15.1.6 Green Public Spaces

Green urban waterfronts are the breathing lungs of a city, a very precious commodity that supports fundamental daily, communal and city-scale functions whereby healthy lifestyles, leisure and recreation are promoted. Greening of high density urban environments is critical and here, 'urban greening' refers to the green spaces within cities including parks, gardens, greenways, children's play areas, outdoor sports facilities and public spaces that have been purposefully developed, as well as the natural areas that have been preserved, protected or enhanced (De Sousa 2014).

Waterfronts can form a central part of the network of such green urban spaces of a city and as such should be recognized and integrated in the wider city planning. Furthermore, tying them to 'broader urban and neighbourhood sustainability, restoration, and economic development initiatives is becoming more widely accepted, thus providing an opportunity to further transform the urban landscape and enhance urban ecologies' (De Sousa 2014).

Gospodini (2001) highlighted that 'in the era of globalization, the relationship between urban economy and urban design, as established throughout history of urban forms, is getting reversed: while for centuries the quality of urban environment has been an outcome of economic growth of cities, nowadays the quality of urban space has become prerequisite for economic development of cities; and urban design is consciously used as a means of enhancing the development prospects of cities'.

Toronto's waterfront developments considered economic aspects alongside environmental and led to the creation of thriving, livable communities that ensured provision of green and accessible waterfronts that became central public spaces for the city. Public access, immediacy to water, pedestrianised environments, common in major waterfront cities (Hong Kong, Singapore, Barcelona, Paris, London, Washington DC, Toronto) promote health and well-being, and support cleaner more sustainable living.

Whilst the importance of public access to waterside land has been well established (Dong 2004; Timur 2013; Shrestha and Shrestha 2008; Moretti 2008), there are inherent challenges of safeguarding the waterfront biodiversity which require considerable attention in terms of environmental planning and urban design. These efforts are reflected in the prominence in urban policies over the years as noted by many researchers (Petrillo and Grenell 1985; Breen and Rigby 1991; Sairinen and Kumpulainen 2006).

Hamilton in Ontario, Canada developed the western section of its harbour into a 3.4 km waterfront trail. This is particularly significant if you consider that this development increased public access to the city's waterfront significantly from 5%

in 1990 to 25% by 2006 (Remedial Action Plan for Hamilton Harbour 2006). Most importantly, the waterfront was instrumental in transforming the city from 'a dirty steel town to a green and healthy city' offering plenty of opportunities for leisure, recreation and proximity with nature (Wakefield 2007).

The influential non-profit organization, Project for Public Spaces, has developed an adaptable framework that is used worldwide (http://www.pps.org/waterfronts/). It promotes waterside public spaces to be people-friendly and designed for multiple uses and activities and avoid single-use unsustainable development, thus creating diversity and attracting more users. The East Coast Park in Singapore follows this model and incorporates a range of accessible public spaces that embrace differing activities (walking, cycling and other leisure activities), catering to the needs of the local community as well as the city at large, whilst also connecting with other public waterfronts of the city. HafenCity, Hamburg is a mixed-use waterfront development that encourages clean modes of transportation, enabled through an extensive network of footpaths and cycle routes.

15.1.7 Social Environments

Sairinen and Kumpulainen (2006) emphasize the social impact assessment of urban waterfront planning, the various means of using and experiencing water edges and their qualitative impacts on the community. They state that 'as a whole, social dimensions provide information about the social effects but also understanding of the social significance, values and meanings of waterfront areas, as well as of the appropriate ways of conserving, preserving and changing these environments for mixed use'. They categorize social dimension of urban waterfront regeneration into resources and identity, social status, access and activities, and waterfront experience. A good example of this would be The Central Waterfront 2 Plan for Toronto's waterfront, which focused on 'Emphasising public access; Building a network of waterfront parks and public spaces; Ensuring high levels of environmental health; and Creating new communities' (Chandran and Gowda 2014).

Having discussed the range of issues related to waterfront developments, the following section examines the approaches adopted with respect to environmental and social sustainability in two comprehensive contemporary cases, namely, HafenCity, Hamburg in Germany and Waterfront Toronto, Toronto in Canada.

15.2 Case Study: HafenCity, Hamburg, Germany

HafenCity Hamburg (HafenCity), located along the River Elbe in Hamburg, Germany, is being transformed from a predominantly derelict, former dockland site into a lively city with a maritime environment that combines work and living, culture and leisure, tourism and shopping. The development, covering 157 ha, will have nearly 7000 homes for 12,000 residents along with commercial developments that will offer more than 45,000 jobs (HafenCity Hamburg 2016a). By the early 2020s, when the HafenCity development is completed, the built area of Hamburg's city centre will have been enlarged by around 40% with a mix of uses including housing, services, arts, recreation, tourism and commerce. While HafenCity will contribute towards new jobs and economic growth the development will also contribute towards environmental protection and health by encouraging both public and private developers to take responsibility in managing the natural environment and its resources sustainably (HafenCity Hamburg 2010).

15.2.1 HafenCity's Environmental Sustainability Goals

HafenCity aims to become an example of sustainability around the world through the redevelopment of former industrial sites (brownfield development) rather than outward expansion (greenfield development). The development aims to encourage the construction of environmentally significant and award winning buildings that promote resource conservation and low carbon emissions, as well as provide numerous green spaces that mitigate the risks from storm surges.

15.2.1.1 Encouraging Sustainable Construction

In 2007, the HafenCity Ecolabel and Eco Award were created to reward developers for sustainable management of energy, public goods and materials in construction as well as for delivering a healthy and comfortable environment for the building's users to work, rest or play in. By certifying the developer's sustainable innovations, it also attracts public awareness to the individual building projects, which in turn increases the status of the HafenCity project as a whole.

15.2.1.2 HafenCity Ecolabel

The Ecolabel is awarded in 'silver' or 'gold' by HafenCity GmbH to buildings that have achieved 'special' or 'excellent' rating in at least three out of five categories of sustainable construction, with all criteria within the assessment categories having to be fulfilled. In all cases it is mandatory to fulfil the conditions of category 1 (Table 15.1), which deals with the sustainable management of energy resources. An application for Ecolabel can be made early in the planning phase, by the site's purchaser or developer. A preliminary certificate is issued after a contractual agreement has been signed to implement the requirements. This means that the Ecolabel can be used for publicity purposes from the point at which the marketing of the project gets underway, before construction commences. The two awards are:

Table 15.1 HafenCity's Ecolabel categories (HafenCity Hamburg 2010)

Ecolabel category	Silver	Gold
1. Sustainable management of energy resources	Lower total primary energy demand than that of the reference building stipulated in the German regulations for energy savings and undercutting permissible transmission heat loss or the permissible heat transfer coefficient	Markedly lower total primary energy demand and markedly undercutting permissible transmission heat loss or the permissible heat transfer coefficient
2. Sustainable management of public goods	Architectural competition, no heavy metal contamination of water bodies, modest demand for potable water, space efficiency, family-orientated design, bicycle parking	Public access to ground floors and basement floors or housing with low car ownership, limited potable water demand, use of roof area, increased space efficiency
3. Use of eco-friendly construction materials	Compliance with requirements regarding construction materials containing halogen, biocides, heavy metals, organic solvents and construction materials described as sensitising, ecologically harmful	Lifecycle analysis of building materials used and undercutting of reference values for selected global impact parameters, wide use of renewable sources
4. Special consideration of health and well-being	Target values for indoor air quality, thermal comfort, acoustic comfort and user influence (regarding indoor climate, lighting and glare protection)	Lower target values for indoor air quality, thermal comfort, acoustic comfort and use influence (indoor climate, lighting and glare protection), plus 20% of area suitable for people with allergies
5. Sustainable facility operations	Optimisation of durability, flexibility of use and building operational costs, modularity of building and variability of design, barrier-free access to all floors	Special solutions for variable use of building components. Compilation of product documentation and its inclusion into building operational manuals or room data sheets

- Special achievement (Silver): Covers ecological building qualities that generate either no extra costs if stipulated at an early stage of the process or additional construction costs if they are seen to be economically justified. Investments are deemed justifiable if they can be largely amortised through operations in the short-term.
- Excellent achievement (Gold): This rating is concerned with the ecological qualities of a building that can be accomplished through innovative measures if some extra costs are incurred at the planning stage and during implementation and construction of the building project. To achieve gold, the criteria of both silver and gold must be met in the chosen categories.

15.2.1.3 Awarding a HafenCity Ecolabel

For new developments to be awarded a HafenCity Ecolabel there is a four-step process that must be followed (HafenCity Hamburg 2016b):

- HafenCity Hamburg GmbH and the purchaser of the land make an agreement to build a certifiable building. The investor commissions an expert to set down the exact criteria for sustainability. Concrete objectives have to be fulfilled in at least three out of five categories including the mandatory category 'sustainable management of energy resources'
- Following an approved building contract and planning documentation, an evaluation by an independent institute commissioned by HafenCity Hamburg GmbH is initiated. If the building meets the requirements, preliminary certification is awarded
- Once the building is completed and is fully operational, independent specialists prepare an inspection report. If the result is positive the final certificate is awarded
- An energy inspection is then conducted one year later. If the specified target levels are not achieved, improvements must be made. If the building fails to reach the target levels after improvement, the certificate may be withdrawn (HafenCity 2016b).

15.2.1.4 HafenCity Eco Award

The HafenCity Eco Award is awarded to buildings, of every type of use, in recognition of their outstanding ecological quality. Projects that are shortlisted will have met the criteria of the Ecolabel (Silver or Gold) as well as have been evaluated on the integration of ecological features into the overall concept and form of the building. Overall, the Eco Award is designed to bolster the positive public impact of outstanding ecological buildings (HafenCity Hamburg 2010).

15.2.1.5 Adapting to Rising Sea Levels

As HafenCity lies to the south of the main Hamburg dike there is no protection for the district from storm surges. It was agreed in the planning phase that surrounding HafenCity with dikes would have created disadvantages for the district, including depriving residents and visitors of sight lines to the water as well as numerous technical and economic challenges of constructing a dike before any construction of buildings: hampering successive development of the district (HafenCity Hamburg 2016c). The result has been the elevation of buildings on plinths made out of compacted fill. Specifically, all new buildings in the district stand on artificial bases eight metres above sea level: out of reach of the most extreme flooding. On the exposed windward side the exterior perimeter will be 8–9 m above sea level. Internally, the flood-secure plinths provide numerous spaces for underground car garages for accommodation of stationary traffic. This reduces the number of above ground parking spaces required, contributing to more effective ground surfaces as a resource. Roads and bridges will also be built above the flood-line: at least 7.5 m above sea level meaning that traffic between HafenCity and the rest of Hamburg can keep flowing even during a storm surge. Regarding responsibility of implementation of flood-secure infrastructure, it is the responsibility of the private developers of buildings to build the artificial compacted bases below the buildings and the responsibility of HafenCity Hamburg GmbH to ensure the roads and bridges are accessible during storm events. Overall, HafenCity can continue to function without restriction even during periods of flooding (HafenCity Hamburg 2016c) (Fig. 15.1).

Fig. 15.1 HafenCity's Philharmonic hall (Pixabay 2016)

15.2.2 HafenCity's Social Sustainability Goals

Achieving social sustainability is an important goal of HafenCity with the development ensuring adequate opportunities for work, leisure, educational and residential spaces that are interlinked by low-carbon transportation options.

15.2.2.1 Public Spaces for Leisure

A key aspect of the HafenCity development is that each neighbourhood is enriched with new urban spaces on the water and beside it. It is projected that 25% of its land —as much as 28 ha—will be public open space. All open spaces, whether plazas, parks or promenades—are on the waterside. In addition, publicly accessible private open spaces will account for a further 13% of the development's area, ensuring plenty of leisure opportunities for residents and visitors (HafenCity Hamburg 2016d). A broad strip up to 15 m-wide along the edges of HafenCity's restored historic quays has been created along the existing land 4.5–5.5 m above sea level to provide 10.5 km of waterside walks, creating additional green space for residents and visitors to enjoy (HafenCity Hamburg 2016c).

15.2.2.2 A Green Core Promoting Relaxation and Recreation

HafenCity's Lohsepark has been re-developed as a green core to encourage recreation and sports as well as provide a place for residents to meet. The park— 100 m-wide and 550 m-long—has been landscaped with lawns and meadows and has four levels: a historical level on the site of the old original area; a park level with broad sweeps of grass; a new city level with its terraces; and the landscaped hills above them. For relaxation the park features benches and seating areas and has both quiet sections and action areas offering games for all age groups including a basketball court (HafenCity Hamburg 2016e).

15.2.2.3 Unique Living Spaces: One District, Ten Neighbourhoods

To ensure HafenCity is not one homogeneous development, each neighbourhood has its own individual profile that utilises the different typographies found across the area including canals, harbour basins or the River Elbe forming the boundaries of the neighbourhood. In addition, each neighbourhood is designed for mixed-use, ensuring no neighbourhood in HafenCity is solely devoted to homes, offices, shops or leisure (HafenCity Hamburg 2016f).

15.2.2.4 Community and Education Spaces

HafenCity has numerous types of social infrastructure to support local residents and provide recreational opportunities. The primary school is integrated with an after-school centre for schoolchildren and is located next door to a nursery and kindergarten for HafenCity's youngest population group. In addition, there is a family service centre that provides back-up childcare for employees of local companies. In fact, the back-up childcare centre has no closing time nor closes over the holiday periods, enabling childcare around the clock if required. The gym located in Katherinenschule is not only used for school sport but also by a local sports club, with courses on offer for children including dancing and karate and for adult karate and recreational football (HafenCity Hamburg 2016g).

15.2.2.5 Low-Carbon Transportation

HafenCity has been designed to promote 'local mobility' in which a large proportion of daily journeys can be taken on foot or bicycle. The development has a comprehensive network of cycle ways and footpaths enabling residents to decrease the number of car journeys (HafenCity Hamburg 2016h). To encourage non-motorised transport, HafenCity has a cycle rental system run by Deutsche Bahn AG while HafenCity itself will offer four bike stations, the first three of which have been installed. The development is also connected to a new underground subway station that is projected to be used by around 35,000 people a day, the equivalent of 26,000 fewer car journeys per day (HafenCity Hamburg 2016i).

15.3 Case Study: Waterfront Toronto, Toronto, Canada

In 2000, the Government of Canada, the Province of Ontario and the City of Toronto each committed $500 million to renew Toronto's waterfront over 30 years. In 2001, the three partners created the Toronto Waterfront Revitalization Corporation (TWRC), now known as Waterfront Toronto, with a mandate to lead and oversee waterfront renewal activities in the designated waterfront area (DWA). The overall corporate objectives of Waterfront Toronto include: (1) implementing a plan that enhances the economic, social and cultural value of the land in the DWA and creates an accessible and active waterfront for living, working and recreation, and to do so in a fiscally and environmentally responsible manner; (2) ensuring the ongoing development in the DWA can continue in a financially self-sustaining manner; and (3) promoting and encouraging the involvement of the private sector in the development of the DWA (City of Toronto 2016).

15.3.1 Waterfront Toronto's Environmental Sustainability Goals

Waterfront Toronto aims to develop the most sustainable waterfront communities in the world that serve as a model for future community development. Initiatives to ensure the development is environmentally sustainable include ensuring minimum green building requirements are met; implementing water conservation measures; developing green infrastructure; installing green roofs; reducing waste and tracking carbon emissions, the development of which are guided by Environmental Management Plans (Waterfront Toronto 2010).

15.3.1.1 Minimum Green Building Requirements

Waterfront Toronto established Minimum Green Building Requirements (MGBR) to help the city realise the revitalisation plan's objective of positioning Toronto as a world leader in creating sustainable communities, with buildings and neighbourhoods that are among the greenest in the world. The MGBR mandate high performance buildings, smart technologies and passive designs (Waterfront Toronto 2016). The MGBR were first introduced by Waterfront Toronto in 2006 and since then has been updated in 2011 and revised in 2014 to better reflect market conditions, changes in the regulatory environment and the wishes of Toronto's residents. The revised MGBR include requirements that:

- Buildings achieve superior levels of energy efficiency
- Developments provide renewable energy generation on-site
- Water conservation measures for buildings, suites and exterior landscaping need to be incorporated
- All parking garages include electric vehicle (EV) infrastructure and EV charging stations within residential and commercial buildings
- Every suite includes energy and water sub-meters
- Adequate bicycle parking and storage spaces in residential buildings at convenient and easily accessible locations.

15.3.1.2 Building on LEED

The MGBR include and build upon the Canada Green Building Council's LEED rating system. In particular, Waterfront Toronto requires a minimum of LEED Gold under the MGBR as well as five LEED certifications that are normally optional. To ensure compliance, Waterfront Toronto obliges all developers engaged in waterfront projects to adhere to the MGBR and compliance agreements are embedded within all developer contracts (Waterfront Toronto 2016).

15.3.1.3 Water Conservation

Waterfront Toronto's buildings must meet MGBR, which include:

- Water efficient landscaping: 100% of the water used for landscaping must come from non-potable sources
- Water use reduction: 40% reduction in potable water use for indoor flow and flush fixtures.

Waterfront Toronto has implemented water conservation and reuse in its major parks too, including:

- Sherbourne Common: The first park in Canada to integrate a storm water management system in its design, where collected storm water and water drawn from Lake Ontario is treated by a UV facility located at the park before being discharged over a series of art sculptures, following which the water is discharged into a wetland feature before release into Lake Ontario. The treated water is also used for irrigation and for washroom facilities in the park, reducing demand for potable water
- Corktown Common: A storm water recycling system collects the park's storm water for UV treatment before being moved to underground storage cells located in the central lawn, from which the water is used for irrigation, park maintenance and to flush the marsh.

15.3.1.4 Green Infrastructure

Waterfront Toronto aims to integrate green infrastructure into every element of urban design, where green infrastructure is defined as natural systems and human-made vegetative technologies that provide ecological and hydrological functions to enhance healthy and sustainable living. Green infrastructure implemented across the DWA to minimise stormwater runoff and enhance water quality include green roofs, community gardens, rain gardens and bio-swales as well as permeable paving, cisterns and soil cells (Waterfront Toronto 2015a).

15.3.1.5 Partnerships to Protect and Restore Aquatic Habitats

Waterfront Toronto is a founding member of Aquatic Habitat Toronto—a consensus-based partnership between agencies, including Fisheries and Oceans Canada, Ministry of Natural Resources, Toronto and Region Conservation, the City of Toronto, and Environment Canada: all of which have a vested interest in improving the aquatic habitat on Toronto's waterfront. The partnership aims to design suitable aquatic habitat in appropriate locations to offset potential impacts of waterfront projects (Waterfront Toronto 2015b).

15.3.1.6 Green Roofs

To capture the multiple benefits of green roofs, including reducing the urban heat island effect, minimising stormwater runoff, improving air quality and building efficiency—Waterfront Toronto requires all developers to install a green roof for 60% of available roof space, or the percent required by the City of Toronto's Green Roof Bylaw, whichever is greater. In addition, all low-sloped roofs need to be designed to accommodate the loads that would be imposed by an intensive green roof, ensuring there is the possibility of installing a green roof in the future without undertaking costly major structural modifications (Waterfront Toronto 2015c).

15.3.1.7 Soil Recycling

Due to the revitalisation of Toronto's waterfront being one of the largest urban brownfield remediation projects in the world, the development is projected to manage around two million cubic metres of contaminated soil over the next 20 years. Typically, contaminated soil is transported to landfills with new soil brought into replace it: a process known as 'dig and dump'. As part of the development's drive to be sustainable, Waterfront Toronto established in 2010 a pilot soil recycling facility in the Port Lands to decontaminate soil on-site to an environmental condition that allows for its reuse (Waterfront Toronto 2015d). Following the pilot's success—including a reduction of greenhouse gas emissions of 36 kg/tonne from reduced truck traffic—Waterfront Toronto supported the development of a privately operated facility on the pilot site. Overall, the projected environmental as well as social cost savings, including reduced traffic accidents, less noise and less congestion, to Waterfront Toronto by using the recycling facility is estimated to be $65 million over 10 years (Waterfront Toronto 2015d).

15.3.1.8 Waste Reduction

To divert waste away from landfills, the MGBR mandate that all kitchens are to have a separate space for the segregation of recyclables, organics and waste. All residential buildings over three storeys must have tri-sorting or separate chutes for each waste stream on each floor, while a collection area for household hazardous waste must be provided in all buildings (Waterfront Toronto 2015e).

15.3.1.9 Construction Waste

To reduce the amount of construction-related waste entering landfills, Waterfront Toronto requires that all construction and demolition projects divert at least 50% of waste, with a target of 75%. This requirement is included in the Environmental Management Plan, which is a plan that describes the processes and procedures designed to mitigate environmental effects that may result from project-related activities in the Waterfront Toronto development area. In addition, the requirement achieves a credit as part of the Waterfront Toronto's LEED for Neighbourhood Development Gold certification (Waterfront Toronto 2010, 2015e).

15.3.1.10 Tracking Carbon Emissions

Waterfront Toronto aims to reduce carbon emissions from the construction and operation of the buildings, parks and neighbourhoods in the revitalised waterfront area. In collaboration with the C40-Climate Positive Development Program, along with funding support from the Ontario Power Authority, Waterfront Toronto has developed a Carbon Tool that analyses and compares the sustainability performance of projects at the design and planning phase over a baseline. This enables Waterfront Toronto to understand how well a project is expected to perform over a build-as-usual scenario and what additional strategies can be implemented to achieve carbon reductions. The tool in particular quantifies how sustainability strategies in energy, water, waste, transport and materials impact carbon reductions, enabling planners to develop ways of improving performance (Waterfront Toronto 2015f).

15.3.2 Waterfront Toronto Social Sustainability Goals

Waterfront Toronto aims to create functional, sustainable and beautiful communities that offer a high quality of life for all who live and work there. The waterfront neighbourhoods are designed to connect with the rest of the city through access to transit, parks, open spaces, pedestrian promenades and cycling lanes.

15.3.2.1 Waterfront Toronto's Public Consultation Strategy

In 2002, Waterfront Toronto created a public consultation strategy to formalise its commitment to public engagement and the principles that would guide the outreach process (Table 15.2).

15.3.2.2 Public Waterfront Access

The waterfront renewal project ensures that both residents and visitors have opportunities to enjoy and connect with the lakefront. Over 13 km of trails and promenades have been created in key areas of the waterfront.

Table 15.2 Principles of Waterfront Toronto's (2003) public consultation strategy

Principle	Description
Accountability	Accurate and timely information will be provided through public consultation processes
Clarity	Well-defined objectives for, and limits to, information, consultation and active participation during planning
Timeliness	Consultations will begin as early as possible to ensure opportunities and issues can emerge and to increase the chance of successful issue resolution and implementation
Openness and inclusivity	Participation will be open to any member of the public or other stakeholder groups that wish to be involved
Flexibility	The consultation process will accommodate the needs of participants taking into consideration different areas of expertise, geographic distribution and availability
Coordination	Initiatives to inform, request feedback and consult citizens will be coordinated between Waterfront Toronto and individual projects to enhance knowledge management, ensure cohesion in decision-making and reduce 'consultation fatigue' among citizens and stakeholders
Evaluation	Waterfront Toronto will evaluate its performance in providing information, conducting consultation and engaging citizens
Commitment	There will be strong leadership and commitment from Waterfront Toronto, politicians, senior managers and public officials to these principles

15.3.2.3 Employee Service Awards

To recognise the contributions made by long-serving employees Waterfront Toronto presents Service Awards to employees at five-year intervals of service. Eligible employees are presented with a certificate and a Service Award at quarterly Town hall meetings, which are attended by all employees. The award presented to eligible employees corresponds in value with the years of service and is chosen by the employee from a catalogue of items. To date, Service Awards have been given to 32 employees with five years of service and 8 employees with 10 years of service (Waterfront Toronto 2015g).

15.3.2.4 Waterfront Toronto Employment Initiative

The Waterfront Toronto Employment Initiative (WTEI) connects unemployed Toronto residents to the employment and training opportunities generated by waterfront revitalisation projects. The Initiative is a collective that includes the City of Toronto's Employment and Social Services Division, the City's Youth Employment Partnership as well as non-governmental organisations including the YMCA of Greater Toronto. WTEI increases access to employment opportunities by working with both employers and job seekers. Waterfront Toronto encourages its partners, including developers, contractors and consultants to participate in the programme by sharing job opportunities as they arise and work with WTEI when customised hiring strategies are appropriate. The key initiatives by WTEI for employers and jobseekers are listed below (Waterfront Toronto 2015h)

WTEI supports employers by:

- Understanding their hiring requirements and preferences, which in turn will inform the development and implementation of customised hiring and training plans
- Providing support to employers to maximise their participation in WTEI by for example referring pre-screen candidates for available opportunities
- Creating a centralised access point for connecting employers with a qualified candidate pool that has been tailored for the employers' needs.

For jobseekers WTEI provides:

- Access to diverse job opportunities generated through the projects on the waterfront
- Advice on effectively finding and competing in job competitions (e.g. resume development, interviewing skills, networking, company researching etc.)
- Support by building partnerships with related unions and/or professional affiliations; and post-hire retention support including on-going career development services.

15.3.2.5 Affordable Housing

To ensure that Waterfront Toronto is an inclusive, diverse and equitable environment for all, the development ensures a minimum of 20% of residential units are designated as Affordable Rental Housing. In addition, 5% of units will be low-end-of-market units. As Waterfront Toronto will have ultra-fast Internet connections and neighbourhood-wide WiFi the development has also developed a digital inclusion strategy to create an accessible and inclusive digital environment for all. The Strategy involves a cross-subsidy model that all residents, including those in affordable housing developments, can access the broadband network. The model is designed as a capital offset so market units subsidise the connection of affordable units (Waterfront Toronto 2015i, j).

15.3.2.6 Public Art Spaces

Waterfront Toronto is installing art in public spaces for the enjoyment of all. This is part of the development's belief that art and creativity are vital aspects of urban life. The goal is to create a contemporary collection of public art that reflects each neighbourhood's character and is accessible to all. During the neighbourhood-planning phase a public art strategy is created and prominent locations are selected for the artwork. While the City of Toronto has a policy that requires developers to set aside 1% of gross construction costs for art—that is often placed in lobbies of buildings making it publicly inaccessible—Waterfront Toronto in contrast will pool all of the public art monies and strategically place art in high profile locations, providing the greatest opportunities for public engagement (Waterfront Toronto 2015k).

15.4 Analysis of Case Studies

HafenCity and Waterfront Toronto both have similar aims of becoming model examples of environmental sustainability, with HafenCity showcasing the sustainable benefits of redeveloping former industrial sites while Waterfront Toronto aiming to be an exemplar of community development that embraces resource efficiency and low-carbon technologies. Achieving social sustainability is an important goal for both the developments, with HafenCity ensuring adequate opportunities for work, leisure, and residential living, while Waterfront Toronto creating functional, sustainable and desirable communities that offer a high quality of life for all who live and work there. In addition, both developments aim to be interlinked by low-carbon transportation options that promote healthy active lifestyles.

15.4.1 Environmental Sustainability

With waterfront developments typically disturbing soil in the event of land reclamation, Waterfront Toronto conducted a pilot study to determine whether it was economically feasible to recycle contaminated soil onsite without having to haul contaminated soil to a nearby landfill only to be replaced with new soil. Following a successful pilot that saw reductions in greenhouse gas emissions as well as reduced traffic congestion, Waterfront Toronto supported the development of a privately-operated soil recycling facility on the pilot site, providing the development with not only economic benefits but environmental as well as social benefits too. Furthermore, to mitigate the damaging effects of waterfront development on the oceanic environment and ecology, Waterfront Toronto along with other stakeholders has formed a consensus-based partnership that aims to improve the lake's aquatic habitat by designing appropriate habitat-spaces.

It has been become more widespread around the world that waterfront development plans have become increasingly integrated with their cities' respective storm water management systems and urban water cycles, as well as other environmental issues. HafenCity for instance, has developed an Ecolabel and Eco Award to reward developers for sustainable management of water and other resources in addition to ensuring a comfortable environment for the building's users to work, rest or play in. Meanwhile, Waterfront Toronto's Minimum Green

Building Requirements include the need for developments to be energy efficient, provide renewable energy as well as reduce potable water consumption.

With respect to flood-resilience, HafenCity requires that all its buildings are elevated on plinths made out of compacted fill. In addition, the development's roads and bridges are also constructed above the flood-line ensuring that traffic between HafenCity and the rest of Hamburg can keep flowing during a storm surge.

Green infrastructure has been described as the essential breathings lungs of a city that offer numerous environmental as well as social benefits including healthy lifestyles, leisure and recreation opportunities. To capitalize on the multiple benefits that green infrastructure provides, HafenCity has transformed an old park into a new green core that also encourages recreation and sports as well as provides a place for residents to meet. In particular, it has terraces with landscaped hills above them with spaces for seating and resting while other sections of the park have been created for sports.

15.4.2 Social Sustainability

Both HafenCity and Waterfront Toronto ensure there are adequate work, educational and leisure opportunities for their residents. For example, the Waterfront Toronto's Employment Initiative connects unemployed Toronto residents to employment and training opportunities generated by waterfront revitalisation projects. As part of the initiative, Waterfront Toronto also encourages its partners, including developers, contractors and consultants to share job opportunities as they arise. Regarding educational and leisure opportunities, HafenCity provides workers with after-school child care services as well as a nursery and kindergarten. In addition, the development has a round-the-clock childcare service if necessary. To provide recreational opportunities, HafenCity has a gym for children as well as local sports clubs to use with classes for both children and adults.

HafenCity and Waterfront Toronto are both committed to promoting low-carbon transportation options that also increase physical health and well-being of both residents and visitors. In HafenCity, there is a comprehensive network of cycle ways and footpaths enabling residents to decrease the number of car journeys. Meanwhile, Waterfront Toronto has developed a series of trails that connect residents and visitors to the waterfront.

Recognizing the value of citizen and stakeholder participation, Waterfront Toronto has a public consultation strategy that formalises its commitment to public engagement and the principles of accountability, clarity, openness and inclusivity that guide its outreach to various stakeholders.

To avoid the world-wide phenomenon of homogenized and artificial environments, HafenCity has been designed so that each of its 10 neighbourhoods have their own individual profile that utilises the various features found in the area. In addition, each neighbourhood is mixed-use ensuring a variety of homes, offices, shops and leisure activities. On the other hand, Waterfront Toronto will pool all of the public art monies and strategically place art in high profile locations to ensure each neighbourhood has a unique character and identity.

15.5 Conclusion

Globally, waterfronts have transitioned from traditional port and industrial activities towards places that in addition to offering residential and commercial spaces, also provide leisure, recreational, heritage, art and entertainment opportunities for both residents and visitors.

Waterfronts form part of the fundamental intersection between the natural and manmade environment and vitally connect the city and its inhabitants to water. A city's waterfront is its prime land, a limited and precious resource with immense pressures and opportunities. Waterfronts as public spaces form part of the vital network of city's breathing lungs that have become all the more important in the context of high density urban environments as they have the potential to bring a range of economic, aesthetic, social, environmental and ecological benefits. They are essentially, humane and democratic spaces that foster community health, wellbeing and social relations through the diverse uses they encompass, and support education, employment and leisure opportunities.

To move towards a more sustainable future in the context of competing world cities, urban waterfront renewal must balance the economic goals with aspirations of a city, its people and environment whilst also respecting its culture and heritage. Gordon (1996) defined successful urban waterfront redevelopments as those that lead to 'improving image; adapting and reusing existing built form; improving public accessibility; integrating waterfronts with their urban surroundings and with the water; thinking small and planning in increments'.

Waterfronts are now recognized as a powerful determinant in the development of urban form and pattern and are integrated in citywide schemes and masterplans. Waterfront regeneration defined by flexible and incremental planning approaches to accommodate the changing values and pace are recommended. Public participatory processes for waterfront regeneration are also strongly encouraged as they engender an improved appreciation of this delicate interface and consequently lead to responsible use, longevity and reduced maintenance.

Increasingly the ecological impacts of waterfront developments and their related activities on natural habitats and the delicate waterside environments are also being considered. Clean up operations, infrastructure improvements for flood resilience and storm water management, greening of waterfronts, creation of new public spaces and improvements to the public realm for urban renewal, city promotion, attracting inward investments, and the creation of more sustainable urban environments are increasingly common place.

Waterfronts offer enormous multi-dimensional potentials for a city and are recognized as important to city's social and environmental sustainability, resilience to climate change impacts, urban ecology and biodiversity, and aquatic ecosystems.

The global waterfronts of the 21st century will increasingly have to respond to the demands and opportunities placed by climate change adaptations, low and zero carbon developments, renewables and their associated infrastructures, that will be of increasing importance and feature prominently affecting land use, waterfront ecology, design and character. Lastly, stakeholder and citizen engagement and participatory design processes will be critical to the shaping of waterfront environments that are distinctive and authentic.

References

Breen A, Rigby D (1991) Waterfronts cities reclaim their edge. McGraw-Hill, Singapore
Chandran ACK, Gowda K (2014) The blue–green edge in Thiruvananthapuram, India: an initiative for urban waterways and waterfront developments. Environ Qual Manage 24(2):45–59
Chang TC, Huang S (2005) Recreating place, replacing memory: creative destruction at the Singapore river. Asia Pacific Viewpoint 46(3):267–280
Cheung DM, Tang B (2015) Social order, leisure, or tourist attraction? The changing planning missions for waterfront space in Hong Kong. Habitat Int 47:231–240
City of Toronto (2016) Waterfront Toronto. http://www1.toronto.ca/wps/portal/contentonly?vgnextoid=8aa5b6a6731d1410VgnVCM10000071d60f89RCRD&vgnextchannel=8c48e26b1af51410VgnVCM10000071d60f89RCRD. Accessed 26 Sept 2016
De Sousa C (2014) The greening of urban post-industrial landscapes: past practices and emerging trends. Local Environ 19(10):1049–1067. doi:10.1080/13549839.2014.886560
Dessai S, van der Sluijs J (2007) Uncertainty and climate change adaptation—a scoping study. Copernicus Institute for Sustainable Development and Innovation, Department of Science Technology and Society, Utrecht, The Netherlands, Report NWS-E-2007-198
Dong L (2004) Waterfront development: a case study of Dalian, China (Master's thesis). Retrieved from http://uwspace.uwaterloo.ca/handle/10012/988
Dovey K (2005) Fluid city: transforming Melbourne's urban waterfront. Routledge, Abingdon
Dyson K, Yocom K (2015) Ecological design for urban waterfronts. Urban Ecosyst 18:189–208
Gersonius B (2012) The resilience approach to climate adaptation applied for flood risk. Ph.D. thesis, TU Delft
Giblett G, Samant S (2011) An investigation of sustainable strategies for the revival of waterfronts in high density urban environments. J Urban Regeneration Renewal 5(2)
Gordon D (1996) Planning, design and managing change in urban waterfront redevelopment. Town Plann Rev 67:261–290
Gospodini A (2001) Urban waterfront redevelopment in Greek cities: a framework for redesigning space. Cities 18(5):285–295
Haasnoot M (2013) Anticipating change: sustainable water policy pathways for an uncertain future. Ph.D. thesis, TU Delft
Hafencity Hamburg (2010) Sustainable construction in HafenCity. http://www.hafencity.com/upload/files/files/Sustainable_Construction_1.4.pdf. Accessed 23 Sept 2016
HafenCity Hamburg (2016a) FlyerInfocenter Kesselhaus. http://www.hafencity.com/upload/files/artikel/151223vK_KesselhausFlyer_final_web.pdf. Accessed 23 Sept 2016
HafenCity Hamburg (2016b) How is the HafenCity Ecolabel awarded? http://www.hafencity.com/en/faq-concepts-planning/how-is-the-hafencity-ecolabel-awarded-.html. Accessed 23 Sept 2016
HafenCity Hamburg (2016c) Flood-secure bases instead of dikes: safe from high water in Hamburg. http://www.hafencity.com/en/concepts/flood-secure-bases-instead-of-dikes-safe-from-high-water-in-hafencity.html. Accessed 23 Sept 2016

HafenCity Hamburg (2016d) The city of plazas, parks and promenades. http://www.hafencity.com/en/concepts/the-city-of-plazas-parks-and-promenades.html.Accessed 23 Sept 2016

HafenCity Hamburg (2016e) Lohsepark, HafenCity's green people's park, opens to the public. http://www.hafencity.com/en/news/lohsepark-hafencity-s-green-people-s-park-opens-to-the-public-.html. Accessed 23 Sept 2016

HafenCity Hamburg (2016f) One district, ten neighbourhoods. http://www.hafencity.com/en/quarters/one-district-ten-neighborhoods.html. Accessed 23 Sept 2016

HafenCity Hamburg (2016g) Something for everyone: social infrastructure in the locality. http://www.hafencity.com/en/living/something-for-everyone-social-infrastructure-in-the-locality.html. Accessed 23 Sept 2016

HafenCity Hamburg (2016h) HafenCity as ecological sustainable city. https://newsroom.cisco.com/dlls/2008/ekits/or_Ecologically_Sustainable_HafenCity_CUD.pdf. Accessed 23 Sept 2016

HafenCity Hamburg (2016i) Many routes to a sustainable city. http://www.hafencity.com/en/concepts/many-routes-to-a-sustainable-city.html. Accessed 23 Sept 2016

IPCC (2007) Climate change 2001 mitigation of climate change. https://www.ipcc.ch/pdf/assessment-report/ar4/wg3/ar4_wg3_full_report.pdf. Accessed 28 Oct 2016

Lukens R, Selberg C (eds) (2004) Guidelines for marine artificial reef materials. 2nd edn. Artificial Reef Subcommittees of the Atlantic and Gulf States Marine Fisheries Commissions

MacLeod G, Ward K (2002) Spaces of utopia and dystopia: landscaping the contemporary city. Geogr Ann Series B Hum Geogr 84(3):153–170

Moretti M (2008) Cities on water and waterfront regeneration: a strategic challenge for the future. In: Proceedings from Grundtvig, II meeting. Rivers of change—river/cities, Warsaw, Poland, 24–27 July. Retrieved from http://www.river-cities.nazwa.pl/www/download/m.moretti_warsaw2008.pdf

Petrillo J, Grenell P (eds) (1985) The urban edge. Where the city meets the sea. William Kaufman, Los Altos, CA

Pixabay (2016) Hamburg Elbe Philharmonic hall. https://pixabay.com/en/hamburg-elbe-philharmonic-hall-1669634/. Accessed 29th Sept 2016

Remedial Action Plan for Hamilton Harbour (2006) Hamilton harbour remedial action plan 2000–2005 highlights and summary. Jan 16

Sairinen R, Kumpulainen S (2006) Assessing social impacts in urban waterfront regeneration. Environ Impact Assess Rev 26:120–135

Samant SR (2014) Waterfront regeneration: experiences in city-building. J Urban Design 19 (2):255–257

Shrestha BK, Shrestha S (2008) Urban waterfront development patterns: water as a structuring element of urbanity. In Feyen J, Shannon K, Neville M (eds) Water and urban development paradigms. Towards an integration of engineering, design and management approaches. CRC Press, London p 105–113

Stevens Q (2009) Artificial waterfronts. Urban Design Int 14:3–21

Timur UP (2013) Urban waterfront regenerations. In Özyavuz M (ed) Advances in landscape architecture. Retrieved from http://www.intechopen.com/books/advances-in-landscape-architecture/urban-waterfront–regenerations. p 169–206

Veelen V, Stone K, Jeuken A (2015) Planning resilient urban waterfronts using adaptive pathways. In: Proceedings of the institution of civil engineers-water management, Feb 2015, vol 168. no 2, Thomas Telford Ltd, pp 49–56

Vollmer D (2009) Urban waterfront rehabilitation: can it contribute to environmental improvements in the developing world? Environ Res Lett 4(2), 024003

Wakefield S (2007) Great expectations: waterfront redevelopment and the Hamilton harbour waterfront trail. Cities 24(4):298–310

Wardekker A, De Jong A, Knoop JM, Van Der Sluijs JP (2010) Operationalising a resilience approach to adapting an urban delta to uncertain climate changes. Technol Forecast Soc Change 77:987–998

Waterfront Toronto (2003) Waterfront revitalization corporation public consultation and participation. http://sr.waterfrontoronto.ca/en/social/resources/public_consultation_strategy.pdf. Accessed 26 Sept 2016

Waterfront Toronto (2010) Environmental management plan for project-related activities. http://www.waterfrontoronto.ca/uploads/documents/wt_emp_final_march_2010_1.pdf. Accessed 26 Sept 2016

Waterfront Toronto (2015a) Water conservation. Waterfront Toronto corporate social responsibility report. http://sr.waterfrontoronto.ca/en/environment/pm5waterconservation.asp. Accessed 26 Sept 2016

Waterfront Toronto (2015b) Habitat creation and restoration. Waterfront Toronto corporate social responsibility report. http://sr.waterfrontoronto.ca/en/environment/pm1aquatichabitat.asp. Accessed 26 Sept 2016

Waterfront Toronto (2015c) Urban green space. Waterfront Toronto corporate social responsibility report. http://sr.waterfrontoronto.ca/en/environment/pm10urbangreenspace.asp. Accessed 26 Sept 2016

Waterfront Toronto (2015d) Contaminated soil management. Waterfront Toronto corporate social responsibility report. http://sr.waterfrontoronto.ca/en/environment/pm6contaminatedsoilmanagement.asp. Accessed 26 Sept 2016

Waterfront Toronto (2015e) Waste diversion. Waterfront Toronto corporate social responsibility report. http://sr.waterfrontoronto.ca/en/environment/pm7wastediversion.asp. Accessed 26 Sept 2016

Waterfront Toronto (2015f) Carbon emissions. Waterfront Toronto corporate social responsibility report. http://sr.waterfrontoronto.ca/en/environment/pm4carbonemissions.asp. Accessed 26 Sept 2016

Waterfront Toronto (2015g) Planning for success. Waterfront Toronto corporate social responsibility report. http://sr.waterfrontoronto.ca/en/social/pm6planningforsuccess.asp. Accessed 26 Sept 2016

Waterfront Toronto (2015h) Community engagement initiatives. Waterfront Toronto corporate social responsibility report. http://sr.waterfrontoronto.ca/en/social/pm7communityengagementinitiatives.asp. Accessed 26 Sept 2016

Waterfront Toronto (2015i) Affordable housing. Waterfront Toronto corporate social responsibility report. http://sr.waterfrontoronto.ca/en/social/AffordableHousing.asp. Accessed 26 Sept 2016

Waterfront Toronto (2015j) Intelligent communities. Waterfront Toronto corporate social responsibility report. http://sr.waterfrontoronto.ca/en/social/Intelligent-Communities.asp. Accessed 26 Sept 2016

Waterfront Toronto (2015k) Public art. Waterfront Toronto corporate social responsibility report. http://sr.waterfrontoronto.ca/en/social/PublicArt.asp. Accessed 26 Sept 2016

Waterfront Toronto (2016) Minimum green building requirements. http://www.waterfrontoronto.ca/our_waterfront_vision/our_future_is_green/green_building_requirements. Accessed 26 Sept 2016

Part IV
Conclusion

Chapter 16
Concluding Remarks

Puay Yok Tan and Chi Yung Jim

The greening of cities is a pursuit that has been a long standing and integral part of the urban development history of human settlements. More than any other urban elements, urban vegetation and the fauna associated with it, underscore the flawed notion that nature is only pristine and has no place in cities. Urban vegetation indeed, represents the most powerful and visible expression that nature has been, and should continue to be an inseparable part of the way cities are planned and managed.

As Y.Q. Feng and P.Y. Tan highlighted in Chap. 3, the greening of cities is driven by a wide range of reasons, from meeting basic existential needs such as urban agriculture for sustenance, to fulfilling higher order human needs, including meeting the innate and perhaps universal dependence on meaningful connections with nature for our well-being. It is also clear that our enhanced understanding of the ecological functions of urban greenery, such as in moderating urban climate (Chap. 4), supporting biodiversity conservation (Chaps. 7 and 13), and urban hydrology (Chap. 10), has provided added impetus in more recent years for greening of cities to enhance the biophysical conditions of the built environment for human well-being. Importantly too, urban vegetation in the form of ancient and heritage trees, as highlighted by C.Y. Jim in Chap. 13, are 'cherished bequests of a bygone era' and serve to remind us that humans and most of the modern infrastructures that we build, are outlived by nature, both literally and metaphorically. It is literal in the sense that remnant patches of woodlands, or individual trees can survive longer

P.Y. Tan (✉)
National University of Singapore, Singapore, Singapore
e-mail: akitpy@nus.edu.sg

C.Y. Jim
University of Hong Kong, Pokfulam, Hong Kong
e-mail: hragjcy@hku.hk

P.Y. Tan and C.Y. Jim (eds.), *Greening Cities: Forms and Functions*,
Advances in 21st Century Human Settlements,
DOI 10.1007/978-981-10-4113-6_16

than generations of urban dwellers. There is consequently an ethical need to protect and conserve precious remnants of vegetation, especially within the confines of cities for posterity. It is metaphorical in the sense that nature is timeless, as the urban environment will always be intimately influenced by biophysical and geochemical flows of materials and energy occurring at the city scale and coarser regional and global scales. Nature is omnipresent and perpetually exerts an external influence on our well-being, and the better we recognise and design the built environment to manage flows of water, nutrients, carbon and energy, etc., the more likely are we going create a resilient environment that responds and adapts to the ebbs and flows of natural forces. This is an *ecology for city* approach of designing cities for human well-being and environmental quality (Childers et al. 2015) and urban greenery, as a medium for introducing nature's services, or ecological services to the built environment (Chap. 2 and 9), should be a critical component of this approach.

The impetus for an emphasis on greening our built environment is also increasingly intensified because of sheer increase in urban populations over the past decades. Increase in urban populations inevitably implies increased land developments, reduction in natural areas, and the attendant environmental consequences of increased pollution and resource consumption. Such changes are especially severe in developing countries in Asia and Africa where projected urban population growth will be the highest. Understanding the benefits, forms, challenges, and means of greening cities thus become important considerations in the overall as well as site-specific approach towards urban planning and design. We also contend that in developed regions of the world, i.e. the Global North, be it in shrinking cities in the American rust belt, or in cities that are still experiencing rapid and very rapid growth in most of Europe (Fons 2012), increasing the knowledge for the methods for urban greening as well as the management of urban greenery, especially one that incorporates the human dimensions, are also equally relevant and valuable.

Various chapters in this book have contributed to enhancing our collective knowledge of urban greening and ecology in the multiple dimensions discussed in the book. However, they also highlight that knowledge gaps still exists. These pertain to techniques and approaches of greening, means for integrating scientific knowledge with practice, social methods for negotiating tradeoffs between benefits of greenery, etc. The multi-disciplinary and the interconnectedness of these issues and challenges will inevitably mean that plugging these knowledge gaps will require further research that incorporates interdisciplinary collaboration not just among scientists, but also with practitioners, policy makers and other stakeholders. In fact, the range of disciplines which our contributors represent: ecologists, geographers, urban climatologists, landscape architects, architects, urban planners, natural resource scientists, etc., reflect the intent of the book to provide multiple perspectives of the relevance of urban greening to urban development. In the remaining sections of this chapter, we summarise the key points that we drew from different contributions by our colleagues in three key strands of thoughts: (1) adopt a multi-functional approach to urban greening; (2) integrate science and design in urban greening practices; and (3) innovate and integrate greenery with the built elements of the city.

16.1 Adopt a Multi-functional Approach to Urban Greening

One of the key knowledge domains that has burgeoned in the last decade is the field of ecosystem services. The exponential growth in publications on ecosystem services (Hubacek and Kronenberg 2013; Potschin and Haines-Young 2011) highlight the attractiveness and utility of the concept to underscore human's dependence on services provided by functions and processes in natural ecosystems for their needs and well-being. In more recent years, this concept has been extended to 'urban ecosystem services', which are the services provided by vegetated areas, both spontaneous or designed, in urbanized areas. As highlighted by W.Y. Chen in Chap. 9, a wide range of ecosystem services has been identified to be also relevant to urban areas. In other words, urban greenery is capable of providing valuable ecological and socio-cultural services to improve urban environmental conditions and support human well-being. This multitude of services has to be recognized over and beyond urban greening as a means to create visually pleasing cityscapes. Urban greenery are often sights of beauty, but beautification alone as argued in Chap. 2, fails to harness the potential of urban greenery to deliver much more benefits needed in cities. These benefits in fact, form the basis for why urban greening directly and indirectly supports urban development goals of urban sustainability, liveability, as well as resilience.

Two interrelated questions ensue from this recognition. Firstly, given that urban greenery provides a suite of urban ecosystem services, which service(s) should be emphasized in the implementation of urban greenery; secondly, what should be the process for determining such services. We suggest that a *multi-functional approach* is a useful basis to manage these challenges. A multi-functional approach in our view addresses two aspects of urban greenery implementation. The first is that urban greenery can be implemented to achieve multi-functionality, that is, to simultaneously provide more than one service. The concept of multi-functional landscape is not new, and has been advocated for agriculture for more than two decades (for instance, see Huang et al. 2015). The concept should also apply to urban landscapes (Lovell and Taylor 2013). Multi-functionality encourages the layering of multiple functions over space and time, and is a means for optimal land use in cities which are perpetually plagued by a contest for space.

The second aspect is that achieving multi-functionality by its very nature, entails engagement of multiple stakeholders in deciding which type of ecosystem services should be provided in a site. The concept of ecosystem services is primarily an anthropocentric view on nature's services, and therefore, who are the beneficiaries of the services, and the process of determining benefits are therefore necessary considerations in the provision of ecosystem services. Inevitably, tradeoffs are also involved as providing one benefit of urban greening, such as a playing field for active recreation, often conflicts with other benefits, such as creating a woodland for biodiversity conservation. When green spaces become increasing reduced or scarce as city intensifies through infill developments, contest over spaces and the types of functions these spaces should provide will increase, and the need to have a socially inclusive process of negotiating tradeoffs will gain in importance. Community engagement tools, such as participatory design, focus groups, online forums, etc. will increasingly need to be used. Scientists alone will not be able to fulfil such expectations as the process of negotiating tradeoffs and forging community agreement is fundamentally social. It is thus crucial that urban greening must involve scientists who provide the scientific facts, the community, social scientists and policy makers who understand needs and communicate scenarios, and design professionals to translate such facts and needs into spatial plans (Sect. 16.2).

16.2 Integrate Science and Design in Urban Greening Practices

The importance of the scientific basis to inform urban greening practices is a recurring theme in this book. Science is needed to evaluate, understand and optimize the benefits from urban greening, such as in urban climate modification (Chap. 4), application of green roofs (Chap. 11), and blue-green infrastructure (Chap. 10). Scientific understanding of landscape ecology and conservation is also crucial to apply ecological principles to design and management of urban greenery (Chap. 2), enhancement of urban biodiversity (Chap. 7), and regeneration and rehabilitation of urban woodlands (Chap. 14). Scientific knowledge of the types and extent of disservices provided by urban greenery is also needed to provide a more complete view of net benefits provided by greenery, identify synergies and tradeoffs to society and environment (Chap. 9). Urban vegetation can also create negative impacts, or ecosystem disservices. Examples of ecosystem disservices include those arising from urban agriculture, highlighted by Lin et al. (Chap. 8), such as the possibility of genetic introgression of natural ecosystems by cultivated urban plants, chemical spill-over from urban farming practices. Other types of ecosystem disservices, such as the release of biogenic volatile organic carbons and their interactions with nitrogen oxides, pollen spores and fungal spores that could negatively affect urban air quality (Leung et al. 2011) also need to be assessed and understood more completely.

Thus, despite several decades of progress in the science behind urban greening, it is also clear that our knowledge is incomplete. This stems from both uncertainties in biophysical and biogeochemical processes mediated by urban vegetation (Pataki et al. 2011), as well as the fact the city is itself a complex system that exerts strong influences on natural processes from anthropogenic activities and the dominance of the built components (buildings and infrastructure). In other words, how much benefits can be provided by urban vegetation is not just about the biological processes, but also about the influence of built and social components, and the complex feedback loops between these interacting components. This is well-illustrated by the high level of uncertainties in assessing the benefits of urban trees on health of urban dwellers through its effects on air quality (see Whitlow et al. 2014a, b). Even though individual trees do indeed sequester air pollutants or trap particulate matter, the overall effect of urban trees at the neighbourhood scale and the outcome on human dwellers is influenced by a suite of confounding factors such as urban form, anthropogenic activities and larger environmental context. To suggest unequivocally that urban trees provide impactful health outcomes is not feasible with current level of uncertainties, and reflects the need for more empirical evidence. Similarly, E. Erell suggested that the magnitude of vegetation on air temperature is often over-stated (Chap. 4), as it depends on the climatic context, urban geometry and species and arrangement of trees.

We suggest here that there are three levels of knowledge gaps that should direct research: (1) at the species level, which species can provide high efficacy in provision of benefits, such as evaporative cooling and pollution uptake, and which species tend to be more adapted to urban conditions under climate change; (2) at the parcel scale, how does the size and composition of urban green spaces affect their performance, especially considering vegetation-urban elements and morphological interactions; (3) at the regional or city scale, how does the distribution of urban green spaces affect their ability to deliver benefits for urban dwellers and the environment. As scale increases, the number of considerations also increases, such as coarser scale environmental variables, and economic and social objectives; therefore, complexity involving urban greening provision also increases. One of the classic illustration of effects scale and distribution is the question of whether single large, or several small (SLOSS) green spaces are more effective urban biodiversity conservation. A similar question can also be asked of how large green spaces need to be provided and how they should be distributed to increase usage, equity and accessibility of urban green spaces under the constraints of space limitation in cities.

How could urban greenery be provided in cities despite such knowledge gaps and uncertainties is thus an important question that should dominate our efforts in greening cities. In addition, a key focus of our efforts should be on applying knowledge that has been gained from science. As design is fundamentally the means to translate our scientific knowledge, such as those covering climate, biodiversity and broader objectives of resilience and sustainability into spatially explicit forms, a focus on the science-design collaboration should be also be priority in our quest for producing greener and more ecological cities. While knowledge gaps exist, it is still possible to use generally applicable strategies to pursue different objectives in urban greening, such as suggested for urban biodiversity (Chap. 7) and for microclimate mitigation (Chap. 4). In addition, as scientific input alone is not adequate to deal with issues requiring normative judgement on topics like economic development and broad societal goals, transdisciplinary approaches involving stakeholders should also be promoted.

16.3 Innovate and Integrate Urban Greenery with Built Components of Cities

In cities, particularly in compact cities like Singapore and Hong Kong, where built densities are high and space for adequate greening is always in competition with other land uses, one critical success factor for greening such cites is to develop methods for integrating greenery with the built elements. The most substantial and visible built component of cities is the buildings, but cities are also criss-crossed by infrastructures such as at-grade and elevated roads, elevated rail lines, vehicular passes, and a dendritic network of canals for urban drainage. These built forms also offer the surfaces and spaces where vegetation can be inserted. Compact cities are also continuously exploiting underground developments, and these spaces, together with the living or working spaces in the interior of buildings, also represent another frontier of bringing vegetation closer to the daily living of urban dwellers.

A good example of integrating greenery with built component of cities is the world-wide momentum in rooftop greening (see Chap. 11) and greening of facades of building walls. In Singapore for instance, considerable progress has been made within a relatively short span of time from the mid-1990s to present time (Tan 2013: 62). These results and elsewhere in the world is an outcome of growing awareness of the benefits provided, policy implementation, and importantly, innovation in products and design that enable safe and effective implementation. In Singapore, other than rooftops and facades of buildings, greenery is also increasingly incorporated into intermediate floors of buildings, which are unenclosed void spaces planted with vegetation, and known locally as 'sky terraces' (Tan 2013: 49). This is a relatively new form of highrise greenery, and its implementation is also supported by floor-area-ratio incentives, which is in itself a policy innovation that promotes rapid implementation of sky terraces in the city.

Another example of greenery and infrastructure integration is the growing push in developing blue-green infrastructure (BGI) (Chap. 10) to manage urban stormwater, which in essence uses vegetation integrated with stormwater receiving basins or channels. Various forms of BGI are designed to convey, detain, retain, infiltrate and remove pollutants in stormwaters to varying extents. For instance, conventional concrete drains can be converted to vegetated bioswales or bioretention systems, which in addition to stormwater conveyance and cleansing functions, also support biodiversity (Kazemi et al. 2010) and education on nature awareness (Church 2015). Given the large footprint of buildings in cities, it is also logical to look at the large areas represented by rooftops spaces as spaces to receive rainwater, treat and store water for reuse. Innovations in combining current rooftop greenery technology with BGI technology will allow rooftop spaces to not only be green, but also function as an avenue to manage water as a resource. Technologies are also currently available to treat grey and black water using vegetated systems that can be designed as landscapes in or around buildings.[1]

Thus, even though cities are dominated by built elements, innovations in products and design can enable vegetation to be integrated with built structures. Thus far, rooftop greenery, green walls, and BGI represent the 'low-hanging fruits' of such types of integration. Future innovations can focus on optimizing performance of existing types of greenery-built components integration, and combining functions of stormwater treatment, heat mitigation, biodiversity support, etc. As cities undergo in-fill development, redevelopment, or upgrade infrastructures, new opportunities are created to insert greenery integrated with the built components. Similarly, with more cities developing plans to reclaim the waterfronts in the post-industrial era (Chap. 15), new opportunities are created for defining new forms of edges between land and water. In pushing new frontiers for inserting greenery, success is really only dependent on our ability to imagine and design a more desirable state for cities. In this state, greenery will play a more integral role in the quality of life for urban dwellers, as well as to create a future in which we exert less pressures on natural resources and respond more sensitively to the forces of nature.

References

Childers DL, Cadenasso ML, Morgan Grove J, Marshall V, McGrath B, Pickett STA (2015) An ecology for cities: a transformational nexus of design and ecology to advance climate change resilience and urban sustainability. Sustainability 7(4):3774–3791. doi:10.3390/su7043774

Church SP (2015) Exploring green streets and rain gardens as instances of small scale nature and environmental learning tools. Landsc Urban Plann 134:229–240. doi:10.1016/j.landurbplan.2014.10.021

Fons J (2012) European land use patterns, vol IV. The urban dimension in the EU-LUPA project. ESPON & TECHALIA Research & Innovation, Luxumbourg

Huang J, Tichit M, Poulot M, Darly S, Li S, Petit C, Aubry C (2015) Comparative review of multifunctionality and ecosystem services in sustainable agriculture. J Environ Manage 149:138–147. doi:10.1016/j.jenvman.2014.10.020

Hubacek K, Kronenberg J (2013) Synthesizing different perspectives on the value of urban ecosystem services. Landsc Urban Plann 109(1):1–6. doi:10.1016/j.landurbplan.2012.10.010

Kazemi F, Beecham S, Gibbs J (2010) Bioretention swales as multifunctional landscapes and their influence on australian urban biodiversity: hymenoptera as biodiversity indicators. Acta Horticulturae 881

Leung DYC, Tsui JKY, Chen F, Yip W-K, Vrijmoed LLP, Liu C-H (2011) Effects of urban vegetation on urban air quality. Landsc Res 36(2):173–188. doi:10.1080/01426397.2010.547570

[1]One example of emerging technology is 'Living Machines' (see https://www3.epa.gov/npdes/pubs/living_machine.pdf).

Lovell ST, Taylor JR (2013) Supplying urban ecosystem services through multifunctional green infrastructure in the United States. Landsc Ecol 28(8):1447–1463. doi:10.1007/s10980-013-9912-y

Pataki DE, Carreiro MM, Cherrier J, Grulke NE, Jennings V, Pincetl S, Pouyat RV, Whitlow TH, Zipperer WC (2011) Coupling biogeochemical cycles in urban environments: ecosystem services, green solutions, and misconceptions. Front Ecol Environ 9(1):27–36. doi:10.1890/090220

Potschin MB, Haines-Young RH (2011) Ecosystem services: exploring a geographical perspective. Prog Phys Geogr 35(5):575–594. doi:10.1177/0309133311423172

Tan PY (2013) Singapore a vertical garden city. Straits Times Press, Singapore

Whitlow TH, Pataki DA, Alberti M, Pincetl S, Setala H, Cadenasso M, Felson A, McComas K (2014a) Comments on "modeled PM2.5 removal by trees in ten U.S. cities and associated health effects" by Nowak et al. (2013). Environ Pollut 191:256. doi:10.1016/j.envpol.2014.03.033

Whitlow TH, Pataki DA, Alberti M, Pincetl S, Setala H, Cadenasso M, Felson A, McComas K (2014b) Response to authors' reply regarding "modeled PM2.5 removal by trees in ten U.S. cities and associated health effects" by Nowak et al. (2013). Environ Pollut 191:258–259. doi:10.1016/j.envpol.2014.03.035

Index

P.Y. Tan and C.Y. Jim (eds.), *Greening Cities: Forms and Functions*, Advances in 21st Century Human Settlements,
DOI 10.1007/978-981-10-4113-6

Kolibri 360 GmbH
Ferdinand-Jühlke-Straße 7
99095 Erfurt, Deutschland
produktsicherheit@kolibri360.de